Lecture Notes in Computational Science and Engineering

96

Editors:

For further volumes:
http://www.springer.com/series/3527

Frank Graziani • Michael P. Desjarlais •
Ronald Redmer • Samuel B. Trickey
Editors

Frontiers and Challenges in Warm Dense Matter

Springer

Editors
Frank Graziani
Lawrence Livermore National Laboratory
Livermore, CA, USA

Michael P. Desjarlais
Sandia National Laboratories
Albuquerque, NM, USA

Ronald Redmer
Institute of Physics
University of Rostock
Rostock, Germany

Samuel B. Trickey
Department of Physics
University of Florida
Gainesville, FL, USA

ISSN 1439-7358 ISSN 2197-7100 (electronic)
ISBN 978-3-319-04911-3 ISBN 978-3-319-04912-0 (eBook)
DOI 10.1007/978-3-319-04912-0
Springer Cham Heidelberg New York Dordrecht London

Library of Congress Control Number: 2014938627

Mathematics Subject Classification (2010): 81-08, 81V45, 81V70, 82C10, 82C22, 82D10, 82D15

Cover illustration: Snapshot of a DFT-MD simulations box with a hydrogen-helium mixture at warm dense matter conditions (4000 K, 1 Mbar). Red: protons, blue: helium nuclei, grey: electron density. (Author: Winfried Lorenzen; for details, see Lorenzen, Becker and Redmer, Chapter 8 of this book)

Printed on acid-free paper

Springer is part of Springer Science+Business Media (www.springer.com)

Contents

Introduction

In Spring 2012, the Institute for Pure and Applied Mathematics (IPAM) of the University of California, Los Angeles, sponsored a Long Program devoted to the computational and theoretical challenges in high energy density physics. A major component was a series of workshops, one of which focused on the exciting field of warm dense matter (WDM). WDM was described in the 2009 Fusion Energy Sciences Advisory Committee's report [1] and the Basic Research Needs for High Energy Density Laboratory Physics report [2] as an extreme state of matter characterized as intermediate "...between condensed matter (solids and liquids), gases, and ideal plasmas. It exists in the lower-temperature portion of the high energy density regime, under conditions where the assumptions of both condensed-matter theory and ideal-plasma theory break down, and where quantum mechanics, particle correlations, and electric forces are all important".

Typically WDM conditions span the eV range ($1\,\text{eV} = 1.16 \times 10^4\,\text{K}$) with densities typical of solids up to very high compressions. For both theoretical methods and for their computational implementation, such magnitudes and large ranges pose immense but fascinating challenges. This volume is devoted to expositions of approaches to those WDM challenges from experts who attended the IPAM workshop. Each chapter is meant to review a specific topic. Hence the chapters are extended and provide plenty of references for the interested reader. We, the editors, feel this volume captures the breadth and depth, along with the excitement of WDM physics.

What makes WDM such an exciting topic is its unique position in density-temperature space. WDM is a meeting point of several distinct physical regimes. Figure 1 (adapted from lectures on WDM given by Richard More at Lawrence Berkeley Laboratory) schematically demonstrates the WDM challenge.

As the meeting point of dense chemistry, condensed matter physics, super-dense matter, and ideal plasmas, WDM exposes severe problems for attempts at straightforward extension of well-developed models. For example, the kinetic equations for ideal plasmas rely on the fact that the number density times the Debye length cubed is a large number. This statement is equivalent to stating that all particle correlations are weak. This is the basis of the kinetic theory description of plasmas

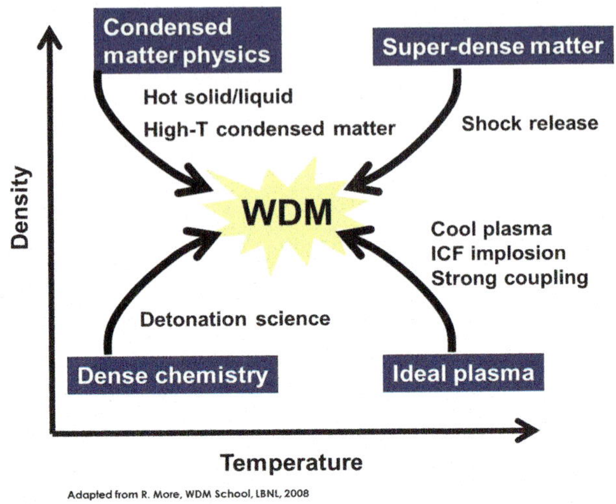

Fig. 1 Warm dense matter is at the confluence of several areas of high energy density physics

expressed in the Vlasov, Fokker-Planck, and Lenard-Balescu equations. For WDM, this approximation breaks down since particle correlations tend to be large and the number density times the cube of the Debye length is not large. Conversely, the approaches of condensed matter physics, specifically the so-called electronic-structure methods and ab initio molecular dynamics, are well-developed for systems in which the electrons can be treated as being at zero-temperature, but not for WDM temperatures. Those methods also have computational costs which scale unfavorably with both temperature and number of electrons.

The FESAC report identified four WDM issues, relevant for this workshop, for which theoretical, computational, and experimental opportunities would exist in the next 10 years. The methodological and computational implications are numerous. The list, with our commentary, is

1. Phase transitions in WDM: Developing advanced theoretical methods and precision experimental techniques that produce validation quality data is sorely needed. Specifically, the challenges include melting, liquid-liquid phase transitions, plasma phase transitions, the location of the liquid-vapor critical point, and the metal-insulator transition.
2. A comprehensive theory connecting WDM regimes: Developing advanced theoretical and computational tools for WDM is key. In addition, the theoretical methods should provide for connectivity between models. Theoretical and computational methods such as density functional theory (DFT) require further developments in the area of orbital-free DFT, advanced exchange-correlation functionals that include temperature effects consistently. In addition, DFT research in going beyond the Born-Oppenheimer approximation and also the inclusion of magnetic fields is needed. Particle simulations methods based on molecular dynamics coupled to quantum hydrodynamics are offering interesting inroads to a comprehensive theory of WDM.

3. Equations of state (EOS) and their dependence on formation history: Validation data for equations of state requires reliance on experimental techniques for accurate measurements in the WDM regime. In addition, a deep question is whether or not WDM is satisfactorily described by a set of state variables, independent of its formation history. The computation of EOS without decomposition of ionic and electronic contributions is a high priority. Another is reliable computation of the EOS of mixtures. In addition, the validity of equilibrium phase diagrams for WDM needs to be explored more satisfactorily, along with the issue of whether two-temperature equations of state are needed (and, if so, what is their physical basis). Finally, our knowledge of EOS in the WDM regime is impacted directly by experiments and theoretical/computational tools that can provide insights into chemistry at high density. Clear understanding of chemical bonds at high density is particularly needed.
4. Transport properties of WDM: Application of current computational and theoretical tools to the calculation of transport coefficients in WDM is key. Improvement also is needed. The coefficients include viscosity, diffusivity, and electric, ionic, and thermal conductivities. Key data to guide theoretical approaches continues to come from experiments conducted at small and large facilities worldwide.

This volume focuses on the theoretical and computational research progress in the four areas described above. Each of the chapters presents an in-depth review of various theoretical and computational approaches currently being developed and applied to problems in WDM physics. Numerous application areas in the WDM regime also are treated.

This volume does not deal with the wide variety of experimental capabilities that exist worldwide for studying WDM. The spectrum of techniques and instruments ranges from small scale, "table-top" to very large multi-billion dollar facilities. Data from them appear throughout this volume. For context therefore, we close with a cursory summary. There are small-scale facilities world-wide that include well-controlled, intense, short-pulse laser sources. Small scale pulsed-power facilities allow the WDM regime to be reached readily. Large pulsed-power facilities, such as the Z machine at Sandia National Laboratory, can provide ramp loading of materials up to 5 Mbar and shock loading up to 40 Mbar, in experimental configurations that permit long time scales and large sample sizes. The Z machine can address EOS and optical properties of elements and mixtures in WDM, strength, shock and release to WDM states, phase boundaries and X-ray scattering studies of WDM. The Linac Coherent Light source (LCLS) at Stanford University probes WDM states in a controllable manner. It employs a rapid heating source that creates WDM conditions and measures it prior to disassembler. LCLS can address dynamic structure factors with spatially resolved elastic Bragg and spectrally resolved Thomson scattering. Inertial confinement fusion facilities, such as Omega at the University of Rochester and the National Ignition Facility (NIF) at Lawrence Livermore National Laboratory, open new areas of high energy density science. NIF in particular can probe matter in the hundreds of Mbar to Gbar pressure regime. These facilities can address high pressure strength and EOS and the transition from

WDM to hot dense matter. Important large scale facilities also exist outside of the United States. The European X-ray free electron laser being built in Germany and the GSI Helmholtz Centre for Heavy Ion Research are examples. An expansion of GSI is now underway, namely, the Facility for Antiproton and Ion Research (FAIR), which is an international accelerator facility under construction which will use antiprotons and ions to perform research in high density plasma physics. Besides Germany, France has invested heavily in WDM experimental facilities. This includes the Laser Mégajoule(LMJ)/PETAL program and LULI (Laboratoire d'Utilisation des Lasers Intenses). Like NIF, LMJ intends to use the "indirect drive" approach. Rutherford Laboratory in the United Kingdom and various laboratories in Russia and China also have significant experimental facilities devoted to obtaining data in the WDM regime.

The editors thank the authors of the chapters for making this volume possible. We are grateful to the staff and leadership of IPAM for supporting dialogue about research which engages mathematics and physical science. The Long Program on Computational Challenges in High Energy Density Plasmas was of immense benefit to all of us, as well as being extremely well-organized and under-girded operationally.

References

1. Advancing the science of high energy density laboratory plasmas, United States Department of Energy Report, 2009
2. Basic research needs for high energy density laboratory physics, United States Department of Energy Report, 2009

Time-Dependent Density-Functional Theory: Features and Challenges, with a Special View on Matter Under Extreme Conditions

Carsten A. Ullrich

Abstract Time-dependent density-functional theory (TDDFT) is a universal quantum mechanical approach to the dynamical many-body problem which can be used to describe matter that is driven out of equilibrium by arbitrary time-dependent perturbations. TDDFT is often applied in the linear-response regime to obtain information about electronic excitations and spectral properties, but it also holds in the strongly nonlinear regime, where the perturbations compete with or even override the internal interactions that provide the structure and stability of matter. The purpose of this article is to give a brief overview of the basic formalism of TDDFT, and then to discuss the advantages, successes, and challenges of TDDFT for describing matter under extreme conditions of pressure and external fields. Two questions will be particularly emphasized: what are "easy" and what are "tough" problems for TDDFT (both from a fundamental and practical point of view), and how can TDDFT deal with dissipation? Some answers will be given, and needs and directions for future research will be pointed out.

1 Introduction and Overview

Time-dependent density-functional theory (TDDFT) is a universal approach to the dynamical many-body problem. This means that one can use TDDFT to describe nonstationary situations in matter such as atoms, molecules, solids, or nanostructures. Such situations can come about in essentially two ways: a system is initially in a stationary state (often, the ground state) and is then acted upon by a perturbation that drives it out of equilibrium, or the system is in a non-eigenstate and propagates freely in time. The external perturbations, and the resulting deviations

C.A. Ullrich (✉)
Department of Physics and Astronomy, University of Missouri, Columbia, MO 65211, USA
e-mail: ullrichc@missouri.edu

F. Graziani et al. (eds.), *Frontiers and Challenges in Warm Dense Matter*, Lecture Notes in Computational Science and Engineering 96,
DOI 10.1007/978-3-319-04912-0_1,

from equilibrium, can be arbitrarily strong: TDDFT can be applied in the linear-response regime (where it provides information about excitation energies and spectral properties) or in the nonlinear regime, where the external perturbations can be strong enough to compete with or even override the internal interactions that provide the structure and stability of matter. This article deals, predominantly, with the latter situation.

The purpose of this article is to give a brief overview of the basic formalism of TDDFT, and then to discuss specifically the advantages, successes, and challenges of TDDFT for describing matter under extreme conditions of pressure and external fields.[1] Two questions will be particularly emphasized: (1) What are "easy" and what are "tough" problems for TDDFT, from a fundamental and practical point of view? (2) How can TDDFT deal with dissipation? Some answers will be given, and needs and directions for future research will be pointed out.

The intended audience for this article has some background in electronic-structure methods, and ideally some prior knowledge of static density-functional theory (DFT), but not necessarily much experience with time-dependent, dynamical phenomena. Interested readers can find a pedagogical introduction and a detailed, up-to-date coverage of TDDFT in two recent books [2, 3].

2 What Do We Mean by Extreme Conditions?

The question of what constitutes an "extreme" condition can be answered in many different ways, depending on the context. For instance, a low-temperature scientist studying superconductivity will consider room temperature to be extremely hot; on the other hand, for a plasma physicist working on a fusion experiment, room temperature is extremely cold. For the rest of us, room temperature is just right.

But temperature is not the only environmental variable that can be controlled; pressure is another example. In materials science, any change in the environment, imposed condition, or perturbation which significantly disrupts and alters the structure of matter can be called extreme. Under such circumstances, perturbation theory ceases to be applicable, and fully nonperturbative solutions of the governing equations need to be sought.

In the following, we shall be interested in situations in which matter is rapidly and suddenly exposed to external influences (i.e., potentials or fields) which are strong enough to lead to extreme conditions. The goal is to study the dynamics in the material following such strong excitations. We thus consider N-electron systems that are governed by the nonrelativistic, time-dependent many-body Schrödinger

[1] Other discussions at this IPAM workshop on warm dense matter involved thermal DFT [1], which is rooted in equilibrium quantum statistical mechanics, and hence is used for static systems at high temperature and pressure. At present, no rigorous extension of thermal DFT to time-dependent and/or non-equilibrium systems exists.

equation:

$$i\frac{\partial}{\partial t}\Psi(\mathbf{r}_1,\ldots,\mathbf{r}_N,t) = \hat{H}(t)\Psi(\mathbf{r}_1,\ldots,\mathbf{r}_N,t)\,, \tag{1}$$

where

$$\hat{H}(t) = \sum_{j=1}^{N}\left\{\frac{1}{2}\left[\frac{\nabla_j}{i} + \mathbf{A}(\mathbf{r}_j,t)\right]^2 + V(\mathbf{r}_j,t)\right\} + \frac{1}{2}\sum_{i\neq j}^{N}\frac{1}{|\mathbf{r}_j - \mathbf{r}_i|}\,. \tag{2}$$

Notice that we have chosen atomic (Hartree) units, where $e = m = \hbar = 1$, and we have ignored the spin of the electrons for notational simplicity. Equations (1) and (2) describe the time evolution of an interacting N-electron system under the influence of a time-dependent scalar potential, $V(\mathbf{r},t)$ and a time-dependent vector potential, $\mathbf{A}(\mathbf{r},t)$. Gauge transformations can be found whereby the scalar potential is transformed away and absorbed into the vector potential; in general, however, it is convenient to keep both V and $\mathbf{A}$ in the formalism.

A suitable measure for the strength of a perturbation is comparison with the appropriate atomic unit. For instance, the atomic unit of the electric field strength is given by

$$E_0 = 5.14221 \times 10^{11}\ \mathrm{V/m}\,. \tag{3}$$

E_0 is that electric field which an electron experiences in the first Bohr orbit of a hydrogen atom. Clearly, if matter is exposed to electric fields of this order of magnitude, field ionization processes are going to become possible. The atomic unit of intensity is given by

$$I_0 = 3.50945 \times 10^{20}\ \mathrm{W/cm^2}\,. \tag{4}$$

I_0 refers to a plane electromagnetic wave with electric-field amplitude E_0. If matter is exposed to radiation with such intensities, the resulting forces on the electrons are comparable to those Coulomb forces that cause electronic binding and cohesion in matter. Under these circumstances, perturbation theory ceases to be applicable, and we need to treat internal Coulomb fields and external fields on the same footings.

Notice that for intensities in excess of about $10^{18}\,\mathrm{W/cm^2}$, the electronic motion in the focus of a laser becomes relativistic. The Schrödinger equation then needs to be replaced by a fully relativistic approach. For even higher intensities, pair production and other quantum electrodynamics effects can take place. In this article we will not be concerned with such intensity regimes (for a review, see [4]).

3 The Basic Features of TDDFT

In this Section we will give a brief review of the basic formalism of time-dependent density-functional theory (TDDFT), which, as of today, is the most promising approach for describing the dynamics of interacting many-electron systems [2].

We begin with the full time-dependent Schrödinger equation without any external vector potentials,

$$i\frac{\partial}{\partial t}\Psi(\mathbf{r}_1,\ldots,\mathbf{r}_N,t) = \left\{\sum_{j=1}^{N}\left[-\frac{\nabla_j^2}{2}+V(\mathbf{r}_j,t)\right]+\frac{1}{2}\sum_{i\neq j}^{N}\frac{1}{|\mathbf{r}_j-\mathbf{r}_i|}\right\}\Psi(\mathbf{r}_1,\ldots,\mathbf{r}_N,t) \tag{5}$$

(we will come back to vector potentials a bit later). Equation (5) is, in general, impossible to solve, even with the most advanced computational resources. At present, this can be done only for small systems, consisting of not more than two interacting electrons in three dimensions (e.g., [5]), or a few more interacting electrons in lower dimensions [6]. Any realistic time-dependent calculation of larger systems requires approximations. The key point is to cast the time-dependent many-body problem into a different form which is still formally exact but which makes it easier to develop approximations that are both accurate and computationally efficient.

TDDFT is based on the Runge-Gross theorem [7], which establishes a one-to-one correspondence, $V(\mathbf{r},t) \leftrightarrow n(\mathbf{r},t)$, between time-dependent potentials and time-dependent single-particle densities: for a given initial many-body state Ψ_0, two different potentials $V(\mathbf{r},t)$ and $V'(\mathbf{r},t)$ (where "different" means that they are not just shifted by a purely time-dependent function) will always cause the system to evolve in time such that the respective time-dependent densities, $n(\mathbf{r},t)$ and $n'(\mathbf{r},t)$, are different. This one-to-one correspondence implies that the time-dependent density,

$$n(\mathbf{r},t) = N\int d^3r_2\ldots\int d^3r_N|\Psi(\mathbf{r},\mathbf{r}_2,\ldots,\mathbf{r}_N,t)|^2\,, \tag{6}$$

formally carries the same information as the potential. But the potential determines the time evolution of the system! Thus, the Hamiltonian, and hence the wave function, formally become a functional of the density, $\Psi[n](\mathbf{r}_1,\ldots,\mathbf{r}_N,t)$. As a consequence, all physical observables are, at least formally, expressible as density functionals:

$$O[n](t) = \langle\Psi[n](t)|\hat{O}(t)|\Psi[n](t)\rangle. \tag{7}$$

(Strictly speaking, there is also a functional dependence on the initial state Ψ_0, but we will ignore this in the following; when starting from the ground state, the initial-state dependence can be absorbed into the density dependence.) In the next Section we will discuss several examples of such observables.

For now, though, the remaining problem is to obtain the time-dependent density (6) without actually having to solve the full time-dependent Schrödinger equation (otherwise, nothing would have been gained!). The idea is the following: instead of solving the time-dependent Schrödinger equation of the *interacting* system, we solve the time-dependent Schrödinger equation of a *noninteracting* system that produces the same density as the interacting system. Such a system is much easier to deal with, since its many-body wave function is simply a Slater determinant,

$$\Phi(\mathbf{r}_1, \ldots, \mathbf{r}_N, t) = \frac{1}{\sqrt{N}} \det\{\varphi_j(\mathbf{r}_j, t)\} , \tag{8}$$

and we only need to solve a single-particle Schrödinger equation to obtain the time-dependent orbitals $\varphi_j(\mathbf{r}, t)$:

$$i \frac{\partial}{\partial t} \varphi_j(\mathbf{r}, t) = \left[-\frac{\nabla^2}{2} + V_s[n](\mathbf{r}, t) \right] \varphi_j(\mathbf{r}, t) . \tag{9}$$

Equation (9) is called the time-dependent Kohn-Sham equation.

Here is a very important point to keep in mind: the time-dependent Kohn-Sham orbitals are designed to reproduce the exact density of the interacting system, i.e.,

$$n(\mathbf{r}, t) = \sum_{j=1}^{N} |\varphi_j(\mathbf{r}, t)|^2 , \tag{10}$$

where $n(\mathbf{r}, t)$ is the same density that one would have obtained from Eq. (6).[2] However, the respective many-body wave functions themselves are different:

$$\Psi(\mathbf{r}_1, \ldots, \mathbf{r}_N, t) \neq \Phi(\mathbf{r}_1, \ldots, \mathbf{r}_N, t) . \tag{11}$$

In other words, the Kohn-Sham system is not meant to reproduce the full many-body wave function of the interacting system, only its single-particle density.

The time-dependent Kohn-Sham equation (9) features an effective single-particle potential whose task is to cause the noninteracting system to reproduce the exact density. It is usually written in the following form:

$$V_s(\mathbf{r}, t) = V(\mathbf{r}, t) + \int d^3r' \frac{n(\mathbf{r}', t)}{|\mathbf{r} - \mathbf{r}'|} + V_{xc}(\mathbf{r}, t) , \tag{12}$$

[2]We assume here that all Kohn-Sham orbitals are either fully occupied or empty. For simplicity, we disregard the possibility of fractional orbital occupation numbers, which would be associated with degeneracies.

that is, as sum of the given potential V of the physical, interacting system, plus the time-dependent Hartree potential V_H plus the time-dependent exchange-correlation (xc) potential V_{xc}. The latter two potentials are functionals of the density, so the time-dependent Kohn-Sham equation has to be solved self-consistently.

Readers who are familiar with ground-state density-functional theory (DFT) [8] will immediately be familiar with TDDFT as well, because both theories—DFT and TDDFT—are based on very similar premises. Indeed, the parallels are obvious: the Runge-Gross theorem [7] is the time-dependent counterpart of the Hohenberg-Kohn theorem [9], and the time-dependent Kohn-Sham equation is a generalization of the static Kohn-Sham approach [10]. However, there are important differences as well, and there are features of TDDFT that are unique to the time-dependent case:

- Ground-state DFT is based on the variational minimum principle. In the time-dependent case, there is no analogous minimum principle. It is possible to derive the formal framework of TDDFT from a stationary-action principle [11, 12]. But in contrast with DFT, where the ground-state energy E_0 is the quantity of central importance, the action is practically of no interest in itself.
- Mathematically, the Kohn-Sham equation of ground-state DFT,

$$\left[-\frac{\nabla^2}{2} + V_s^0[n_0](\mathbf{r})\right] \varphi_j^0(\mathbf{r}) = \varepsilon_j \varphi_j^0(\mathbf{r}) , \tag{13}$$

  is a boundary-value problem, where the static effective potential V_s^0 depends self-consistently on the ground-state density $n_0(\mathbf{r})$. By contrast, the time-dependent Kohn-Sham equation (9) represents an initial-value problem. One proceeds in three steps:

  1. Prepare the initial state, which is usually the ground state (technically, one can start from any initial state, but non-ground-state or non-stationary initial states are rarely considered). This means that one needs to carry out a static Kohn-Sham calculation for the system of interest, i.e., solve Eq. (9), to get a set of ground-state Kohn-Sham orbitals $\varphi_j^0(\mathbf{r})$ and orbital energies ε_j.
  2. Solve the time-dependent Kohn-Sham equation (9) from the initial time t_0 to the desired final time t_1, where $\varphi_i(\mathbf{r}, t_0) = \varphi_j^0(\mathbf{r})$ are the initial orbitals. The time propagation of the orbitals gives the time-dependent density $n(\mathbf{r}, t)$ via Eq. (10).
  3. Calculate the desired observable(s) as functionals of $n(\mathbf{r}, t)$.

- The static xc potential is defined as

$$V_{xc}^0[n_0](\mathbf{r}) = \left.\frac{\delta E_{xc}[n]}{\delta n(\mathbf{r})}\right|_{n_0(\mathbf{r})} . \tag{14}$$

  The common strategy for deriving approximations of V_{xc}^0 is to start with an approximation for the xc energy E_{xc} and then obtain the xc potential via Eq. (14). By contrast, the time-dependent xc potential is approximated directly.

- The time-dependent xc potential $V_{xc}[n](\mathbf{r},t)$ has many features and exact properties that are analogous to those of the static xc potential: for instance, it must be self-interaction free, it must have the correct asymptotic $-1/r$ behavior, and it must exhibit an overall discontinuity upon change of particle number [13]. However, there are also features that have no counterpart in static DFT and are truly dynamic. The most important one is that the time-dependent xc potential has a memory: $V_{xc}[n](\mathbf{r},t)$ at time t depends on densities $n(\mathbf{r}',t')$ at earlier times, where $t' \leq t$. This memory is, in principle, infinitely long-ranged.
- The most common approximation for the time-dependent xc potential is the adiabatic approximation, which ignores all memory effects:

$$V_{xc}^{\text{adia}}(\mathbf{r},t) = V_{xc}^{0,\text{approx}}[n_0](\mathbf{r})\big|_{n_0(\mathbf{r})\to n(\mathbf{r},t)} . \tag{15}$$

Here, an approximate xc potential from static DFT, $V_{xc}^{0,\text{approx}}[n_0](\mathbf{r})$, is evaluated with the instantaneous time-dependent density. This is, obviously, very convenient, because this means that we can simply and straightforwardly use our favorite approximation from static DFT in TDDFT. A widely used example is the adiabatic local-density approximation (ALDA).

But how can we be sure that the adiabatic approximation is justified? Are there instances in which the memory of the xc potential is important? This is one of the key questions in TDDFT, and we will address it in several places below.

4 Practical Aspects of TDDFT

Practitioners of TDDFT may face two kinds of challenges, formal and computational ones. What constitutes a computational challenge of course depends on the resources and facilities at one's disposal. Today, there exist several computer codes for propagating the time-dependent Kohn-Sham equations in molecules and materials, most notably, the `octopus` [14–16] and the `SIESTA` [17, 18] packages. With these codes, it is possible to study the real-time dynamics of systems with hundreds of electrons.

But let us now take a more fundamental point of view. While TDDFT is formally exact and universal, and can thus in principle be used for any system and any type of electron dynamics (as long as the nonrelativistic time-dependent Schrödinger equation applies), one has to work with approximate functionals in practice. Given the available approximations, there are some situations that are easy, and others that are tough (which means, it is difficult to get good results). Let us now discuss those cases and give some examples. Notice that whenever we say "TDDFT" in the following, we mean using state-of-the-art approximations, and not the formally exact (but practically inaccessible) theory.

4.1 What TDDFT Can Do Well

There are several distinct types of "easy" dynamics that can be successfully treated with TDDFT:

- *When the dynamics of the interacting system is qualitatively similar to the corresponding noninteracting system.*

  Whenever this is the case, the adiabatic approximation (15) for the xc potential works well, which means that it is possible to use the same approximation in the time-dependent case that is used to prepare the initial ground state of the system (i.e., the LDA, or a GGA, or a hybrid functional). Here are two very different scenarios that are well described by the adiabatic approximation:

  1. In linear-response calculations of the excitation spectrum of the system, whenever the spectral features are dominated by single-particle excitation processes. This is because such processes will find a counterpart in the spectrum of the noninteracting Kohn-Sham system. The dynamical xc potential then does not need to create any new spectral features, and merely adjusts and corrects the features that are already there.
  2. In calculations of strongly excited systems under the influence of high-intensity fields, whenever the external field dominates over the electron-electron interactions. In this case, highly nonlinear multiphoton processes are prevalent, such as multiple ionization or high-harmonic generation.

- *When the electron dynamics is highly collective, and the density flows in a "hydrodynamic" manner, without much compression, deformation, or abrupt changes, and with small gradients in the velocity field.*

  An important class of this type of dynamics is plasmon modes in (quasi-) metallic systems such as bulk metals, metal surfaces, nanoparticles, clusters, quantum dots, or doped semiconductor heterostructures. Plasmons—collective charge-density oscillations—are essentially a phenomenon of classical electrodynamics, and dynamical many-body effects only give relatively small corrections to the plasmon dispersions. We will discuss plasmon damping below in Sect. 5.

4.2 Easy Observables

In Sect. 3 we showed that all physical observables are formally functionals of the time-dependent density, see Eq. (7). TDDFT gives, in principle, the exact density, but nothing else; all quantities of interest must be obtained from $n(\mathbf{r},t)$. Fortunately, there are several important quantities that are easy to calculate in this way.

The easiest observable is the density itself, which shows how electrons move during a dynamical process. The time propagation is unitary, so the total norm is conserved; but to describe ionization or charge transfer processes, it is often of interest to obtain the number of electrons that escape from a given spatial region $\mathcal{V}$:

$$N_{\rm esc}(t) = N - \int_{\mathscr{V}} d^3\, n(\mathbf{r},t)\ . \tag{16}$$

Here, $\mathscr{V}$ can be thought of as a "box" that surrounds the entire system (in case we wish to calculate ionization rates of atoms or molecules), or it could be a part of a larger molecule or part of a unit cell of a periodic solid.

Another easy class of observables is moments of the density, such as the dipole moment:

$$\mathbf{d}(t) = \int d^3r\ \mathbf{r}n(\mathbf{r},t)\ . \tag{17}$$

The dipole moment can be considered directly, i.e., in real time, to study the behavior of charge-density oscillations. Alternatively, it can be Fourier transformed to yield the dipole power spectrum $|d(\omega)|^2$ or related observable quantities such as the photoabsorption cross section.

Higher moments of the density, such as the quadrupole moment, can be calculated just as easily, but are less frequently considered.

4.3 Where TDDFT Faces Challenges

There are various dynamical situations and phenomena that can be tough to describe with TDDFT, using present-day approximations.

- *When the electron dynamics of the interacting system is highly correlated.*
  This is the case whenever one considers excitation processes—in the linear or in the nonlinear regime—that do not have a counterpart in the Kohn-Sham system. In particular, multiple excitations (e.g., double excitations) are notoriously difficult to capture with TDDFT, because they require xc functionals beyond the adiabatic approximation [19, 20]. At present, it appears that the most promising avenue towards treating double excitations with TDDFT is by constructing xc functionals using many-body techniques [21].
  What we just discussed seems to be in contradiction with the earlier statement (Sect. 4.1) that TDDFT does very well when collective plasmon excitations are concerned. After all, there are no plasmons in a noninteracting system! The explanation has to do with two things. First of all, charge plasmons are already obtained on the level of the RPA (random phase approximation), which has no xc at all.[3] Secondly, the most popular xc functional in TDDFT, the ALDA, has the

[3]The RPA in linear response is equivalent to the time-dependent Hartree approximation, and has no dynamical xc contributions. It is not to be confused with the RPA in ground-state DFT, which is a method to calculate correlation energies using the so-called adiabatic connection fluctuation-dissipation approach [22].

interacting homogeneous electron gas as a reference system, which can of course sustain a plasmon.

But even for plasmons, TDDFT faces serious difficulties when it comes to describing the damping. The reason is, ultimately, the same as the difficulties to describe double excitations in finite systems: both originate from nonadiabatic correlation effects. We will come back to dissipation in the next section.

Another notorious, correlated problem is that of direct double ionization of atoms in strong fields, also known as the "helium knee" problem [23, 24]. Here, one electron gets ionized by the strong laser field, but rather than escaping from the ion, it gets driven back during the second half of the laser cycle, returns to the ion and knocks out another electron [25]. As a result, the double ionization rate is much higher than what one would expect in the absence of such rescattering. From the perspective of TDDFT, it turns out that the double ionization problem can now be considered solved: the key is to use xc functionals that have a discontinuity whenever the particle number changes by an integer [13].

Notice that we have used the terminology "highly correlated" in the strictly dynamical context. Problems that arise in so-called "strongly-correlated" materials, which are already present in ground-state DFT, will of course carry over into TDDFT as well. A typical class of examples is Mott insulators, which are incorrectly described in DFT using most standard xc functionals. DFT usually predicts a metallic ground state in such systems, and TDDFT will thus predict a qualitatively wrong dynamical behavior.

- *When highly delocalized, long-ranged excitation processes take place.*

  Single excitation processes in small and medium-sized molecules are well described with TDDFT using standard local and semilocal xc functionals such as LDA and GGA, or hybrid functionals (typical errors are of the order of 0.3 eV or less). But for larger systems one encounters two basic classes of excitation processes where these xc functionals are inadequate:

  1. In large molecular complexes there exist so-called *charge-transfer excitations*, in which excited electrons get redistributed between different spatial regions, or fragments, of the total system. Standard xc functionals typically underestimate charge-transfer excitation energies; an accurate description requires xc functionals with a long spatial range. So-called range-separated hybrid functionals [26, 27] have recently emerged as a promising solution for this problem.
  2. In organic or inorganic extended semiconducting or insulating systems, interband optical excitations are dominated by *excitonic* effects. In a simple picture, excitons are bound electron-hole pairs whose mean distance can extend over many lattice constants of the system. The standard many-body approach for describing such excitonic effects is via the Bethe-Salpeter equation, but they can also be captured with TDDFT (since TDDFT is in principle an exact theory). However, for the electron and the hole to form a bound state, again a long-ranged xc functional is needed [28–30]. LDA or GGA, even most of the standard hybrid functionals, do not give excitons.

Let us emphasize that in the case of delocalized excitation processes it is the long-rangedness of the xc effects that is crucial, not the adiabatic approximation.

- *When the electron dynamics is extremely "non-hydrodynamic" or extremely fast.*

As we have seen, the adiabatic approximation works best for dynamical processes that are slow compared to characteristic intrinsic frequency scales of the system, and for excitations (individual or collective) of the many-body system that have an analog in the Kohn-Sham system. When these conditions are not met, nonadiabatic effects occur and the adiabatic approximation can break down. The failure of the adiabatic approximation for very rapid and high-frequency dynamics has been observed in numerical simulations [31, 32].

From a hydrodynamic perspective, the breakdown of the adiabatic approximation is associated with large gradients of the velocity field of the electrons: in other words, with rapid compression and rarefaction of the electron density. This may happen, for instance, during tunneling processes through barriers or constrictions, or during sudden switching, fast collisions, or violent shake-up processes driven by intense high-frequency fields.

In general, it is not easy to predict just by looking at the external potential $V(\mathbf{r}, t)$ whether or when such extreme conditions will occur in the system; even if the potential undergoes large and sudden changes, the density may be quite sluggish in responding. In general, there is no alternative but calculate the actual time evolution of the system and see what happens. For finite systems, Thiele and Kümmel [33] have given a simple criterion involving the total noninteracting kinetic energy obtained from the time-dependent Kohn-Sham orbitals,

$$T_s(t) = \sum_j \int d^3r |\nabla \varphi_j(t)|^2 . \tag{18}$$

They suggest that the adiabatic approximation remains valid until the dynamics leads to situations in which the kinetic energy varies too rapidly, with a critical value

$$\left. \frac{\partial}{\partial t} T_s(t) \right|_{\text{crit}} \approx \frac{T_s(t=0)}{\tau} . \tag{19}$$

Here τ is a characteristic time scale that is of the order of the period associated with the lowest excitation energy of the system.

For a more detailed hydrodynamic analysis of the electron dynamics, local measures of the deformation of the electron liquid can be considered [31]. However, such analysis is quite involved in three dimensions, since the deformation then needs to be expressed as a tensorial quantity [34].

4.4 Difficult Observables

The easy observables we discussed in Sect. 4.2 could all be calculated directly and in elementary ways from the time-dependent density $n(\mathbf{r}, t)$, see Eqs. (16) and (17). But there are many quantities and observables of great physical interest that cannot be easily obtained from the density, even though they can formally be defined as density functionals. Let us consider an example.

Equation (16) gives the total number of escaped electrons, which in general can be nonintegral. For instance, if we consider a helium atom in a laser field, a value of $N_{\rm esc} = 0.5$ would indicate that on average half an electron has been removed. In reality there are of course no "half-electrons", so we have to interpret this result in a probabilistic sense: it could for instance mean that there is 50 % probability that the helium atom is singly ionized, and 50 % probability that it is not ionized; but other scenarios, involving doubly ionized helium, are also possible. The probabilities to find an atom or molecule in a certain charge state $+m$ are defined as [35]

$$P^0(t) = \int_{\mathscr{V}} d^3r_1 \ldots \int_{\mathscr{V}} d^3r_N \, |\Psi(\mathbf{r}_1, \ldots, \mathbf{r}_N, t)|^2 \tag{20}$$

$$P^{+1}(t) = \int_{\overline{\mathscr{V}}} d^3r_1 \int_{\mathscr{V}} d^3r_2 \ldots \int_{\mathscr{V}} d^3r_N \, |\Psi(\mathbf{r}_1, \ldots, \mathbf{r}_N, t)|^2 \tag{21}$$

and similarly for all other $P^{+m}(t)$. Here $\overline{\mathscr{V}}$ denotes all space outside the integration box $\mathscr{V}$ surrounding the system. The ion probabilities are defined in terms of the full many-body wave function $\Psi(t)$, which is a density functional according to the Runge-Gross theorem; but it is not possible to extract the ion probabilities $P^{+m}(t)$ directly from the density in an elementary way. Since dealing with the full wave function is prohibitively expensive, one may be tempted to resort to approximating $\Psi(t)$ by the Kohn-Sham Slater determinant $\Phi(t)$, in spite of the discussion of Sect. 3. One then obtains the Kohn-Sham ion probabilities

$$P_s^0(t) = N_1(t) N_2(t) \ldots N_N(t) \tag{22}$$

$$P_s^{+1}(t) = \sum_{j=1}^{N} N_1(t) \ldots N_{j-1}(t) \big(1 - N_j(t)\big) N_{j+1}(t) \ldots N_N(t) \tag{23}$$

and similarly for all other $P_s^{+m}(t)$, where

$$N_j(t) = \int_{\mathscr{V}} d^3r |\varphi_j(\mathbf{r}, t)|^2 \,. \tag{24}$$

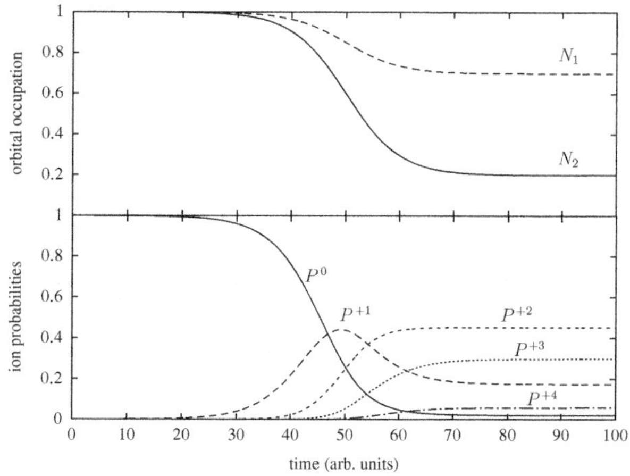

Fig. 1 Schematic illustration of the ionization dynamics of a 4-electron system with two doubly occupied orbitals (e.g., the Be atom). *Top panel*: time-dependent orbital occupation numbers of the first and second orbital. The decrease in N_1 and N_2 is typical for a short-pulse excitation around $t = 50$. *Bottom panel*: associated Kohn-Sham ion probabilities

The Kohn-Sham ion probabilities are easily obtained from the orbitals; but apart from certain limiting cases [35], this remains essentially an uncontrolled approximation [36, 37].

Figure 1 presents a simple illustration to show how these formulas work. We consider a 4-electron system (such as, e.g., the Be atom), with two initially doubly occupied orbitals. The orbital populations N_1 and N_2 are assumed to decrease as $N_j(t) = 1 - a_j(1 + e^{-(t-t_p)/b})^{-1}$, where we take $a_1 = 0.3$, $a_2 = 0.8$, $b = 0.2$, and $t_p = 50$. This behavior mimics the ionization of an atom by a short intense laser pulse with peak at t_p. As the orbital populations decrease (which indicates that density is removed from the system), the ion probabilities go through various stages. At the end of the pulse, the system can be found in any of the possible ionized states (from zero to fourfold) with different probabilities.

The ion probabilities are a typical example of a "difficult" observable in TDDFT. There are several other examples:

- *Photoelectron spectra.* The photoelectron kinetic energy distribution spectrum is formally defined as

$$P(E)dE = \lim_{t\to\infty} \sum_{k=1}^{N} |\langle \Psi_E^k | \Psi(t)\rangle|^2 dE\,, \tag{25}$$

  where $|\Psi_E^k\rangle$ is a many-body eigenstate with k electron in the continuum and total kinetic energy E of the continuum electrons. There are approximate ways of calculating photoelectron spectra from the density or from the Kohn-Sham orbitals [38–40], but the rigorous method involves the full wave functions.
- *State-to-state transition probabilities.* The S-matrix describes the transition between two states:

$$S_{i,f} = \lim_{t\to\infty} \langle \Psi_f | \Psi(t)\rangle\,, \tag{26}$$

where $\Psi(t)$ starts out from the initial state Ψ_i, and Ψ_f is some final state. Getting the S-matrix from the density can be accomplished using a complicated implicit read-out procedure [41].

- *Momentum distributions*. Ion recoil momenta are of great interest in high-intense field or scattering experiments. In the context of calculating Compton profiles using DFT, momentum distributions have been of interest for a long time [42,43]. In TDDFT, the problem is formally similar to the problem of calculating ion probabilities from the density, and in principle requires the full wave function in momentum space. The Kohn-Sham momentum distributions can be taken as approximation, but this is not always well justified [44].
- *Transition density matrix*. The transition density matrix is a quantity that is defined in the linear response regime. As the name indicates, it refers to a specific excitation of the system (typically, a large molecular system), and maps the distribution and coherences of the excited electron and the associated hole. In particular, the transition density matrix is useful to visualize excitonic effects. There is no easy way to obtain it directly from the density; the best we can do is to construct the transition density matrix from Kohn-Sham orbitals [45].

All the foregoing examples have in common that they are explicit expressions of the many-body wave function, or of the N-body density matrix of the system, and therefore can be only implicitly expressed as density functionals. One can get approximate results by replacing the full many-body wave function $\Psi(t)$ with the Kohn-Sham wave function $\Phi(t)$, but there is no guarantee that this will give good results.

5 TDDFT and Dissipation

Dissipation is a phenomenon that occurs whenever a system interacts with its environment; energy flows away from the system and is transferred into the available degrees of freedom of the environment, thus effectively getting "lost" insofar as the system under study is concerned. There exists no system in nature that is fully and absolutely shielded from any environmental influence, although there are many systems under study that are extremely well isolated. Therefore, dissipation can often be neglected when describing electron dynamics. On the other hand, there exist many situations in which dissipative effects are prominent, or at least cannot be neglected. In particular, dissipation plays an important role in the response of materials to very strong perturbations or extreme conditions: it determines the rate at which energy can be absorbed and redistributed, and converted into structural changes such as fragmentation or melting.

All occurrences of dissipation have in common that they involve a coupling between the dynamics of a system (which may be finite or extended) and degrees of freedom that are not part of the system. Often one refers to the latter as a "bath". There are many varieties of system-bath coupling: for instance, the coupling of electrons and phonons in a molecule or crystal, the coupling between an atom and the photons in a cavity, or the coupling between different kinds of electronic degrees of freedom within the same system.

TDDFT, as we have presented it here, is defined for electronic systems governed by the many-body Schrödinger equation (5). This means that TDDFT can describe purely electronic dissipation processes in principle exactly. But to treat dissipation into nonelectronic degrees of freedom such as motion of the ions, TDDFT needs to be extended. In the following, we will briefly review the current status of TDDFT for dissipation.

5.1 Coupling Between Electron and Nuclear Dynamics

From a formal point of view, electronic and nuclear degrees of freedom need to be treated on an equal footing, by starting from a many-body Schrödinger equation for the coupled electronic-nuclear system. It is possible to formulate a multicomponent TDDFT treatment for this problem [46–48]. However, finding suitable xc functionals for the nuclear degrees of freedom is difficult, and thus multicomponent electron-nuclear DFT and TDDFT have not yet found widespread application.

In practice, the coupling between ionic and electronic degrees of freedom is usually carried out within a molecular dynamics framework. In the regime of strong excitations, non-Born-Oppenheimer molecular dynamics is required. This can be done in various ways; the so-called TDDFT-Ehrenfest approach can be formulated with the following coupled set of equations:

$$i\frac{\partial}{\partial t}\varphi_j(\mathbf{r},t) = \left[-\frac{\nabla^2}{2} + V_e(\mathbf{r},t) + V_{\rm Hxc}[n](\mathbf{r},t) + W_{en}(\{\mathbf{r}\},\{\mathbf{R}(t)\})\right]\varphi_j(\mathbf{r},t)\,, \tag{27}$$

$$M_j\frac{\partial^2}{\partial t^2}\mathbf{R}_j(t) = -\nabla_{\mathbf{R}_j}\left[V_{n,j}(\mathbf{R}_j,t)+W_{nn}(\mathbf{R}(t))+\int d^3r\, n(\mathbf{r},t)W_{en}(\{\mathbf{r}\},\{\mathbf{R}(t)\})\right]. \tag{28}$$

Here, Eq. (27) is the time-dependent Kohn-Sham equation featuring a potential acting on the electrons only, V_e, the combined Hartree and xc potentials, $V_{\rm Hxc}$, and the electron-nuclear interaction W_{en}, which depends on all electronic and nuclear coordinates $\{\mathbf{r}\}$ and $\{\mathbf{R}\}$. The second equation is the classical equation of motion of the nuclei, with an external potential $V_{n,j}$ acting on the jth nucleus and the nucleus-nucleus interaction W_{nn}. The last term on the right-hand side of Eq. (28) is a mean-field interaction between the instantaneous time-dependent electronic density and the nuclei.

The TDDFT-Ehrenfest approach can thus be characterized as a mixed quantum-classical approach, with classical nuclear dynamics in the average field caused by the electrons. This approximate treatment of the coupled dynamics works well under the following circumstances:

- If a single nuclear path is dominant;
- For ultrafast processes, and at the initial state of an excitation, before significant of potential energy surfaces can occur;
- When a large number of electronic excitations are involved, so that the nuclear dynamics is governed by the average forces (for instance, in metals or during strong-field processes, if a large amount of energy is absorbed).

The TDDFT-Ehrenfest approach thus appears to be the method of choice for strongly excited coupled electron-nuclear systems: it allows a computational treatment of relaxation and energy dissipation processes into nuclear degrees of freedom, and can be used to describe processes such as laser-induced melting and Coulomb explosion.

But the TDDFT-Ehrenfest approach also has clear limitations, for instance for processes where a branching of ionic trajectories occurs and the excited-state dynamics takes place on multiple paths of different Born-Oppenheimer potential energy surfaces. Such processes are important in many chemical reactions, and obviously extend beyond the scope of a mean-field approach; to deal with them, so-called surface hopping schemes have been developed [49, 50]. A detailed presentation of the various TDDFT/molecular dynamics approaches, together with many applications and references, can be found in Chap. 17 of Ref. [2].

5.2 *Electronic Dissipation: Time-Dependent Current-DFT*

In an isolated system such as a single atom or molecule, the electron dynamics is free of dissipation within the time-dependent many-body Schrödinger equation, and hence within TDDFT. This implies that electronic excitation energies are real, spectral linewidths are zero, and associated lifetimes of excited states are infinite.

However, in extended systems the situation is different: there exist excitations that have finite lifetimes, even if impurity or disorder or phonon scattering effects are not included. A prime example is plasmons in metals. In the simplest possible point of view of a homogeneous system, plasmons are collective oscillations of the electron liquid against the uniform positive jellium background of the metal.

Plasmon damping can be nicely visualized in the case of intersubband plasmons in doped semiconductor quantum wells. Such excitations can be realized, e.g., in $GaAs/Al_xGa_{1-x}As$ square potential wells [51]. If, say, the direction of growth is the z direction, then electrons in this quantum well are essentially confined in a box along this direction, and behave as free particles in the $x - y$ plane, with in-plane wave vectors $\mathbf{q}_{||}$. Figure 2 (top left) illustrates the possible single-particle transitions between the lowest two subbands. The shape of the plasmon is shown in the bottom left panel, where the red line is the profile of the charge-density distribution which

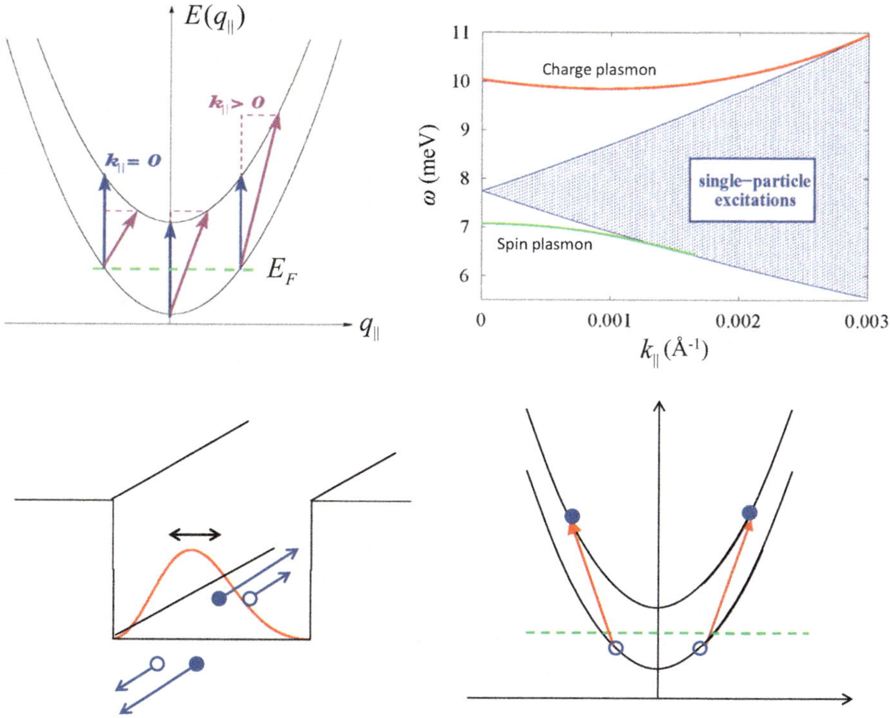

Fig. 2 Intersubband plasmon damping in quantum wells. *Top left*: Single-particle intersubband excitations between the first subband, occupied up to the Fermi energy E_F, to the second subband. The transitions can take place vertically (zero in-plane momentum transfer) or nonvertically. *Top right*: intersubband particle-hole continuum, and charge and spin plasmon dispersions. *Bottom left*: schematic illustration of the decay of an intersubband charge plasmon (with $k_{||} = 0$) into two in-plane particle-hole pairs. *Bottom right*: the two particle hole pairs have opposite momentum, as dictated by in-plane momentum conservation

oscillates back and forth in the quantum well (along the z direction), as indicated by the two-sided arrow above it.

If all intersubband excitations take place collectively and in phase, then we have an intersubband plasmon mode. This is a collective excitation of the many-body system in which the charge density oscillates along the direction of growth of the well (i.e., along the z-direction). As long as the plasmon dispersion is away from the single particle-hole continuum (see Fig. 2 top right), the plasmon mode is rather robust. Once it merges with the particle-hole continuum, the so-called Landau damping leads to a rapid dephasing of the plasmon mode. If plasmons are calculated with TDDFT using xc functionals that do not have a memory, such as the ALDA, then they are characterized by an infinite lifetime away from the particle-hole continuum.

However, intersubband plasmons can dissipate into multiple particle-hole pairs, as illustrated in the bottom part of Fig. 2. Physically, this means that the collective degrees of freedom associated with motion perpendicular to the plane of the quantum well couple to a "bath" of in-plane degrees of freedom [52]. This coupling is mediated via Coulomb scattering; it involves two-particle-hole processes, which are of higher order than the single-particle processes that give rise to Landau damping, and is therefore rather weak.

How does TDDFT account for plasmon damping via two-particle (or multiple) excitation processes? The answer lies in the xc potential $V_{xc}[n](\mathbf{r},t)$: there are no multiple excitations in the adiabatic approximation (15), so it is necessary to use nonadiabatic xc functionals. Plasmons are usually discussed within linear response theory; the quantity of interest is then the linearized xc potential, given by

$$\delta V_{xc}(\mathbf{r},t) = \int dt' \int d^3r' f_{xc}(\mathbf{r},t,\mathbf{r}',t')\delta n(\mathbf{r}',t') , \tag{29}$$

which defines the xc kernel $f_{xc}(\mathbf{r},t,\mathbf{r}',t')$ and its Fourier transform, $f_{xc}(\mathbf{r},\mathbf{r}',\omega)$ [53]. In general, the xc kernel is a complex, frequency-dependent object; within the adiabatic approximation, f_{xc} is frequency-independent and real. This immediately explains the absence of any plasmon damping in the adiabatic approximation [54].

The construction of nonadiabatic xc functionals is far from a trivial task. The underlying reason has become known as the "ultranonlocality problem" of TDDFT [55], which states that it is impossible to construct a memory-dependent xc potential $V_{xc}(\mathbf{r},t)$ that depends only on the local density at the same point $\mathbf{r}$. The reason for this is not hard to appreciate [56]: the electron dynamics in any kind of system necessarily involves the flow and rearrangement of probability density, and a density volume element at a given space-time point $(\mathbf{r},t)$ must have originated from a different spatial location $\mathbf{r}'$ at an earlier time $t' < t$. Hence, the time-dependent density $n(\mathbf{r},t)$ is not the most suitable variable for an easy construction of memory-dependent xc potentials, because it does not allow a local approximation.

It turns out that it is possible instead to construct a local, memory-dependent xc *vector* potential as a functional of the current density[4] $\mathbf{j}(\mathbf{r},t)$; in other words, if we "upgrade" from TDDFT to time-dependent current-DFT (TDCDFT) [59, 60]. The time-dependent Kohn-Sham equation in TDCDFT can be written like this:

$$i\frac{\partial}{\partial t}\varphi_j(\mathbf{r},t) = \left[\frac{1}{2}\left(\frac{\nabla}{i} + \mathbf{A}(\mathbf{r},t) + \mathbf{A}_{xc}(\mathbf{r},t)\right)^2 + V(\mathbf{r},t) + V_H(\mathbf{r},t)\right]\varphi_j(\mathbf{r},t) . \tag{30}$$

Here, $V(\mathbf{r},t)$ and $\mathbf{A}(\mathbf{r},t)$ are external scalar and vector potentials, and all xc effects have been gauge transformed into the xc vector potential $\mathbf{A}_{xc}(\mathbf{r},t)$, which is a

[4] Note that $\mathbf{j}(\mathbf{r},t)$ is the *physical* current density. This is in contrast with ground-state CDFT, where the basic variable is the *paramagnetic* current density [57, 58].

functional of the current density. TDCDFT allows one to describe systems governed by the time-dependent many-body Schrödinger equation (2) in principle exactly.

A memory-dependent xc vector potential can be formulated as follows [61, 62]:

$$\frac{\partial \mathbf{A}_{\mathrm{xc}}(\mathbf{r},t)}{\partial t} = -\nabla V_{\mathrm{xc}}^{\mathrm{ALDA}}(\mathbf{r},t) + \frac{\nabla \cdot \boldsymbol{\sigma}_{\mathrm{xc}}(\mathbf{r},t)}{n(\mathbf{r},t)}, \tag{31}$$

where the viscoelastic xc stress tensor is given by

$$\begin{aligned}\sigma_{\mathrm{xc},\mu\nu}(\mathbf{r},t) = &\int_{-\infty}^{t} \eta_{\mathrm{xc}}(\mathbf{r},t,t')\left[\nabla_\mu v_\nu(\mathbf{r},t') - \nabla_\nu v_\mu(\mathbf{r},t') - \frac{2}{3}\nabla\cdot\mathbf{v}(\mathbf{r},t')\delta_{\mu\nu}\right] \\ &+ \int_{-\infty}^{t} \zeta_{\mathrm{xc}}(\mathbf{r},t,t')\nabla\cdot\mathbf{v}(\mathbf{r},t')\delta_{\mu\nu}\,. \end{aligned}\tag{32}$$

Here, $\mathbf{v}(\mathbf{r},t) = \mathbf{j}(\mathbf{r},t)/n(\mathbf{r},t)$ is the time-dependent velocity field, and η_{xc} and ζ_{xc} are Fourier transforms of the complex, frequency-dependent viscosity coefficients of the homogeneous electron liquid [62]. The xc functional (31) is valid up to second order in the spatial derivatives of the velocity field; notice that the *gradients* have to be small, but the velocities themselves can be large.

The explicit memory dependence of the xc vector potential (31) causes the dynamics to be dissipative, as long as the velocity field $\mathbf{v}(\mathbf{r},t)$ is nonuniform. There have been several applications of TDCDFT in the linear-response regime for dissipative electron dynamics in the areas of quantum transport [63, 64], the stopping power of metallic conduction electrons for slow ionic impact [65], and plasmon damping in doped quantum wells [66, 67]. The xc vector potential (31) has also been applied in the real-time, nonlinear regime [68, 69].

In general, TDCDFT performs well if the system behaves in a "hydrodynamic" way, that is, the electronic velocity fields are relatively smooth, without too much local compression or rarefaction. In other words, plasmonics is the ideal field of application. Bulk insulators or finite systems such as atoms or molecules are generally not well described with the xc functional (31) [70–72].

A subtle issue arises in the damping of plasmons in low-dimensional systems such as the intersubband plasmons discussed in Fig. 2. Doped quantum wells are quasi-two-dimensional systems, and electronic viscosity coefficients that come from the three-dimensional homogeneous electron liquid may lead to overdamping, since the two-dimensional in-plane particle-hole excitations offer fewer degrees of freedom to couple into. Some recent efforts have been dedicated to treat electronic dissipation in low-dimensional systems beyond the local approximation [73, 74].

5.3 Other Approaches to Treat Dissipation Within TDDFT

Let us briefly mention some other methodologies to describe dissipation in combination with TDDFT.

- *TDDFT for weakly disordered systems.* In this approach, TDDFT in the linear-response regime is combined with a microscopic, ab-initio description of disorder scattering (arising from neutral, charged, or magnetic impurities, defects, or surface roughness) [75, 76]. Applications are in the calculations of plasmon linewidths [67] and transport properties in doped or magnetic semiconductors [77]. No extension into the nonlinear or real-time regime exists at present.
- *Master equation approach.* This formalism is based on the following equation of motion of the statistical density matrix $\boldsymbol{\rho}$ of the system:

$$\frac{\partial}{\partial t}\boldsymbol{\rho} = -[H, \boldsymbol{\rho}] - \mathscr{R}[\boldsymbol{\rho}] \,. \tag{33}$$

Microscopic expressions for the relaxation matrix $\mathscr{R}[\boldsymbol{\rho}]$ can be derived from first principles under some simplifying assumptions such as perturbative, time-averaged treatment of collisions, and no memory (Markov approximation). It is possible to prove a generalized Runge-Gross theorem for N-electron systems that evolve under a master equation [78]. Some applications of the TDDFT-master equation approach exist in molecular transport under weak bias.

In the nonlinear regime of strongly driven systems, it is convenient and instructive to work with a simplified treatment of dissipation in the form of phenomenological parameters T_1 and T_2. Specifically, for a two-level system the density matrix has a simple 2×2 form. Assuming that the time evolution of the system is described by the wave function $\phi(t) = c_1(t)\phi_1 + c_2(t)\phi_2$, where ϕ_1 and ϕ_2 are the wave functions of the two levels, the components of the density matrix are $\rho_{ij}(t) = c_i^*(t)c_j(t)$, where $i, j = 1, 2$. We then have

$$\mathscr{R}[\boldsymbol{\rho}] = \begin{pmatrix} \dfrac{\rho_{11} - \rho_{11}^0}{T_1} & \dfrac{\rho_{12} - \rho_{12}^0}{T_2} \\ \dfrac{\rho_{21} - \rho_{21}^0}{T_2} & \dfrac{\rho_{22} - \rho_{22}^0}{T_1} \end{pmatrix}, \tag{34}$$

where the superscript "0" indicates the value at the initial time. Here, T_1 is the population relaxation time and T_2 is the dephasing time. In the weak excitation limit, $1/T_2$ shows up as the linewidth of an excitation. But the approach can also be applied in the nonlinear, strongly driven regime [79].

- *TDDFT for open and stochastic systems.* Existence theorems have been proved by D'Agosta and Di Ventra for interacting many-particle systems interacting with arbitrary external baths [80, 81], which extends the applicability of TD(C)DFT to open quantum systems. In practice, a generalized time-dependent Kohn-Sham equation of motion is solved multiple times in random stochastic potentials, and the system property of interest is obtained through an averaging procedure. The applicability of the method has been demonstrated for dissipative quantum molecular dynamics [82, 83]. A similar approach by Aspuru-Guzik and coworkers [84] has been applied to study the dynamics of atoms coupled to heat baths.

6 Concluding Remarks

Today, the main area in which TDDFT is used is in theoretical chemistry, for calculating excitation energies and optical properties of large molecular systems. It has become a standard feature of many popular computer codes in quantum chemistry and materials science, and as such has reached a similar status for electron dynamics as DFT has for ground-state properties.

In this article, we have presented the TDDFT formalism and its features and capabilities in a nutshell, with a special emphasis on successes and challenges in the nonlinear regime and for matter under extreme conditions. Even though it represents an approach that is in principle exact, in practice there are some things that TDDFT (with the currently available, state-of-the-art approximations and implementations) can do well, and other situations in which it faces difficulties. In particular, observables that are implicit functionals of the density remain a challenge, and continue to be the subject of active research and development. The description of dissipation in TDDFT is a subject of active research as well, although much progress has been made in our understanding and in constructing the appropriate formal framework.

In spite of these challenges, there can be no doubt that TDDFT is the only method that is capable of dealing with the nonlinear, real-time dynamics of many interacting electrons driven by arbitrary external fields, or coupled to the ions via adiabatic or nonadiabatic molecular dynamics. Large-scale TDDFT numerical simulations of materials are now within reach, such as the recent study by Yabana and coworkers [85] who simulated high-intensity fs laser pulses acting on crystalline solids and leading to dielectric breakdown. As our computational capabilities continue to increase, we can expect many more such applications of TDDFT in the area of materials science under a broad variety of extreme (or not so extreme) conditions.

Acknowledgements This work was supported by NSF Grant No. DMR-1005651.

References

1. N.D. Mermin, Phys. Rev. **137**, A1441 (1965)
2. C.A. Ullrich, *Time-Dependent Density-Functional Theory: Concepts and Applications* (Oxford University Press, Oxford/New York, 2012)

3. M.A.L. Marques, N.T. Maitra, F.M.S. Nogueira, E.K.U. Gross, A. Rubio (eds.), *Fundamentals of Time-Dependent Density Functional Theory* (Springer, Berlin, 2012)
4. A. Di Piazza, C. Müller, K.Z. Hatsagortsyan, C.H. Keitel, Rev. Mod. Phys. **84**, 1177 (2012)
5. K. Harumiya, H. Kono, Y. Fujimura, I. Kawata, A. D. Bandrauk, Phys. Rev. A **66**, 043403 (2002)
6. N. Helbig, J.I. Fuks, M. Casula, M.J. Verstraete, M.A.L. Marques, I.V. Tokatly, A. Rubio, Phys. Rev. A **83**, 032503 (2011)
7. E. Runge, E.K.U. Gross, Phys. Rev. Lett. **52**, 997 (1984)
8. E. Engel, R.M. Dreizler, *Density Functional Theory: An Advanced Course* (Springer, Berlin, 2012)
9. P. Hohenberg, W. Kohn, Phys. Rev. **136**, B864 (1964)
10. W. Kohn, L.J. Sham, Phys. Rev. **140**, A1133 (1965)
11. R. van Leeuwen, Phys. Rev. Lett. **80**, 1280 (1998)
12. G. Vignale, Phys. Rev. A **77**, 062511 (2008)
13. M. Lein, S. Kümmel, Phys. Rev. Lett. **94**, 143003 (2005)
14. M.A.L. Marques, A. Castro, G.F. Bertsch, A. Rubio, Comput. Phys. Commun. **151**, 60 (2003)
15. A. Castro, H. Appel, M. Oliveira, C.A. Rozzi, X. Andrade, F. Lorenzen, M.A.L. Marques, E.K.U. Gross, A. Rubio, Phys. Stat. Sol. B **243**, 2465 (2006)
16. X. Andrade, J. Alberdi-Rodriguez, D.A. Strubbe, M.J.A. Oliveira, F. Nogueira, A. Castro, J. Muguerza, A. Arruabarrena, S.G. Louie, A. Aspuru-Guzik, A. Rubio, M.A.L. Marques, J. Phys. Condens. Matter **24**, 233202 (2011)
17. J.M. Soler, E. Artacho, J.D. Gale, A. García, J. Junquera, P. Ordejón, D. Sánchez-Portal, J. Phys. Condens. Matter **14**, 2745 (2002)
18. E. Artacho, E. Anglada, O. Diéguez, J.D. Gale, A. García, J. Junquera, R.M. Martin, P. Ordejón, J.M. Pruneda, D. Sánchez-Portal, J. M. Soler, J. Phys. Condens. Matter **20**, 064208 (2008)
19. N.T. Maitra, F. Zhang, R.J. Cave, K. Burke, J. Chem. Phys. **120**, 5932 (2004)
20. C.A. Ullrich, J. Chem. Phys. **125**, 234108 (2006)
21. D. Sangalli, P. Romaniello, G. Onida, A. Marini, J. Chem. Phys. **134**, 034115 (2011)
22. D.C. Langreth, J.P. Perdew, Phys. Rev. B **15**, 2884 (1977)
23. D.N. Fittinghoff, P.R. Bolton, B. Chang, K.C. Kulander, Phys. Rev. Lett. **69**, 2642 (1992)
24. B. Walker, B. Sheehy, L.F. DiMauro, P. Agostini, K.J. Schafer, K.C. Kulander, Phys. Rev. Lett. **73**, 1227 (1994)
25. P.B. Corkum, Phys. Rev. Lett. **71**, 1994 (1993)
26. T. Yanai, D.P. Tew, N.C. Handy, Chem. Phys. Lett. **393**, 51 (2004)
27. R. Baer, E. Livshits, U. Salzner, Annu. Rev. Phys. Chem. **61**, 85 (2010)
28. G. Onida, L. Reining, A. Rubio, Rev. Mod. Phys. **74**, 601 (2002)
29. Z. Yang, Y. Li, C.A. Ullrich, J. Chem. Phys. **137**, 014513 (2012)
30. Z. Yang, C.A. Ullrich, Phys. Rev. B **87**, 195204 (2013)
31. C.A. Ullrich, I.V. Tokatly, Phys. Rev. B **73**, 235102 (2006)
32. M. Thiele, E.K.U. Gross, S. Kümmel, Phys. Rev. Lett. **100**, 153004 (2008)
33. M. Thiele, S. Kümmel, Phys. Rev. A **79**, 052503 (2009)
34. I.V. Tokatly, Phys. Rev. B **71**, 165104 (2005); Ibid. 165105 (2005)
35. C.A. Ullrich, J. Mol. Struct. (Theochem) **501–502**, 315 (2000)
36. D. Lappas, R. van Leeuwen, J. Phys. B: At. Mol. Opt. Phys. **31**, L249 (1998)
37. F. Wilken, D. Bauer, Phys. Rev. Lett. **97**, 203001 (2006)
38. A. Pohl, P.-G. Reinhard, E. Suraud, Phys. Rev. Lett. **84**, 5090 (2000)
39. H.S. Nguyen, A.D. Bandrauk, C.A. Ullrich, Phys. Rev. A **69**, 063415 (2004)
40. V. Véniard, R. Taïeb, A. Maquet, Laser Phys. **13**, 465 (2003)
41. N. Rohringer, S. Peter, J. Burgdörfer, Phys. Rev. A **74**, 042512 (2004)
42. L. Lam, P.M. Platzman, Phys. Rev. B **9**, 5122 (1974)
43. G.E.W. Bauer, Phys. Rev. B **27**, 5912 (1983)
44. F. Wilken, D. Bauer, Phys. Rev. A **76**, 023409 (2007)
45. Y. Li, C.A. Ullrich, Chem. Phys. **391**, 157 (2011)
46. T. Kreibich, E.K.U. Gross, Phys. Rev. Lett. **86**, 2984 (2001)

47. O. Butriy, H. Ebadi, P.L. de Boeij, R. van Leeuwen, E.K.U. Gross, Phys. Rev. A **76**, 052514 (2007)
48. T. Kreibich, R. van Leeuwen, E.K.U. Gross, Phys. Rev. A **78**, 022501 (2008)
49. J.C. Tully, J. Chem. Phys. **93**, 1061 (1990)
50. J.C. Tully, J. Chem. Phys. **137**, 22A301 (2012)
51. P. Harrison, *Quantum Wells, Wires and Dots* (Wiley, Chichester, 2005)
52. R. D'Agosta, G. Vignale, Phys. Rev. Lett. **96**, 016405 (2006)
53. E.K.U. Gross, W. Kohn, Phys. Rev. Lett. **55**, 2850 (1985); Erratum: Ibid. **57**, 923 (1986)
54. J.F. Dobson, G.H. Harris, A.J. O'Connor, J. Phys. Condens. Matter **2**, 6461 (1990)
55. G. Vignale, in *Time-Dependend Density Functional Theory*, ed. by M.A.L. Marques, C.A. Ullrich, F. Nogueira, A. Rubio, K. Burke, E.K.U. Gross. Lecture Notes in Physics, vol 706 (Springer, Berlin, 2006), p. 75
56. J.F. Dobson, M.J. Bünner, E.K.U. Gross, Phys. Rev. Lett. **79**, 1905 (1997)
57. G. Vignale, M. Rasolt, Phys. Rev. Lett. **59**, 2360 (1987)
58. G. Vignale, M. Rasolt, Phys. Rev. B **37**, 10685 (1988)
59. S.K. Ghosh, A.K. Dhara, Phys. Rev. A **38**, 1149 (1988)
60. G. Vignale, Phys. Rev. B **70**, 201102 (2004)
61. G. Vignale, W. Kohn, Phys. Rev. Lett. **77**, 2037 (1996)
62. G. Vignale, C.A. Ullrich, S. Conti, Phys. Rev. Lett. **79**, 4878 (1997)
63. N. Sai, M. Zwolak, G. Vignale, M. Di Ventra, Phys. Rev. Lett. **94**, 186810 (2005)
64. D. Roy, G. Vignale, M. Di Ventra, Phys. Rev. B **83**, 075428 (2011)
65. V.U. Nazarov, J.M. Pitarke, Y. Takada, G. Vignale, Y.-C. Chang, Phys. Rev. B **76**, 205103 (2007)
66. C.A. Ullrich, G. Vignale, Phys. Rev. B **58**, 15756 (1998)
67. C.A. Ullrich, G. Vignale, Phys. Rev. Lett. **87**, 037402 (2001)
68. H.O. Wijewardane, C.A. Ullrich, Phys. Rev. Lett. **95**, 086401 (2005)
69. C.A. Ullrich, J. Chem. Phys. **125**, 234108 (2006)
70. C.A. Ullrich, K. Burke, J. Chem. Phys. **121**, 28 (2004)
71. M. van Faassen, Int. J. Mod. Phys. B **20**, 3419 (2006)
72. J.A. Berger, P.L. de Boeij, R. van Leeuwen, Phys. Rev. B **75**, 035116 (2007)
73. R. D'Agosta, M.D. Ventra, G. Vignale, Phys. Rev. B **76**, 035320 (2007)
74. I. D'Amico and C.A. Ullrich, Phys. Rev. B **88**, 155324 (2013)
75. C.A. Ullrich, G. Vignale, Phys. Rev. B **65**, 245102 (2002)
76. F.V. Kyrychenko, C.A. Ullrich, J. Phys. Condens. Matter **21**, 084202 (2009)
77. F.V. Kyrychenko, C.A. Ullrich, Phys. Rev. B **83**, 205206 (2011)
78. K. Burke, R. Car, R. Gebauer, Phys. Rev. Lett. **94**, 146803 (2005)
79. H.O. Wijewardane, C.A. Ullrich, Appl. Phys. Lett. **84**, 3984 (2004)
80. M. Di Ventra, R. D'Agosta, Phys. Rev. Lett. **98**, 226403 (2007)
81. R. D'Agosta, M. Di Ventra, Phys. Rev. B **78**, 165105 (2008)
82. H. Appel, M. Di Ventra, Phys. Rev. B **80**, 212303 (2009)
83. F. Biele, R. D'Agosta, J. Phys. Condens. Matter **24**, 273201 (2012)
84. J. Yuen-Zhou, D.G. Tempel, C.A. Rodríguez-Rosario, A. Aspuru-Guzik, Phys. Rev. Lett. **104**, 043001 (2010)
85. K. Yabana, S. Sugiyama, Y. Shinohara, T. Otobe, G.F. Bertsch, Phys. Rev. B **85**, 045134 (2012)

Thermal Density Functional Theory in Context

Aurora Pribram-Jones, Stefano Pittalis, E.K.U. Gross, and Kieron Burke

Abstract This chapter introduces thermal density functional theory, starting from the ground-state theory and assuming a background in quantum mechanics and statistical mechanics. We review the foundations of density functional theory (DFT) by illustrating some of its key reformulations. The basics of DFT for thermal ensembles are explained in this context, as are tools useful for analysis and development of approximations. This review emphasizes thermal DFT's strengths as a consistent and general framework.

1 Introduction

The subject matter of high-energy-density physics is vast [1], and the various methods for modeling it are diverse [2–4]. The field includes enormous temperature, pressure, and density ranges, reaching regimes where the tools of plasma physics are appropriate [5]. But, especially nowadays, interest also stretches down to warm dense matter (WDM), where chemical details can become not just relevant, but vital [6]. WDM, in turn, is sufficiently close to zero-temperature, ground-state electronic structure that the methods from that field, especially Kohn-Sham density

A. Pribram-Jones • K. Burke
Department of Chemistry, University of California, Irvine, CA 92697, USA
e-mail: apribram@uci.edu

S. Pittalis
Department of Chemistry, University of California, Irvine, CA 92697, USA

CNR–Istituto Nanoscienze, Centro S3, Via Campi 213a, I-41125, Modena, Italy

E.K.U. Gross
Max-Planck-Institut für Mikrostrukturphysik, Weinberg 2, D-06120 Halle, Germany

F. Graziani et al. (eds.), *Frontiers and Challenges in Warm Dense Matter*, Lecture Notes in Computational Science and Engineering 96,
DOI 10.1007/978-3-319-04912-0_2,

functional theory (KS DFT) [7, 8], provide a standard paradigm for calculating material-specific properties with useful accuracy.

It is important to understand, from the outset, that the logic and methodology of KS-DFT is at times foreign to other techniques of theoretical physics. The procedures of KS-DFT appear simple, yet the underlying theory is surprisingly subtle. Consequently, progress in developing useful approximations, or even writing down formally correct expressions, has been incredibly slow. As the KS methodology develops in WDM and beyond, it is worth taking a few moments to wrap one's head around its logic, as it does lead to one of the most successful paradigms of modern electronic structure theory [9].

This chapter sketches how the methodology of KS DFT can be generalized to warm systems, and what new features are introduced in doing so. It is primarily designed for those unfamiliar with DFT to get a general understanding of how it functions and what promises it holds in the domain of warm dense matter. Section 2 is a general review of the basic theorems of DFT, using the original methodology of Hohenberg-Kohn [10] and then the more general Levy-Lieb construction [11, 12]. In Sect. 3, we discuss approximations, which are always necessary in practice, and several important exact conditions that are used to guide their construction. In Sect. 4, we review the thermal KS equations [13] and some relevant statistical mechanics. Section 5 summarizes some of the most important exact conditions for thermal ensembles [14, 15]. Last, but not least, in Sect. 6 we review some recent results that generalize ground-state exact scaling conditions and note some of the main differences between the finite-temperature and the ground-state formulation.

2 Density Functional Theory

A reformulation of the interacting many-electron problem in terms of the electron density rather than the many-electron wavefunction has been attempted since the early days of quantum mechanics [16–18]. The advantage is clear: while the wavefunction for interacting electrons depends in a complex fashion on all the particle coordinates, the particle density is a function of only three spatial coordinates.

Initially, it was believed that formulating quantum mechanics solely in terms of the particle density gives only an approximate solution, as in the Thomas-Fermi method [16–18]. However, in the mid-1960s, Hohenberg and Kohn [10] showed that, for systems of electrons in an external potential, all the properties of the many-electron ground state are, in principle, exactly determined by the ground-state particle density alone.

Another important approach to the many-particle problem appeared early in the development of quantum mechanics: the single-particle approximation. Here, the two-particle potential representing the interaction between particles is replaced by some effective, one-particle potential. A prominent example of this approach is the Hartree-Fock method [19, 20], which includes only exchange contributions in its effective one-particle potential. A year after the Hohenberg-Kohn theorem had been proven, Kohn and Sham [21] took a giant leap forward. They took the ground state particle density as the basic quantity and showed that both exchange and correlation effects due to the electron-electron interaction can be treated through an effective single-particle Schrödinger equation. Although Kohn and Sham wrote their paper using the local density approximation, they also pointed out the exactness of that scheme if the exact exchange-correlation functional were to be used (see Sect. 2.3). The KS scheme is used in almost all DFT calculations of electronic structure today. Much development in this field remains focused on improving approximations to the exchange-correlation energy (see Sect. 3).

The Hohenberg-Kohn theorem and Kohn-Sham scheme are the basic elements of modern density-functional theory (DFT) [9, 22, 23]. We will review the initial formulation of DFT for non-degenerate ground states and its later extension to degenerate ground states. Alternative and refined mathematical formulations are then introduced.

2.1 Introduction

The non-relativistic Hamiltonian[1] for N interacting electrons[2] moving in a static potential $v(\mathbf{r})$ reads (in atomic units)

$$\hat{H} = \hat{T} + \hat{V}_{ee} + \hat{V} := -\frac{1}{2}\sum_{i=1}^{N} \nabla_i^2 + \frac{1}{2}\sum_{\substack{i,j=1 \\ i \neq j}}^{N} \frac{1}{|\mathbf{r}_i - \mathbf{r}_j|} + \sum_{i=1}^{N} v(\mathbf{r}_i). \tag{1}$$

Here, $\hat{T}$ is the total kinetic-energy operator, $\hat{V}_{ee}$ describes the repulsion between the electrons, and $\hat{V}$ is a local (multiplicative) scalar operator. This includes the interaction of the electrons with the nuclei (considered within the Born-Oppenheimer approximation) and any other external scalar potentials.

The eigenstates, $\Psi_i(\mathbf{r}_1, \ldots, \mathbf{r}_N)$, of the system are obtained by solving the eigenvalue problem

$$\hat{H}\Psi_i(\mathbf{r}_1, \ldots, \mathbf{r}_N) = E_i\Psi_i(\mathbf{r}_1, \ldots, \mathbf{r}_N), \tag{2}$$

with appropriate boundary conditions for the physical problem at hand. Equation (2) is the time-independent Schrödinger equation. We are particularly interested in the ground state, the eigenstate with lowest energy, and assume the wavefunction can be normalized.

[1] See Refs. [24] or [25] for quantum mechanical background that is useful for this chapter.

[2] In this work, we discuss only spin-unpolarized electrons.

Due to the interactions among the electrons, $\hat{V}_{ee}$, an explicit and closed solution of the many-electron problem in Eq. (2) is, in general, not possible. But because accurate prediction of a wide range of physical and chemical phenomena requires inclusion of electron-electron interaction, we need a path to accurate approximate solutions.

Once the number of electrons with Coulombic interaction is given, the Hamiltonian is determined by specifying the external potential. For a given $v(\mathbf{r})$, the total energy is a functional of the many-body wavefunction $\Psi(\mathbf{r}_1, \ldots, \mathbf{r}_N)$:

$$E_v[\Psi] = \langle \Psi | \hat{T} + \hat{V}_{ee} + \hat{V} | \Psi \rangle \ . \tag{3}$$

The energy functional in Eq. (3) may be evaluated for any N-electron wavefunction, and the Rayleigh-Ritz variational principle ensures that the ground state energy, E_v, is given by

$$E_v = \inf_{\Psi} E_v[\Psi], \tag{4}$$

where the infimum is taken over all normalized, antisymmetric wavefunctions. The Euler-Lagrange equation expressing the minimization of the energy is

$$\frac{\delta}{\delta \Psi} \{ E_v[\Psi] - \mu \left[\langle \Psi | \Psi \rangle - 1 \right] \} = 0, \tag{5}$$

where the functional derivative is performed over $\Psi \in \mathscr{L}^2(\mathbb{R}^{3N})$ (defined as in Ref. [26]). Relation (5) again leads to the many-body Schrödinger equation and the Lagrangian multiplier μ can be identified as the chemical potential.

We now have a procedure for finding approximate solutions by restricting the form of the wavefunctions. In the Hartree-Fock (HF) approximation, for example, the form of the wave-function is restricted to a single Slater determinant. Building on the HF wavefunction, modern quantum chemical methods can produce extremely accurate solutions to the Schrödinger equation [27]. Unfortunately, wavefunction-based approaches that go beyond HF usually are afflicted by an impractical growth of the numerical effort with the number of particles. Inspired by the Thomas-Fermi approach, one might wonder if the role played by the wavefunction could be played by the particle density, defined as

$$n(\mathbf{r}) := \langle \Psi | \sum_{i=1}^{N} \delta(\hat{r} - \hat{r}_i) | \Psi \rangle = N \int d\mathbf{r}_2 \ldots \int d\mathbf{r}_N \left| \Psi(\mathbf{r}, \mathbf{r}_2, \ldots, \mathbf{r}_N) \right|^2, \tag{6}$$

from which

$$\int d^3r \, n(\mathbf{r}) = N. \tag{7}$$

In that case, one would deal with a function of only three spatial coordinates, regardless of the number of electrons.

2.2 Hohenberg-Kohn Theorem

Happily, the two-part Hohenberg-Kohn (HK) Theorem assures us that the electronic density alone is enough to determine all observable quantities of the systems. These proofs cleverly connect specific sets of densities, wavefunctions, and potentials, exposing a new framework for the interacting many-body problem.

Let **P** be the set of external potentials leading to a *non-degenerate* ground state for N electrons. For a given potential, the corresponding ground state, Ψ, is obtained through the solution of the Schrödinger equation:

$$v \quad \longrightarrow \quad \Psi, \ \text{with} \ v \in \mathbf{P}. \tag{8}$$

Wavefunctions obtained this way are called interacting v-representable. We collect these ground state wavefunctions in the set **W**. The corresponding particle densities can be computed using definition (6):

$$\Psi \quad \longrightarrow \quad n, \ \text{with} \ \Psi \in \mathbf{W}. \tag{9}$$

Ground state particle densities obtained this way are also called interacting v-representable. We denote the set of these densities as **D**.

2.2.1 First Part

Given a density $n \in \mathbf{D}$, the first part of the Hohenberg-Kohn theorem states that the wavefunction $\Psi \in \mathbf{W}$ leading to n is unique, apart from a constant phase factor. The proof is carried out by *reductio ad absurdum* and is illustrated in Fig. 1.

Consider two different wavefunctions in **W**, Ψ_1 and Ψ_2, that differ by more than a constant phase factor. Next, let n_1 and n_2 be the corresponding densities computed by Eq. (6). Since, by construction, we are restricting ourselves to non-degenerate ground states, Ψ_1 and Ψ_2 must come from two different potentials. Name these v_1 and v_2, respectively.

Assume that these different wavefunctions yield the same density:

$$\Psi_1 \neq \Psi_2 \ \text{but} \ n_1(\mathbf{r}) = n_2(\mathbf{r}). \tag{10}$$

Application of the Rayleigh-Ritz variational principle yields the inequality

$$\langle \Psi_1 | \hat{H}_1 | \Psi_1 \rangle < \langle \Psi_2 | \hat{H}_1 | \Psi_2 \rangle, \tag{11}$$

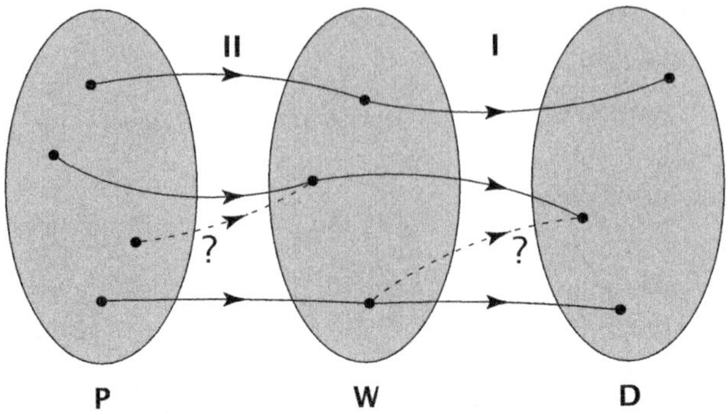

Fig. 1 The Hohenberg-Kohn Theorem proves the one-to-one mappings between potentials and ground-state wavefunctions and between ground-state wavefunctions and ground-state densities. The dotted lines indicated by question marks show the two-to-one mappings disproved by Hohenberg and Kohn [26, 28]

from which we obtain

$$E_1 < \langle \Psi_2 | \hat{H}_2 + (\hat{V}_1 - \hat{V}_2) | \Psi_2 \rangle = E_2 + \int d^3r\, n_1(\mathbf{r})\, [v_1(\mathbf{r}) - v_2(\mathbf{r})]\,. \tag{12}$$

Reversing the role of systems 1 and 2 in the derivation, we find

$$E_2 < \langle \Psi_1 | \hat{H}_1 + (\hat{V}_2 - \hat{V}_1) | \Psi_1 \rangle = E_1 + \int d^3r\, n_2(\mathbf{r})\, [v_2(\mathbf{r}) - v_1(\mathbf{r})]\,. \tag{13}$$

The assumption that the two densities are equal, $n_1(\mathbf{r}) = n_2(\mathbf{r})$, and addition of the inequalities (12) and (13) yields

$$E_1 + E_2 < E_1 + E_2, \tag{14}$$

which is a contradiction. We conclude that the foregoing hypothesis (10) was wrong, so $n_1 \neq n_2$. Thus each density is the ground-state density of, at most, one wavefunction. This mapping between the density and wavefunction is written

$$n \longrightarrow \Psi, \text{ with } n \in \mathbf{D} \text{ and } \Psi \in \mathbf{W}. \tag{15}$$

2.2.2 Second Part

Having specified the correspondence between density and wavefunction, Hohenberg and Kohn then consider the potential. By explicitly inverting the Schrödinger equation,

$$\sum_{i=1}^{N} v(\mathbf{r}_i) = E - \frac{\left(\hat{T} + \hat{V}_{ee}\right) \Psi(\mathbf{r}_1, \mathbf{r}_2, \ldots, \mathbf{r}_N)}{\Psi(\mathbf{r}_1, \mathbf{r}_2, \ldots, \mathbf{r}_N)}, \tag{16}$$

they show the elements Ψ of **W** also determine the elements v of **P**, apart from an additive constant.

We summarize this second result by writing

$$\Psi \longrightarrow v, \text{ with } \Psi \in \mathbf{W} \text{ and } v \in \mathbf{P}. \tag{17}$$

2.2.3 Consequences

Together, the first and second parts of the theorem yield

$$n \longrightarrow v + const, \quad \text{with } n \in \mathbf{D} \text{ and } v \in \mathbf{P}, \tag{18}$$

that the ground state particle density determines the external potential up to a trivial additive constant. This is the first HK theorem.

Moreover, from the first part of the theorem it follows that any ground-state observable is a functional of the ground-state particle density. Using the one-to-one dependence of the wavefunction, $\Psi[n]$, on the particle density,

$$\langle \Psi | \hat{O} | \Psi \rangle = \langle \Psi[n] | \hat{O} | \Psi[n] \rangle = O[n]. \tag{19}$$

For example, the following functional can be defined:

$$E_{v,\mathrm{HK}}[n] := \langle \Psi[n] | \hat{T} + \hat{V}_{ee} + \hat{V} | \Psi[n] \rangle = F_{\mathrm{HK}}[n] + \int d^3r \, n(\mathbf{r}) v(\mathbf{r}), \tag{20}$$

where v is a given external potential and n can be any density in **D**. Note that

$$F_{\mathrm{HK}}[n] := \langle \Psi[n] | \hat{T} + \hat{V}_{ee} | \Psi[n] \rangle \tag{21}$$

is independent of v. The second HK theorem is simply that $F_{\mathrm{HK}}[n]$ is independent of $v(\mathbf{r})$. This is therefore a universal functional of the ground-state particle density. We use the subscript, HK, to emphasize that this is the original density functional of Hohenberg and Kohn.

Let n_0 be the ground-state particle density of the potential v_0. The Rayleigh-Ritz variational principle (4) immediately tells us

$$E_{v_0} = \min_{n \in \mathbf{D}} E_{v_0,\mathrm{HK}}[n] = E_{v_0,\mathrm{HK}}[n_0]. \tag{22}$$

We have finally obtained a variational principle based on the particle density instead of the computationally expensive wavefunction.

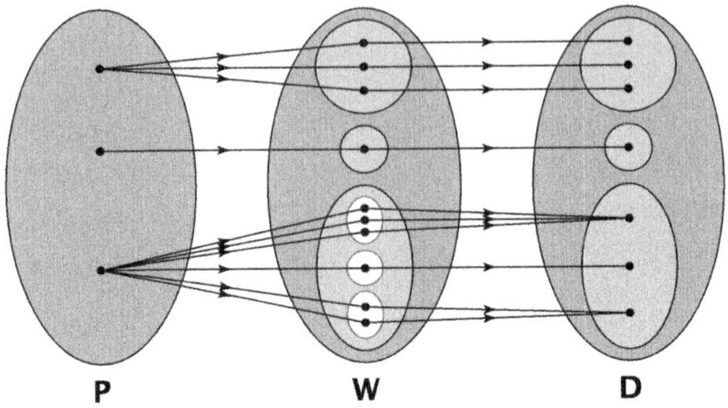

Fig. 2 The mappings between sets of potentials, wavefunctions, and densities can be extended to include potentials with degenerate ground states. This is seen in the one-to-many mappings between **P** and **W**. Note also the many-to-one mappings from **W** to **D** caused by this degeneracy [28, 30]

2.2.4 Extension to Degenerate Ground States

The Hohenberg-Kohn theorem can be generalized by allowing **P** to include local potentials having *degenerate* ground states [11, 28, 29]. This means an entire subspace of wavefunctions can correspond to the lowest eigenvalue of the Schrödinger equation (2). The sets **W** and **D** are enlarged accordingly, to include all the additional ground-state wavefunctions and particle densities.

In contrast to the non-degenerate case, the solution of the Schrödinger equation (2) now establishes a mapping from **P** to **W** which is one-to-many (see Fig. 2). Moreover, different degenerate wavefunctions can have the same particle density. Equation (6), therefore, establishes a mapping from **W** to **D** that is many-to-one. However, any one of the degenerate ground-state densities still determines the potential uniquely.

The first part of the HK theorem needs to be modified in light of this alteration of the mapping between wavefunctions and densities. To begin, note that two degenerate subspaces, sets of ground states of two different potentials, are disjoint. Assuming that a common eigenstate Ψ can be found, subtraction of one Schrödinger equation from the other yields

$$(\hat{V}_1 - \hat{V}_2)\Psi = (E_1 - E_2)\Psi. \tag{23}$$

For this identity to be true, the eigenstate Ψ must vanish in the region where the two potentials differ by more than an additive constant. This region has measure greater than zero. Eigenfunctions of potentials in **P**, however, vanish only on sets of measure zero [31]. This contradiction lets us conclude that v_1 and v_2 cannot have common eigenstates. We then show that ground states from two different potentials always have different particle densities using the Rayleigh-Ritz variational principle as in the non-degenerate case.

However, two or more degenerate ground state wavefunctions can have the same particle density. As a consequence, neither the wavefunctions nor a generic ground state property can be determined uniquely from knowledge of the ground state particle density alone. This demands reconsideration of the definition of the universal F_{HK} as well. Below, we verify that the definition of F_{HK} does not rely upon one-to-one correspondence among ground state wavefunctions and particle densities.

The second part of the HK theorem in this case proceeds as in the original proof, with each ground state in a degenerate level determining the external potential up to an additive constant. Combining the first and second parts of the proof again confirms that any element of **D** determines an element of **P**, up to an additive constant. In particular, any one of the degenerate densities determines the external potential. Using this fact and that the total energy is the same for all wavefunctions in a given degenerate level, we define F_{HK}:

$$F_{\mathrm{HK}}[n] := E\,[v[n]] - \int d^3r\; v[n](\mathbf{r})n(\mathbf{r}). \tag{24}$$

This implies that the value of

$$F_{\mathrm{HK}}[n] = \langle \Psi_0 \to n|\hat{T} + \hat{V}_{ee}|\Psi_0 \to n\rangle \tag{25}$$

is the same for all degenerate ground-state wavefunctions that have the same particle density. The variational principle based on the particle density can then be formulated as before in Eq. (22).

2.3 *Kohn-Sham Scheme*

The exact expressions defining F_{HK} in the previous section are only formal ones. In practice, F_{HK} must be approximated. Finding approximations that yield usefully accurate results turns out to be an extremely difficult task, so much so that pure, orbital-free approximations for F_{HK} are not pursued in most modern DFT calculations. Instead, efficient approximations can be constructed by introducing the Kohn-Sham scheme, in which a useful decomposition of F_{HK} in terms of other density functionals is introduced. In fact, the Kohn-Sham decomposition is so effective that effort on orbital-free DFT utilizes the Kohn-Sham structure, but not its explicitly orbital-dependent expressions.

Consider the Hamiltonian of N non-interacting electrons

$$\hat{H}_s = \hat{T} + \hat{V} := -\frac{1}{2}\sum_{i=1}^{N}\nabla_i^2 + \sum_{i=1}^{N} v(\mathbf{r}_i). \tag{26}$$

Mimicking our procedure with the interacting system, we group external local potentials in the set $\mathbf{P}$. The corresponding non-interacting ground state wavefunctions Ψ_s are then grouped in the set $\mathbf{W}^\mathbf{s}$, and their particle densities n_s are grouped in $\mathbf{D}^\mathbf{s}$. We can then apply the HK theorem and define the non-interacting analog of F_{HK}, which is simply the kinetic energy:

$$T_s[n_s] := E\left[v[n_s]\right] - \int d^3r\ v[n_s](\mathbf{r})n_s(\mathbf{r}). \tag{27}$$

Restricting ourselves to non-degenerate ground states, the expression in Eq. (27) can be rewritten to stress the one-to-one correspondence among densities and wavefunctions:

$$T_s[n_s] = \langle \Psi_s[n_s]|\hat{T}|\Psi_s[n_s]\rangle\ . \tag{28}$$

We now introduce a fundamental assumption: for each element n of $\mathbf{D}$, a potential v_s in $\mathbf{P}^\mathbf{s}$ exists, with corresponding ground-state particle density $n_s = n$. We call v_s the Kohn-Sham potential. In other words, interacting v-representable densities are also assumed to be non-interacting v-representable. This maps the interacting problem onto a non-interacting one.

Assuming the existence of v_s, the HK theorem applied to the class of non-interacting systems ensures that v_s is unique up to an additive constant. As a result, we find the particle density of the interacting system by solving the non-interacting eigenvalue problem, which is called the Kohn-Sham equation:

$$\hat{H}_s\Phi = E\Phi. \tag{29}$$

For non-degenerate ground states, the Kohn-Sham ground-state wavefunction is a single Slater determinant. In general, when considering degenerate ground states, the Kohn-Sham wavefunction can be expressed as a linear combination of several Slater determinants [12, 32]. There also exist interacting ground states with particle densities that can only be represented by an ensemble of non-interacting particle densities [33–37]. We will come back to this point in Sect. 2.5.

Here we continue by considering the simplest cases of non-degenerate ground states. Equation (29) can be rewritten in terms of the single-particle orbitals as follows:

$$\left[-\frac{1}{2}\nabla^2 + v_s(\mathbf{r})\right]\varphi_i(\mathbf{r}) = \epsilon_i\varphi_i(\mathbf{r})\ . \tag{30}$$

The single-particle orbitals $\varphi_i(\mathbf{r})$ are called Kohn-Sham orbitals and Kohn-Sham wavefunctions are Slater determinants of these orbitals. Via the Kohn-Sham equations, the orbitals are implicit functionals of $n(\mathbf{r})$. We emphasize that – although in DFT the particle density is the only basic variable – the Kohn-Sham orbitals are proper fermionic single-particle states. The ground-state Kohn-Sham wavefunction is obtained by occupying the N eigenstates with lowest eigenvalues. The corresponding density is

$$n(\mathbf{r}) = \sum_{i=1}^{N} n_i |\varphi_i(\mathbf{r})|^2, \tag{31}$$

with n_i the ith occupation number.

In the next section, we consider the consequences of introducing the Kohn-Sham system in DFT.

2.3.1 Exchange-Correlation Energy Functional

A large fraction of $F_{\text{HK}}[n]$ can be expressed in terms of kinetic and electrostatic energy. This decomposition is given by

$$F_{\text{HK}}[n] = T_s[n] + U[n] + E_{xc}[n] . \tag{32}$$

The first term is the kinetic energy of the Kohn-Sham system,

$$T_s[n] = -\frac{1}{2} \sum_{i=1}^{N} \int \mathrm{d}^3 r \; \varphi_i^*(\mathbf{r}) \nabla^2 \varphi_i(\mathbf{r}) . \tag{33}$$

The second is the Hartree energy (a.k.a. electrostatic self-energy, a.k.a. Coulomb energy),

$$U[n] = \frac{1}{2} \int \int d^3 r d^3 r' \, \frac{n(\mathbf{r}) n(\mathbf{r}')}{|\mathbf{r} - \mathbf{r}'|} . \tag{34}$$

The remainder is defined as the exchange-correlation energy,

$$E_{xc}[n] := F_{\text{HK}}[n] - T_s[n] - U[n] . \tag{35}$$

For systems having more than one particle, E_{xc} accounts for exchange and correlation energy contributions. Comparing Eqs. (32) and (20), the total energy density functional is

$$E_{v,\text{HK}}[n] = T_s[n] + U[n] + E_{xc}[n] + \int d^3 r \; n(\mathbf{r}) v(\mathbf{r}). \tag{36}$$

Consider now the Euler equations for the interacting and non-interacting system. Assuming the differentiability of the functionals (see Sect. 2.5), these necessary conditions for having energy minima are

$$\frac{\delta F_{\text{HK}}}{\delta n(\mathbf{r})} + v(\mathbf{r}) = 0 \tag{37}$$

and

$$\frac{\delta T_s}{\delta n(\mathbf{r})} + v_s(\mathbf{r}) = 0, \tag{38}$$

respectively. With definition (32), from Eqs. (37) and (38), we obtain

$$v_s(\mathbf{r}) = v_H[n](\mathbf{r}) + v_{xc}[n](\mathbf{r}) + v(\mathbf{r}). \tag{39}$$

Here, $v(\mathbf{r})$ is the external potential acting upon the interacting electrons, $v_H[n](\mathbf{r})$ is the Hartree potential,

$$v_H[n](\mathbf{r}) = \int d^3r' \frac{n(\mathbf{r}')}{|\mathbf{r} - \mathbf{r}'|} = \frac{\delta U}{\delta n(\mathbf{r})}, \tag{40}$$

and $v_{xc}[n](\mathbf{r})$ is the exchange-correlation potential,

$$v_{xc}[n](\mathbf{r}) = \frac{\delta E_{xc}[n]}{\delta n(\mathbf{r})}. \tag{41}$$

Through the decomposition in Eq. (32), a significant part of F_{HK} is in the explicit form of $T_s[n] + U[n]$ without approximation. Though often small, the E_{xc} density functional still represents an important part of the total energy. Its exact functional form is unknown, and it therefore must be approximated in practice. However, good and surprisingly efficient approximations exist for E_{xc}.

We next consider reformulations of DFT, which allow analysis and solution of some important technical questions at the heart of DFT. They also have a long history of influencing the analysis of properties of the exact functionals.

2.4 Levy's Formulation

An important consequence of the HK theorem is that the Rayleigh-Ritz variational principle based on the wavefunction can be replaced by a variational principle based on the particle density. The latter is valid for all densities in the set **D**, the set of v-representable densities. Unfortunately, v-representability is a condition which is not easily verified for a given function $n(\mathbf{r})$. Hence it is highly desirable to formulate the variational principle over a set of densities characterized by simpler conditions. This was provided by Levy [11] and later reformulated and extended by Lieb [12]. In this and the sections that follow, Lebesgue and Sobolev spaces are defined in the usual way [26, 38].

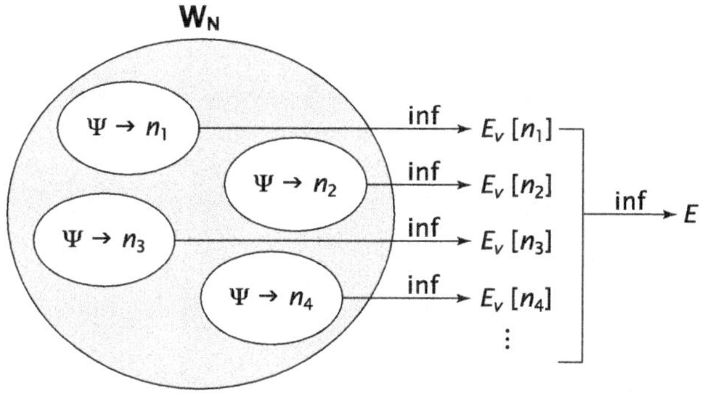

Fig. 3 This diagram shows the two-step minimization of Levy's constrained search. The first infimum search is over all wavefunctions corresponding to a certain density n_i. The second search runs over all of the densities [30, 40]

First, the set $\mathbf{W}$ is enlarged to $\mathbf{W_N}$, which includes all possible antisymmetric and normalized N-particle wavefunctions Ψ. The set $\mathbf{W_N}$ now also contains N-particle wavefunctions which are *not* necessarily ground-state wavefunctions to some external potential v, though it remains in the same Sobolev space [26] as $\mathbf{W}$: $\mathscr{H}^1(\mathbb{R}^{3N})$. Correspondingly, the set $\mathbf{D}$ is replaced by the set $\mathbf{D_N}$. $\mathbf{D_N}$ contains the densities generated from the N-particle antisymmetric wavefunctions in $\mathbf{W_N}$ using Eq. (6):

$$\mathbf{D_N} = \left\{ n \mid n(\mathbf{r}) \geq 0, \int d^3r \, n(\mathbf{r}) = N, n^{1/2}(\mathbf{r}) \in \mathscr{H}^1(\mathbb{R}^3) \right\} . \tag{42}$$

The densities of $\mathbf{D_N}$ are therefore called N-representable. Harriman's explicit construction [39] shows that any integrable and positive function $n(\mathbf{r})$ is N-representable.

Levy reformulated the variational principle in a constrained-search fashion (see Fig. 3):

$$E_v = \inf_{n \in \mathbf{D_N}} \left\{ \inf_{\Psi \to n | \Psi \in \mathbf{W_N}} \langle \Psi | \hat{T} + \hat{V}_{ee} | \Psi \rangle + \int d^3r \, n(\mathbf{r}) v(\mathbf{r}) \right\} . \tag{43}$$

In this formulation, the search inside the braces is constrained to those wavefunctions which yield a given density n – therefore the name "constrained search". The minimum is then found by the outer search over all densities. The potential $v(\mathbf{r})$ acts like a Lagrangian multiplier to satisfy the constraint on the density at each point in space. In this formulation, F_{HK} is replaced by

$$F_{\mathrm{LL}}[n] := \inf_{\Psi \to n} \langle \Psi | \hat{T} + \hat{V}_{ee} | \Psi \rangle, \quad \text{with } \Psi \in \mathbf{W_N} \text{ and } n \in \mathbf{D_N} . \tag{44}$$

The functional $E_{\rm HK}$ can then be replaced by

$$E_{v,\rm LL}[n] := F_{\rm LL}[n] + \int d^3r\; n(\mathbf{r})v(\mathbf{r}), \quad \text{with } n \in \mathbf{D_N}. \tag{45}$$

If, for a given v_0, the corresponding ground-state particle density, n_0, is inserted, then

$$E_{v_0,\rm LL}[n_0] = E_{v_0,\rm HK}[n_0] = E_{v_0}, \tag{46}$$

from which

$$F_{\rm LL}[n] = F_{\rm HK}[n], \quad \text{for all } n \in \mathbf{D}\,. \tag{47}$$

Furthermore, if any other particle density is inserted, we obtain

$$E_{v_0,\rm LL}[n] \geq E_{v_0}, \quad \text{for } n \neq n_0 \ \text{ and } n \in \mathbf{D_N}. \tag{48}$$

In this approach, the degenerate case does not require particular care. In fact, the correspondences between potentials, wavefunctions and densities are not explicitly employed as they were in the previous Hohenberg-Kohn formulation. However, the N-representability is of secondary importance in the context of the Kohn-Sham scheme. There, it is still necessary to assume that the densities of the interacting electrons are non-interacting v-representable as well. We discuss this point in more detail in the next section.

Though it can be shown that the $F_{\rm LL}[n]$ infimum is a minimum [12], the functional's lack of convexity causes a serious problem in proving the differentiability of $F_{\rm LL}$ [12]. Differentiability is needed to define an Euler equation for finding $n(\mathbf{r})$ self-consistently. This is somewhat alleviated by the Lieb formulation of DFT (see below).

2.5 Ensemble-DFT and Lieb's Formulation

In the remainder of this section, we are summarizing more extensive and pedagogical reviews that can be found in Refs. [26, 28], and [41]. Differentiability of functionals is, essentially, related to the convexity of the functionals. Levy and Lieb showed that the set $\mathbf{D}$ is not convex [12]. In fact, there exist combinations of the form

$$n(\mathbf{r}) = \sum_{k=1}^{M} \lambda_k n_k(\mathbf{r}), \; \lambda_k = 1 \;\; (0 \leq \lambda_k \leq 1), \tag{49}$$

where n_k is the density corresponding to degenerate ground state Ψ_k, that are not in **D** [12,42].

A convex set can be obtained by looking at ensembles. The density of an ensemble can be defined through the (statistical, or von Neumann) density operator

$$\hat{D} = \sum_{k=1}^{M} \lambda_k |\Psi_k\rangle\langle\Psi_k|, \text{ with } \sum_{k=1}^{M} \lambda_k = 1 \;\; (0 \leq \lambda_k \leq 1) . \tag{50}$$

The expectation value of an operator $\hat{O}$ on an ensemble is defined as

$$O := \mathrm{Tr}\left\{\hat{D}\hat{O}\right\}, \tag{51}$$

where the symbol "Tr" stands for the trace taken over an arbitrary, complete set of orthonormal N-particle states

$$\mathrm{Tr}\{\hat{D}\hat{O}\} := \sum_{k} \langle\Phi_k|(\hat{D}\hat{O})|\Phi_k\rangle. \tag{52}$$

The trace is invariant under unitary transformations of the complete set for the ground-state manifold of the Hamiltonian $\hat{H}$ (see Eq.(50)). Since

$$\mathrm{Tr}\left\{\hat{D}\hat{O}\right\} = \sum_{k=1}^{M} \lambda_k \langle\Psi_k|\hat{O}|\Psi_k\rangle, \tag{53}$$

the energy obtained from a density matrix of the form (50) is the total ground-state energy of the system.

Densities of the form (49) are called ensemble v-representable densities, or E-V-densities. We denote this set of densities as $\mathbf{D_{EV}}$. Densities that can be obtained from a single wavefunction are said to be pure-state (PS) v-representable, or PS-V-densities. The functional F_{HK} can then be extended as [43]

$$F_{\mathrm{EHK}}[n] := Tr\left\{\hat{D}\left(\hat{T} + \hat{V}_{ee}\right)\right\}, \text{ with } n \in \mathbf{D_{EV}} \tag{54}$$

where $\hat{D}$ has the form (50) and is any density matrix giving the density n. However, the set $\mathbf{D_{EV}}$, just like **D**, is difficult to characterize. Moreover, as for F_{HK} and F_{LL}, a proof of the differentiability of F_{EHK} (and for the non-interacting versions of the same functional) is not available.

In the Lieb formulation, however, differentiability can be addressed to some extent [12,44,45] . In the work of Lieb, **P** is restricted to $\mathbf{P} = \mathscr{L}^{3/2}(\mathbb{R}^3) + \mathscr{L}^{\infty}(\mathbb{R}^3)$ and wavefunctions are required to be in

$$\mathbf{W_N} = \{\Psi \mid ||\Psi|| = 1, T[\Psi] \leq \infty\} . \tag{55}$$

The universal functional is defined as

$$F_{\mathrm{L}}[n] := \inf_{\hat{D} \to n \in \mathbf{D_N}} Tr\left\{\hat{D}\left(\hat{T} + \hat{V}_{ee}\right)\right\}, \tag{56}$$

and it can be shown that the infimum is a minimum [12]. Note that in definition (56), $\hat{D}$ is a generic density matrix of the form

$$\hat{D} = \sum_k \lambda_k |\Psi_k\rangle\langle\Psi_k|, \quad \text{with} \sum_k \lambda_k = 1 \ \ (0 \leq \lambda_k \leq 1), \tag{57}$$

where $\Psi_k \in \mathbf{W_N}$. The sum is not restricted to a finite number of degenerate ground states as in Eq. (50). This minimization over a larger, less restricted set leads to the statements

$$F_{\mathrm{L}}[n] \leq F_{\mathrm{LL}}[n], \quad \text{for } n \in \mathbf{D_N}, \tag{58}$$

and

$$F_{\mathrm{L}}[n] = F_{\mathrm{LL}}[n] = F_{\mathrm{HK}}[n], \quad \text{for } n \in \mathbf{D}. \tag{59}$$

$F_{\mathrm{L}}[n]$ is defined on a convex set, and it is a convex functional. This implies that $F_{\mathrm{L}}[n]$ is differentiable at any ensemble v-representable densities and nowhere else [12]. Minimizing the functional

$$E_{\mathrm{L}}[n] := F_{\mathrm{L}}[n] + \int d^3r\, n(\mathbf{r})v(\mathbf{r}) \tag{60}$$

with respect to the elements of $\mathbf{D_{EV}}$ by the Euler-Lagrange equation

$$\frac{\delta F_{\mathrm{L}}}{\delta n(\mathbf{r})} + v(\mathbf{r}) = 0 \tag{61}$$

is therefore well-defined on the set $\mathbf{D_{EV}}$ and generates a valid energy minimum.

We finally address, although only briefly, some important points about the Kohn-Sham scheme and its rigorous justification. The results for F_{L} carry over to $T_{\mathrm{L}}[n]$. That is, the functional

$$T_{\mathrm{L}}[n] = \inf_{\hat{D} \to n} Tr\left\{\hat{D}\hat{T}\right\}, \quad \text{with } n \in \mathbf{D_N} \tag{62}$$

is differentiable at any non-interacting ensemble v-representable densities and nowhere else. We can gather all these densities in the set $\mathbf{D^s_{EV}}$. Then, the Euler-Lagrange equation

$$\frac{\delta T_{\mathrm{L}}}{\delta n(\mathbf{r})} + v_s(\mathbf{r}) = 0 \tag{63}$$

is well defined on the set $\mathbf{D}^{\mathbf{s}}_{\mathbf{EV}}$ only. One can then redefine the exchange-correlation functional as

$$E_{xc,\mathrm{L}}[n] = F_{\mathrm{L}}[n] - T_{\mathrm{L}}[n] - U[n], \tag{64}$$

and observe that the differentiability of $F_{\mathrm{L}}[n]$ and $T_{\mathrm{L}}[n]$ implies the differentiability of $E_{xc}[n]$ only on $\mathbf{D_{EV}} \cap \mathbf{D}^{\mathbf{s}}_{\mathbf{EV}}$. The question as to the size of the latter set remains. For densities defined on a discrete lattice (finite or infinite) it is known [46] that $\mathbf{D_{EV}} = \mathbf{D}^{\mathbf{s}}_{\mathbf{EV}}$. Moreover, in the continuum limit, $\mathbf{D_E}$ and $\mathbf{D}^{\mathbf{s}}_{\mathbf{E}}$ can be shown to be dense with respect to one another [12,44,45]. This implies that any element of $\mathbf{D_{EV}}$ can be approximated, with an arbitrary accuracy, by an element of $\mathbf{D}^{\mathbf{s}}_{\mathbf{EV}}$. Whether or not the two sets coincide remains an open question.

3 Functional Approximations

Numerous approximations to E_{xc} exist, each with its own successes and failures [9]. The simplest is the local density approximation (LDA), which had early success with solids[21]. LDA assumes that the exchange-correlation energy density can be approximated locally with that of the uniform gas. DFT's popularity in the chemistry community skyrocketed upon development of the generalized gradient approximation (GGA) [47]. Inclusion of density gradient dependence generated sufficiently accurate results to be useful in many chemical and materials applications.

Today, many scientists use hybrid functionals, which substitute a fraction of single-determinant exchange for part of the GGA exchange [48–50]. More recent developments in functional approximations include meta-GGAs [51], which include dependence on the kinetic energy density, and hyper-GGAs [51], which include exact exchange as input to the functional. Inclusion of occupied and then unoccupied orbitals as inputs to functionals increases their complexity and computational cost; the idea that this increase is coupled with an increase in accuracy was compared to Jacob's Ladder [51]. The best approximations are based on the exchange-correlation hole, such as the real space cutoff of the LDA hole that ultimately led to the GGA called PBE [52,53]. An introduction to this and some other exact properties of the functionals follows in the remainder of this section.

Another area of functional development of particular importance to the warm dense matter community is focused on orbital-free functionals [28, 54–57]. These approximations bypass solution of the Kohn-Sham equations by directly approximating the non-interacting kinetic energy. In this way, they recall the original, pure DFT of Thomas-Fermi theory [16–18]. While many approaches have been tried over the decades, including fitting techniques from computer science [58], no general-purpose solution of sufficient accuracy has been found yet.

3.1 Exact Conditions

Though we do not know the exact functional form for the universal functional, we do know some facts about its behavior and the relationships between its components. Collections of these facts are called exact conditions. Some can be found by inspection of the formal definitions of the functionals and their variational properties. The correlation energy and its constituents are differences between functionals evaluated on the true and Kohn-Sham systems. As an example, consider the kinetic correlation:

$$T_c[n] = T[n] - T_s[n]. \tag{65}$$

Since the Kohn-Sham kinetic energy is the lowest kinetic energy of any wavefunction with density $n(\mathbf{r})$, we know T_c must be non-negative. Other inequalities follow similarly, as well as one from noting that the exchange functional is (by construction) never positive [22]:

$$E_x \leq 0, \ E_c \leq 0, \ U_c \leq 0, \ T_c \geq 0. \tag{66}$$

Some further useful exact conditions are found by uniform coordinate scaling [59]. In the ground state, this procedure requires scaling all the coordinates of the wavefunction[3] by a positive constant γ, while preserving normalization to N particles:

$$\Psi_\gamma(\mathbf{r_1}, \mathbf{r_2}, \ldots, \mathbf{r_N}) = \gamma^{3N/2}\Psi(\gamma\mathbf{r_1}, \gamma\mathbf{r_2}, \ldots, \gamma\mathbf{r_N}), \tag{67}$$

which has a scaled density defined as

$$n_\gamma(\mathbf{r}) = \gamma^3 n(\gamma\mathbf{r}). \tag{68}$$

Scaling by a factor larger than one can be thought of as squeezing the density, while scaling by $\gamma < 1$ spreads the density out. For more details on the many conditions that can be extracted using this technique and how they can be used in functional approximations, see Ref. [22].

Of greatest interest in our context are conditions involving exchange-correlation and other components of the universal functional. Through application of the foregoing definition of uniform scaling, we can write down some simple uniform scaling equalities. Scaling the density yields

$$T_s[n_\gamma] = \gamma^2 T_s[n] \tag{69}$$

[3]Here and in the remainder of the chapter, we restrict ourselves to square-integrable wavefunctions over the domain $\mathbb{R}^{3N}$.

for the non-interacting kinetic energy and

$$E_x[n_\gamma] = \gamma\, E_x[n] \tag{70}$$

for the exchange energy. Such simple conditions arise because these functionals are defined on the non-interacting Kohn-Sham Slater determinant. On the other hand, although the density from a scaled interacting wavefunction is the scaled density, the scaled wavefunction is not the ground-state wavefunction of the scaled density. This means correlation scales less simply and only inequalities can be derived for it.

Another type of scaling that is simply related to coordinate scaling is interaction scaling, the adiabatic change of the interaction strength [60]. In the latter, the electron-electron interaction in the Hamiltonian, V_{ee}, is multiplied by a factor, λ between 0 and 1, while holding n fixed. When $\lambda = 0$, interaction vanishes. At $\lambda = 1$, we return to the Hamiltonian for the fully interacting system. Due to the simple, linear scaling of V_{ee} with coordinate scaling, we can relate it to scaling of interaction strength. Combining this idea with some of the simple equalities above leads to one of the most powerful relations in ground-state functional development, the adiabatic connection formula [61, 62]:

$$E_{xc}[n] = \int_0^1 d\lambda U_{xc}[n](\lambda), \tag{71}$$

where

$$U_{xc}[n](\lambda) = V_{ee}[\Psi^\lambda[n]] - U[n] \tag{72}$$

and $\Psi^\lambda[n]$ is the ground-state wavefunction of density n for a given λ and

$$\Psi^\lambda[n](\mathbf{r_1}, \mathbf{r_2}, \ldots, \mathbf{r_N}) = \lambda^{3N/2}\Psi[n_{1/\lambda}](\lambda\mathbf{r_1}, \lambda\mathbf{r_2}, \ldots, \lambda\mathbf{r_N}). \tag{73}$$

Interaction scaling also leads to some of the most important exact conditions for construction of functional approximations, the best of which are based on the exchange-correlation hole. The exchange-correlation hole represents an important effect of an electron sitting at a given position. All other electrons will be kept away from this position by exchange and correlation effects, due to the antisymmetry requirement and the Coulomb repulsion, respectively. This representation allows us to calculate V_{ee}, the electron-electron repulsion, in terms of an electron distribution function.[4]

To define the hole distribution function, we need first to introduce the pair density function. The pair density, $P(\mathbf{r}, \mathbf{r}')$ describes the distribution of the electron pairs. This is proportional to the the probability of finding an electron in a volume d^3r around position $\mathbf{r}$ *and* a second electron in the volume d^3r' around $\mathbf{r}'$. In terms of

[4]For a more extended discussion of these topics, see Ref. [60].

the electronic wavefunction, it is written as follows

$$P(\mathbf{r},\mathbf{r}') = N(N-1)\int d^3r_3\ldots\int d^3r_N\ |\Psi(\mathbf{r},\mathbf{r}',\ldots,\mathbf{r}_N)|^2. \tag{74}$$

We then can define the conditional probability density of finding an electron in d^3r' after having already found one at $\mathbf{r}$, which we will denote $n_2(\mathbf{r},\mathbf{r}')$. Thus

$$n_2(\mathbf{r},\mathbf{r}') = P(\mathbf{r},\mathbf{r}')/n(\mathbf{r}). \tag{75}$$

If the positions of the electrons were truly independent of one another (no electron-electron interaction and no antisymmetry requirement for the wavefunction) this would be just $n(\mathbf{r}')$, independent of $\mathbf{r}$. But this cannot be, as

$$\int d^3r'\ n_2(\mathbf{r},\mathbf{r}') = N-1. \tag{76}$$

The conditional density integrates to one fewer electron, since one electron is at the reference point. We therefore define a "hole" density:

$$n_2(\mathbf{r},\mathbf{r}') = n(\mathbf{r}') + n_{\text{hole}}(\mathbf{r},\mathbf{r}'). \tag{77}$$

which is typically negative and integrates to -1 [60]:

$$\int d^3r'\ n_{\text{hole}}(\mathbf{r},\mathbf{r}') = -1. \tag{78}$$

The exchange-correlation hole in DFT is given by the coupling-constant average:

$$n_{xc}(\mathbf{r},\mathbf{r}') = \int_0^1 d\lambda\ n^{\lambda}_{\text{hole}}(\mathbf{r},\mathbf{r}'), \tag{79}$$

where $n^{\lambda}_{\text{hole}}$ is the hole in Ψ^{λ}. So, via the adiabatic connection formula (Eq. 71), the exchange-correlation energy can be written as a double integral over the exchange-correlation hole:

$$E_{xc} = \frac{1}{2}\int d^3r\ n(\mathbf{r})\int d^3r'\ \frac{n_{xc}(\mathbf{r},\mathbf{r}')}{|\mathbf{r}-\mathbf{r}'|}. \tag{80}$$

By definition, the exchange hole is given by $n_x = n^{\lambda=0}_{\text{hole}}$ and the correlation hole, n_c, is everything *not* in n_x. The exchange hole may be readily obtained from the (ground-state) pair-correlation function of the Kohn-Sham system. Moreover $n_x(\mathbf{r},\mathbf{r}) = 0$, $n_x(\mathbf{r},\mathbf{r}') \leq 0$, and for one particle systems $n_x(\mathbf{r},\mathbf{r}') = -n(\mathbf{r}')$. If the Kohn-Sham state is a single Slater determinant, then the exchange energy assumes the form of the Fock integral evaluated with occupied Kohn-Sham orbitals. It is straightforward to verify that the exchange-hole satisfies the sum rule

$$\int d^3r' \, n_x(\mathbf{r}, \mathbf{r}') = -1 \; ; \tag{81}$$

and thus

$$\int d^3r' \, n_c(\mathbf{r}, \mathbf{r}') = 0 \, . \tag{82}$$

The correlation hole is a more complicated quantity, and its contributions oscillate from negative to positive in sign. Both the exchange and the correlation hole decay to zero at large distances from the reference position **r**.

These and other conditions on the exact hole are used to constrain exchange-correlation functional approximations. The seemingly unreasonable reliability of the simple LDA has been explained as the result of the "correctness" of the LDA exchange-correlation hole [63, 64]. Since the LDA is constructed from the uniform gas, which has many realistic properties, its hole satisfies many mathematical conditions on this quantity [65]. Many of the most popular improvements on LDA, including the PBE generalized gradient approximation, are based on models of the exchange-correlation hole, not just fits of exact conditions or empirical data [52]. In fact, the most successful approximations usually are based on models for the exchange-correlation hole, which can be explicitly tested [66]. Unfortunately, insights about the ground-state exchange-correlation hole do not simply generalize as temperatures increase, as will be discussed later.

4 Thermal DFT

Thermal DFT deals with statistical ensembles of quantum states describing the thermodynamical equilibrium of many-electron systems. The grand canonical ensemble is particularly convenient to deal with the symmetry of identical particles. In the limit of vanishing temperature, thermal DFT reduces to an equiensemble ground state DFT description [67], which, in turn, reduces to the standard pure-state approach for non-degenerate cases.

While in the ground-state problem the focus is on the ground state energy, in the statistical mechanical framework the focus is on the grand canonical potential. Here, the grand canonical Hamiltonian plays an analogous role as the one played by the Hamiltonian for the ground-state problem. The former is written

$$\hat{\Omega} = \hat{H} - \tau \hat{S} - \mu \hat{N} , \tag{83}$$

where $\hat{H}$, $\hat{S}$, and $\hat{N}$ are the Hamiltonian, entropy, and particle-number operators. The crucial quantity by which the Hamiltonian differs from its grand-canonical version is the entropy operator[5]:

$$\hat{S} = -k_B \ln \hat{\Gamma} \ , \tag{84}$$

where

$$\hat{\Gamma} = \sum_{N,i} w_{N,i} |\Psi_{N,i}\rangle\langle\Psi_{N,i}| \ . \tag{85}$$

$|\Psi_{N,i}\rangle$ are orthonormal N-particle states (that are not necessarily eigenstates in general) and $w_{N,i}$ are normalized statistical weights satisfying $\sum_{N,i} w_{N,i} = 1$. $\hat{\Gamma}$ allows us to describe the thermal ensembles of interest.

Observables are obtained from the statistical average of Hermitian operators

$$O[\hat{\Gamma}] = \mathrm{Tr}\,\{\hat{\Gamma}\hat{O}\} = \sum_N \sum_i w_{N,i} \langle\Psi_{N,i}|\hat{O}|\Psi_{N,i}\rangle \ . \tag{86}$$

These expressions are similar to Eq. (53), but here the trace is not restricted to the ground-state manifold.

In particular, consider the average of the $\hat{\Omega}$, $\Omega[\hat{\Gamma}]$, and search for its minimum at a given temperature, τ, and chemical potential, μ. The quantum version of the Gibbs Principle ensures that the minimum exists and is unique (we shall not discuss the possible complications introduced by the occurrence of phase transitions). The minimizing statistical operator is the grand-canonical statistical operator, with statistical weights given by

$$w^0_{N,i} = \frac{\exp[-\beta(E^0_{N,i} - \mu N)]}{\sum_{N,i} \exp[-\beta(E^0_{N,i} - \mu N)]}. \tag{87}$$

$E^0_{N,i}$ are the eigenvalues of N-particle eigenstates. It can be verified that $\Omega[\hat{\Gamma}]$ may be written in the usual form

$$\Omega = E - \tau S - \mu N = -k_B \tau \ln Z_G, \tag{88}$$

where Z_G is the grand canonical partition function; which is defined by

$$Z_G = \sum_N \sum_j e^{-\beta(E^0_{N,i} - \mu N)} \ . \tag{89}$$

[5]Note that, we eventually choose to work in a system of units such that the Boltzmann constant is $k_B = 1$, that is, temperature is measured in energy units.

The statistical description we have outlined so far is the standard one. Now, we wish to switch to a density-based description and thereby enjoy the same benefits as in the ground-state problem. To this end, the minimization of Ω can be written as follows:

$$\Omega^{\tau}_{v-\mu} = \min_{n} \left\{ F^{\tau}[n] + \int d^3r\, n(\mathbf{r})(v(\mathbf{r}) - \mu) \right\} \tag{90}$$

with $n(\mathbf{r})$ an ensemble N-representable density and

$$F^{\tau}[n] := \min_{\hat{\Gamma}\to n} F^{\tau}[\hat{\Gamma}] = \min_{\hat{\Gamma}\to n} \left\{ T[\hat{\Gamma}] + V_{ee}[\hat{\Gamma}] - \tau S[\hat{\Gamma}] \right\}. \tag{91}$$

This is the constrained-search analog of the Levy functional [11, 40], Eq. (44). It replaces the functional originally defined by Mermin [13] in the same way that Eq. (44) replaces Eq. (21) in the ground-state theory.[6]

Equation (91) defines the thermal universal functional. Universality of this quantity means that *it does not depend explicitly on the external potential nor on* μ. This is very appealing, as it hints at the possibility of widely applicable approximations.

We identify $\Gamma^{\tau}[n]$ as the minimizing statistical operator in Eq. (91). We can then define other interacting density functionals at a given temperature by taking the trace over the given minimizing statistical operator. For example, we have:

$$T^{\tau}[n] := T[\hat{\Gamma}^{\tau}[n]] \tag{92}$$

$$V^{\tau}_{ee}[n] := V_{ee}[\hat{\Gamma}^{\tau}[n]] \tag{93}$$

$$S^{\tau}[n] := S[\hat{\Gamma}^{\tau}[n]]. \tag{94}$$

In order to introduce the thermal Kohn-Sham system, we proceed analogously as in the zero-temperature case. We assume that there exists an ensemble of non-interacting systems with same average particle density *and* temperature of the interacting ensemble. Ultimately, this determines the one-body Kohn-Sham potential, which includes the corresponding chemical potential. Thus, the noninteracting (or Kohn-Sham) universal functional is defined as

$$F^{\tau}_{s}[n] := \min_{\hat{\Gamma}\to n} K^{\tau}[\hat{\Gamma}] = K^{\tau}[\hat{\Gamma}^{\tau}_{s}[n]] = K^{\tau}_{s}[n], \tag{95}$$

where $\hat{\Gamma}^{\tau}_{s}[n]$ is a statistical operator that describes the Kohn-Sham ensemble and $K^{\tau}[\hat{\Gamma}] := T[\hat{\Gamma}] - \tau S[\hat{\Gamma}]$ is a combination we have chosen to call the kentropy.

[6]The interested reader may find the extension of the Hohenberg-Kohn theorem to the thermal framework in Mermin's paper.

We can also write the corresponding Kohn-Sham equations at non-zero temperature, which are analogous to Eqs. (30) and (39) [21]:

$$\left[-\frac{1}{2}\nabla^2 + v_s(\mathbf{r})\right]\varphi_i(\mathbf{r}) = \epsilon_i^\tau \varphi_i(\mathbf{r}) \tag{96}$$

$$v_s(\mathbf{r}) = v_H[n](\mathbf{r}) + v_{xc}[n](\mathbf{r}) + v(\mathbf{r}). \tag{97}$$

The accompanying density formula is

$$n(\mathbf{r}) = \sum_i f_i |\varphi_i(\mathbf{r})|^2, \tag{98}$$

where

$$f_i = \left(1 + e^{(\epsilon_i^\tau - \mu)/\tau}\right)^{-1}. \tag{99}$$

Equations (96) and (97) look strikingly similar to the case of non-interacting Fermions. However, the Kohn-Sham weights, f_i, are not simply the familiar Fermi functions, due to the temperature dependence of the Kohn-Sham eigenvalues.

Through the series of equalities in Eq. (95), we see that the non-interacting universal density functional is obtained by evaluating the kentropy on a non-interacting, minimizing statistical operator which, at temperature τ, yields the average particle density n. The seemingly simple notation of Eq. (95) reduces the kentropy first introduced as a functional of the statistical operator to a finite-temperature functional of the density. From the same expression, we see that the kentropy plays a role in this framework analogous to that of the kinetic energy within ground-state DFT. Finally, we spell-out the components of $F_s^\tau[n]$:

$$F_s^\tau[n] = T_s^\tau[n] - \tau S_s^\tau[n] , \tag{100}$$

where $T_s^\tau[n] := T[\hat{\Gamma}_s^\tau[n]]$ and $S_s^\tau[n] := S[\hat{\Gamma}_s^\tau[n]]$.

Now we identify other fundamental thermal DFT quantities. First, consider the decomposition of the interacting grand-canonical potential as a functional of the density given by

$$\Omega_{v-\mu}^\tau[n] = F_s^\tau[n] + U[n] + \mathscr{F}_{xc}^\tau[n] + \int d^3r\, n(\mathbf{r})\,(v(\mathbf{r}) - \mu) . \tag{101}$$

Here, $U[n]$ is the Hartree energy having the form in Eq. (34). The adopted notation stresses that temperature dependence of $U[n]$ enters only through the input equilibrium density. The exchange-correlation free-energy density functional is given by

$$\mathscr{F}_{xc}^\tau[n] = F^\tau[n] - F_s^\tau[n] - U[n] . \tag{102}$$

It is also useful to introduce a further decomposition:

$$\mathscr{F}_{xc}^{\tau}[n] := \mathscr{F}_{x}^{\tau}[n] + \mathscr{F}_{c}^{\tau}[n] . \tag{103}$$

This lets us analyze the two terms on the right hand side along with their components.

The exchange contribution is

$$\mathscr{F}_{x}^{\tau}[n] = V_{\mathrm{ee}}[\Gamma_{s}^{\tau}[n]] - U[n] . \tag{104}$$

Note that the average on the right hand side is taken with respect to the Kohn-Sham ensemble and that kinetic and entropic contributions do not contribute to exchange effects explicitly. Interaction enters in Eq. (104) in a fashion that is reminiscent of (but not the same as) finite-temperature Hartree-Fock theory. In fact, $\mathscr{F}_{x}^{\tau}[n]$ may be expressed in terms of the square modulus of the finite-temperature Kohn-Sham one-body density matrix. Thus $\mathscr{F}_{x}^{\tau}[n]$ has an explicit, known expression, just as does $\mathscr{F}_{s}^{\tau}[n]$. For the sake of practical calculations, however, approximations are still needed.

The fundamental theorems of density functional theory were proven for any ensemble with monotonically decreasing weights [68] and were applied to extract excitations [69–71]. But simple approximations to the exchange for such ensembles are corrupted by ghost interactions [72] contained in the ensemble Hartree term. The Hartree energy defined in Eq. (34) is defined as the electrostatic self-energy of the density, both for ground-state DFT and at non-zero temperatures. But the physical ensemble of Hartree energies is in fact the Hartree energy of each ensemble member's density, added together with the weights of the ensembles. Because the Hartree energy is quadratic in the density, it therefore contains ghost interactions [72], i.e., cross terms, that are unphysical. These must be canceled by the exchange energy, which must therefore contain a contribution:

$$\Delta E_{X}^{GI} = \sum_{i} w_i U[n_i] - U\left[\sum_{i} w_i n_i\right]. \tag{105}$$

Such terms appear only when the temperature is non-zero and so are missed by any ground-state approximation to E_X.

Consider, now, thermal DFT correlations. We may expect correctly that these will be obtained as differences between interacting averages and the noninteracting ones. The kinetic correlation energy density functional is

$$T_{c}^{\tau}[n] := T^{\tau}[n] - T_{s}^{\tau}[n], \tag{106}$$

and similar forms apply to $S_{c}^{\tau}[n]$ and $K_{c}^{\tau}[n]$. Another important quantity is the correlation potential density functional. At finite-temperature, this is defined by

$$U_{c}^{\tau}[n] := V_{\mathrm{ee}}[\Gamma^{\tau}[n]] - V_{\mathrm{ee}}[\Gamma_{s}^{\tau}[n]] . \tag{107}$$

Finally, we can write the correlation free energy as follows

$$\mathscr{F}_c^\tau[n] = K_c^\tau[n] + U_c^\tau[n] = E_c^\tau[n] - \tau S_c^\tau[n] \tag{108}$$

where

$$K_c^\tau[n] = T_c^\tau[n] - \tau S_c^\tau[n] \tag{109}$$

is the correlation kentropy density functional and

$$E_c^\tau[n] := T_c^\tau[n] + U_c^\tau[n] \tag{110}$$

generalizes the expression of the correlation energy to finite temperature. Above, we have noticed that entropic contributions do not enter explicitly in the definition of $\mathscr{F}_x^\tau[n]$. From Eq. (108), on the other hand, we see that the correlation entropy is essential for determining $\mathscr{F}_c^\tau[n]$. Further, it may be grouped together with the kinetic contributions (as in the first identity) or separately (as in the second identity), depending on the context of the current analysis.

In the next section, we consider finite-temperature analogs of the exact conditions described earlier for the ground state functionals. This allow us to gain additional insights about the quantities identified so far.

5 Exact Conditions at Non-Zero Temperature

In the following, we review several properties of the basic energy components of thermal Kohn-Sham DFT [14, 15].

We start with some of the most elementary properties, their signs [14]:

$$\mathscr{F}_x^\tau[n] \leq 0, \ \mathscr{F}_c^\tau[n] \leq 0, \ U_c^\tau[n] \leq 0, \ K_c^\tau[n] \geq 0. \tag{111}$$

The sign of $\mathscr{F}_x^\tau[n]$ is evident from the definition given in terms of the Kohn-Sham one-body reduced density matrix [26]. The others may be understood in terms of their variational properties. For example, let us consider the case for $K_c^\tau[n]$. We know that the Kohn-Sham statistical operator minimizes the kentropy

$$K_s^\tau[n] = K^\tau[\hat{\Gamma}_s^\tau[n]] \ . \tag{112}$$

Thus, we also know that $K_s^\tau[n]$ must be less than $K^\tau[n] = K^\tau[\Gamma^\tau[n]]$, where $\Gamma^\tau[n]$ is the equilibrium statistical operator. This readily implies that

$$K_c^\tau[n] = K^\tau[\hat{\Gamma}^\tau[n]] - K^\tau[\hat{\Gamma}_s^\tau[n]] \geq 0. \tag{113}$$

An approximation for $K_c^\tau[n]$ that does not respect this inequality will not simply have the "wrong" sign. Much worse is that results from such an approximation will suffer from improper variational character.

A set of remarkable and useful properties are the scaling relationships. What follows mirrors the zero-temperature case, but an important and intriguing difference is the relationship between coordinate and temperature scaling.

We first introduce the concept of uniform scaling of statistical ensembles in terms of a particular scaling of the corresponding statistical operators.[7] Wavefunctions of each state in the ensemble can be scaled as in Eq. (67). At the same time, we require that the statistical mixing is not affected, so the statistical weights are held fixed under scaling (we shall return to this point in Sect. 6.1). In summary, the scaled statistical operator is

$$\hat{\Gamma}_\gamma := \sum_N \sum_i w_{N,i} |\Psi_{\gamma,N,i}\rangle\langle\Psi_{\gamma,N,i}|, \tag{114}$$

where (the representation free) Hilbert space element $|\Psi_\gamma\rangle$ is such that $\Psi_\gamma(\mathbf{r}_1,\ldots,\mathbf{r}_N) = \langle\mathbf{r}_1,\ldots,\mathbf{r}_N|\Psi_\gamma\rangle$. For sake of simplicity, we restrict ourselves to states of the type typically considered in the ground-state formalism.

Equation (114) leads directly to scaling relationships for any observable. For instance, we find

$$N[\hat{\Gamma}_\gamma] = N[\hat{\Gamma}], \tag{115}$$

$$T[\hat{\Gamma}_\gamma] = \gamma^2 T[\hat{\Gamma}], \text{ and} \tag{116}$$

$$S[\hat{\Gamma}_\gamma] = S[\hat{\Gamma}]\,. \tag{117}$$

Combining these, we find

$$\hat{\Gamma}_s^\tau[n_\gamma] = \hat{\Gamma}_{\gamma,s}^{\tau/\gamma^2}[n] \text{ and } F_s^\tau[n_\gamma] = \gamma^2 F_s^{\tau/\gamma^2}[n]. \tag{118}$$

Equation (118) states that the value of the non-interacting universal functional evaluated at a scaled density is related to the value of the same functional evaluated on the unscaled density at a scaled temperature. Equation (118) constitutes a powerful statement, which becomes more apparent by rewriting it as follows [14]:

$$F_s^{\tau'}[n] = \frac{\tau'}{\tau} F_s^\tau[n_{\sqrt{\tau/\tau'}}]. \tag{119}$$

This means that, if we know $F_s^\tau[n]$ at some non-zero temperature τ, we can find its value at any other temperature by scaling its argument.

---

[7] Uniform coordinate scaling may be considered as (very) careful dimensional analysis applied to density functionals. Dufty and Trickey analyze non-interacting functionals in this way in Ref. [15].

Scaling arguments allow us to extract other properties of the functionals, such as some of their limiting behaviors. For instance, we can show that in the "high-density" limit, the kinetic term dominates [14]:

$$T_s^{\infty}[n] = \lim_{\gamma \to +\infty} F_s[n_\gamma]/\tau^2 \tag{120}$$

while in the "low-density" limit, the entropic term dominates:

$$S_s^{\infty}[n] = \lim_{\gamma \to 0} F_s[n_\gamma]\tau. \tag{121}$$

Also, we may consider the interacting universal functional for a system with coupling strength equal to λ

$$F^{\tau,\lambda}[n] = \min_{\hat{\Gamma} \to n} \left\{ T[\hat{\Gamma}] + \lambda V_{ee}[\hat{\Gamma}] - \tau S[\hat{\Gamma}] \right\}, \tag{122}$$

and note that in general,

$$\hat{\Gamma}^{\tau,\lambda}[n] \neq \hat{\Gamma}^{\tau}[n]. \tag{123}$$

We can relate these two statistical operators [14]. In fact, one can prove

$$\hat{\Gamma}^{\tau,\lambda}[n] = \hat{\Gamma}_{\lambda}^{\tau/\lambda^2}[n_{1/\lambda}] \text{ and } F^{\tau,\lambda}[n] = \lambda^2 F^{\tau/\lambda^2}[n_{1/\lambda}]. \tag{124}$$

In the expressions above, a single superscript implies full interaction [14]. Equation (124) demands scaling of the coordinates, the temperature, and the strength of the interaction at once. This procedure connects one equilibrium state to another equilibrium state, that of a "scaled" system. Equation (124) may be used to state other relations similar to those discussed above for the non-interacting case.

Scaling relations combined with the Hellmann-Feynman theorem allow us to generate the thermal analog of one of the most important statements of ground-state DFT, the adiabatic connection formula [14]:

$$\mathscr{F}_{xc}^{\tau}[n] = \int_0^1 d\lambda \; U_{xc}^{\tau}[n](\lambda), \tag{125}$$

where

$$U_{xc}^{\tau}[n](\lambda) = V_{ee}[\hat{\Gamma}^{\tau,\lambda}[n]] - U[n] \tag{126}$$

and a superscript λ implies an electron-electron interaction strength equal to λ. The interaction strength runs between zero, corresponding to the noninteracting Kohn-Sham system, and one, which gives the fully interacting system. All this must be done while keeping the density constant. In thermal DFT, an expression like

Eq. (125) offers the appealing possibility of defining an approximation for $\mathscr{F}_{xc}^{\tau}[n]$ *without* having to deal with kentropic contributions explicitly.

Another interesting relation generated by scaling connects the exchange-correlation to the exchange-only free energy [14]:

$$\mathscr{F}_{x}^{\tau}[n] = \lim_{\gamma\to\infty} \mathscr{F}_{xc}^{\gamma^2\tau}[n_\gamma]/\gamma. \tag{127}$$

This may be considered the definition of the exchange contribution in an xc functional, and so Eq. (127) may also be used to extract an approximation for the exchange free energy, if an approximation for the exchange-correlation free energy as a whole is given (for example, if obtained from Eq. (125)).

Despite decades of research [73–76], thermal exchange-correlation GGAs have not been fully developed. The majority of the applications in the literature have adopted two *practical* methods: one uses plain finite-temperature LDA, the other uses ground-state GGAs within the thermal Kohn-Sham scheme. This latter method ignores any modification to the exchange-correlation free energy functional due to its non-trivial temperature dependence. As new approximations are developed, exact conditions such as those above are needed to define consistent and reliable thermal approximations.

6 Discussion

In this section, we discuss several aspects that may not have been fully clarified by the previous, relatively abstract sections. First, by making use of a simple example, we will illustrate in more detail the tie between temperature and coordinate scaling. Then, with the help of another example, we will show how scaling and other exact properties of the functionals can guide development and understanding of approximations. The last subsection notes some complications in importing tools directly from ground-state methods to thermal DFT.

6.1 Temperature and Coordinate Scaling

Here we give an illustration of how the scaling of the statistical operators introduced in the previous section is applicable to thermal ensembles. Our argument applies – with proper modifications and additions, such as the scaling of the interaction strength – to all Coulomb-interacting systems with all one-body external potentials. For sake of simplicity, we shall restrict ourselves to non-interacting fermions in a one-dimensional harmonic oscillator at thermodynamic equilibrium.

Let us start from the general expression of the Fermi occupation numbers

$$n_i(\tau, \mu, \epsilon_i) = \left(1 + e^{\beta(\epsilon_i - \mu)}\right)^{-1}, \tag{128}$$

where ϵ_i is the ith eigenvalue of the harmonic oscillator, $\epsilon_i = \omega(i + 1/2)$. For our system, the (time-independent) Schrödinger equation is:

$$\left\{-\frac{1}{2}\frac{d^2}{dx^2} + v(x)\right\} \phi_i(x) = \epsilon_i \phi_i(x) \ . \tag{129}$$

Now, we multiply the x-coordinates by γ

$$\left\{-\frac{1}{2\gamma^2}\frac{d^2}{dx^2} + v(\gamma x)\right\} \phi_i(\gamma x) = \epsilon_i \phi_i(\gamma x). \tag{130}$$

We then multiply both sides by γ^2:

$$\left\{-\frac{1}{2}\frac{d^2}{dx^2} + \gamma^2 v(\gamma x)\right\} \phi_i(\gamma x) = \gamma^2 \epsilon_i \phi_i(\gamma x). \tag{131}$$

Substituting $\tilde{v}(x) = \gamma^2 v(\gamma x)$, $\tilde{\phi}_i(x) = \sqrt{\gamma}\phi_i(\gamma x)$ (to maintain normalization), and $\tilde{\epsilon}_i = \gamma^2 \epsilon_i$ yields

$$\left\{-\frac{1}{2}\frac{d^2}{dx^2} + \tilde{v}(x)\right\} \tilde{\phi}_i(x) = \tilde{\epsilon}_i \tilde{\phi}_i(x). \tag{132}$$

The latter may be interpreted as the Schrödinger equation for a "scaled" system. In the special case of the harmonic oscillator,

$$\gamma^2 v(\gamma x) = \gamma^4 v(x), \tag{133}$$

the frequency scales quadratically, consistent with the scaling of the energies described just above. Now, let us look at the occupation numbers for the "scaled" system

$$n_i(\tau, \tilde{\mu}, \tilde{\epsilon}_i) = \left(1 + e^{\beta(\tilde{\epsilon}_i - \tilde{\mu})}\right)^{-1}, \tag{134}$$

where $\tilde{\mu} = \gamma^2 \mu$ (in this way, the average number of particle is kept fixed too). These occupation numbers are equal to those of the original system at a temperature τ/γ^2,

$$n_i(\tau, \tilde{\mu}, \tilde{\epsilon}_i) = n_i(\tau/\gamma^2, \mu, \epsilon_i). \tag{135}$$

Thus the statistical weights of the scaled system are precisely those of the original system, at a suitably scaled temperature.

6.2 Thermal-LDA for Exchange Energies

In ground-state DFT, uniform coordinate scaling of the exchange has been used to constrain the form of the exchange-enhancement factor in GGAs. In thermal DFT, a "reduction" factor, R_x, enters already in the expression of a LDA for the exchange energies. This lets us capture the reduction in exchange with increasing temperature, while keeping the zero-temperature contribution well-separated from the modification entirely due to non-vanishing temperatures.

The behavior of R_x can be understood using the basic scaling relation for the exchange free energy. Observe that, from the scaling of $\hat{\Gamma}_s^{\tau}$, U, and $V_{ee}[\Gamma_s^{\tau}[n]]$, one readily arrives at

$$\mathscr{F}_x^{\tau}[n_{\gamma}] = \gamma \mathscr{F}_x^{\tau/\gamma^2}[n]. \tag{136}$$

Since

$$\mathscr{F}_x^{\mathrm{LDA},\tau}[n] = \int d^3r \; f_x^{\tau}(n(r)), \tag{137}$$

Eq. (136) implies that a thermal-LDA exchange free energy density must have the form [14]

$$f_x^{\mathrm{unif},\tau}(n) = e_x^{\mathrm{unif}}(n) R_x(\Theta), \tag{138}$$

where $e_x^{\mathrm{unif}}(n) = -A_x n^{4/3}$, $A_x = (3/4\pi)(3\pi^2)^{1/3}$, and R_x can only depend on τ and n through the electron degeneracy $\Theta = 2\tau/(3\pi^2 n(\mathbf{r}))^{2/3}$.

The LDA is exact for the uniform electron gas and so automatically satisfies many conditions. As such, it also reduces to the ground-state LDA as temperature drops to zero:

$$R_x \to 1 \text{ as } \tau \to 0. \tag{139}$$

Moreover, for fixed n, we expect

$$\mathscr{F}_x^{\tau}/U \to 0 \text{ as } \tau \to \infty \tag{140}$$

because the effect of the Pauli exclusion principle drops off as the behavior of the system becomes more classical. Moreover, since $U[n]$ does not depend explicitly on the temperature, fixing n also fixes U. We conclude that, the reduction factor must drop to zero:

$$R_x \to 0 \text{ as } \tau \to \infty. \tag{141}$$

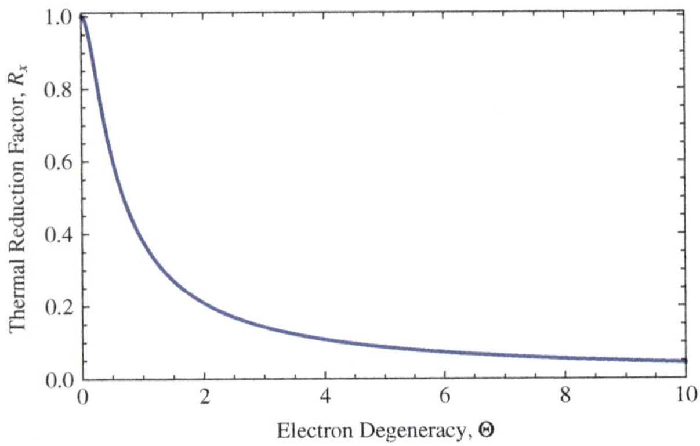

Fig. 4 Perrot and Dharma-Wardana's parameterization [74] of the thermal reduction factor for the exchange free energy for the uniform gas is plotted versus the electron degeneracy parameter

Now, let us consider the parameterization of R_x for the uniform gas by Perrot and Dharma-Wardana [74]:

$$R_x^{\text{unif}}(\Theta) \approx \left(\frac{4}{3}\right) \frac{0.75 + 3.04363\Theta^2 - 0.092270\Theta^3 + 1.70350\Theta^4}{1 + 8.31051\Theta^2 + 5.1105\Theta^4} \tanh \Theta^{-1} \ , \tag{142}$$

Here, $\Theta = \tau/\epsilon_F = 2\tau/k_F^2$ and k_F is the Fermi wavevector. Note the factor of $4/3$ that is not present in their original paper, which arises because we include a factor of $3/4$ in A_x they do not. Figure 4 shows the plot of this reduction factor. From both Fig. 4 and Eq. (142), it is apparent that the parametrization satisfies all the exact behaviors discussed just above.

6.3 Exchange-Correlation Hole at Non-Zero Temperature

Previously, we have emphasized that in ground-state DFT, the exchange-correlation hole function was vital for constructing reliable approximations. Therefore, it is important to reconsider this quantity in the context of thermal DFT. As we show below, this does not come without surprises.

In the grand canonical ensemble, the pair correlation function is a sum over statistically weighted pair correlation functions of each of the states in the ensemble labeled with collective index, ν (in this section, we follow notation and convention of Refs. [77] and [78]). A state $\Psi_{\lambda,\nu}$ has particle number N_ν, energy E_ν, and corresponds to λ-scaled interaction. If its weight in the ensemble is denoted as

$$w_{\lambda,\nu} = \frac{e^{-\beta(E_{\lambda,\nu} - \mu N_\nu)}}{\sum_\nu e^{-\beta(E_{\lambda,\nu} - \mu N_\nu)}}, \tag{143}$$

the ensemble average of the exchange-correlation hole density is

$$\langle n_{xc}^{\lambda}(\mathbf{r},\mathbf{r}')\rangle = \sum_{\nu} w_{\lambda,\nu} n_{xc,\nu}^{\lambda}(\mathbf{r},\mathbf{r}'). \tag{144}$$

However, the exchange-correlation hole function used to obtain $\mathscr{F}_{xc}^{\tau}$ through a λ integration requires the addition of more complicated terms [78]:

$$n_{xc}^{\lambda}(\mathbf{r},\mathbf{r}') = \langle n_{xc}^{\lambda}(\mathbf{r},\mathbf{r}')\rangle + \sum_{\nu} w_{\lambda,\nu} \frac{[n_{\lambda,\nu}(\mathbf{r}) - n(\mathbf{r})]}{n(\mathbf{r})}[n_{\lambda,\nu}(\mathbf{r}') + n_{xc,\nu}^{\lambda}(\mathbf{r},\mathbf{r}')], \tag{145}$$

where $n_{xc,\nu}^{\lambda}$ is the usual exchange-correlation hole corresponding to $\Psi_{\lambda,\nu}$ with particle density $n_{\lambda,\nu}$.

Thus, the sum rule stated in the ground state gets modified as follows [79]

$$\int d^3r'\, n_{xc}^{\lambda}(\mathbf{r},\mathbf{r}') = -1 + \sum_{\nu} w_{\lambda,\nu} \frac{n_{\lambda,\nu}(\mathbf{r})}{n(\mathbf{r})}[N_{\nu} - \langle N\rangle]. \tag{146}$$

The last expression shows that the sum rule for the thermal exchange-correlation hole accounts for an additional term due to particle number fluctuations. Worse still, this term carries along with it state-dependent, and therefore system-dependent, quantities. This is an important warning that standard methodologies for producing reliable ground-state functional approximations must be properly revised for use in the thermal context.

7 Conclusion

Thermal density functional theory is an area ripe for development in both fundamental theory and the construction of approximations because of rapidly expanding applications in many areas. Projects underway in the scientific community include construction of temperature-dependent GGAs, exact exchange methods for non-zero temperatures [80], orbital-free approaches at non-zero temperatures [81], and continued examination of the exact conditions that may guide both of these developments. In the world of warm dense matter, simulations are being performed, often very successfully [82], generating new insights into both materials science and the quality of our current approximations. As discussed above, techniques honed for zero-temperature systems should be carefully considered before being applied to thermal problems. Studying exact properties of functionals may guide efficient progress in application to warm dense matter. In context, thermal DFT emerges as as a clear and solid framework that provides users and developers practical and formal tools of general fundamental relevance.

Acknowledgements We would like to thank the Institute for Pure and Applied Mathematics for organization of Workshop IV: Computational Challenges in Warm Dense Matter and for hosting APJ during the Computational Methods in High Energy Density Physics long program. APJ thanks the U.S. Department of Energy (DE-FG02-97ER25308), SP and KB thank the National Science Foundation (CHE-1112442), and SP and EKUG thank European Community's FP7, CRONOS project, Grant Agreement No. 280879.

References

1. N.R.C.C. on High Energy Density Plasma Physics Plasma Science Committee, *Frontiers in High Energy Density Physics: The X-Games of Contemporary Science* (The National Academies Press, Washington, D.C., 2003)
2. T.R. Mattsson, M.P. Desjarlais, Phys. Rev. Lett. **97**, 017801 (2006)
3. F.R. Graziani, V.S. Batista, L.X. Benedict, J.I. Castor, H. Chen, S.N. Chen, C.A. Fichtl, J.N. Glosli, P.E. Grabowski, A.T. Graf, S.P. Hau-Riege, A.U. Hazi, S.A. Khairallah, L. Krauss, A.B. Langdon, R.A. London, A. Markmann, M.S. Murillo, D.F. Richards, H.A. Scott, R. Shepherd, L.G. Stanton, F.H. Streitz, M.P. Surh, J.C. Weisheit, H.D. Whitley, High Energy Density Phys. **8**(1), 105 (2012)
4. K.Y. Sanbonmatsu, L.E. Thode, H.X. Vu, M.S. Murillo, J. Phys. IV France **10**(PR5), Pr5 (2000)
5. S. Atzeni, J. Meyer-ter Vehn, *The Physics of Inertial Fusion: Beam-Plasma Interaction, Hydrodynamics, Hot Dense Matter* (Oxford University Press, New York, 2004)
6. M.D. Knudson, M.P. Desjarlais, Phys. Rev. Lett. **103**, 225501 (2009)
7. A. Kietzmann, R. Redmer, M.P. Desjarlais, T.R. Mattsson, Phys. Rev. Lett. **101**, 070401 (2008)
8. S. Root, R.J. Magyar, J.H. Carpenter, D.L. Hanson, T.R. Mattsson, Phys. Rev. Lett. **105**(8), 085501 (2010)
9. K. Burke, J. Chem. Phys. **136**, 150901 (2012)
10. P. Hohenberg, W. Kohn, Phys. Rev. **136**(3B), B864 (1964)
11. M. Levy, Proc. Natl. Acad. Sci. USA **76**(12), 6062 (1979)
12. E.H. Lieb, Int. J. Quantum Chem. **24**(3), 243 (1983)
13. N.D. Mermin, Phys. Rev. **137**, A: 1441 (1965)
14. S. Pittalis, C.R. Proetto, A. Floris, A. Sanna, C. Bersier, K. Burke, E.K.U. Gross, Phys. Rev. Lett. **107**, 163001 (2011)
15. J.W. Dufty, S.B. Trickey, Phys. Rev. B **84**, 125118 (2011). Interested readers should note that Dufty and Trickey also have a paper concerning interacting functionals in preparation
16. L.H. Thomas, Math. Proc. Camb. Phil. Soc. **23**(05), 542 (1927)
17. E. Fermi, Rend. Acc. Naz. Lincei **6**, 602–607 (1927)
18. E. Fermi, Z. für Phys. A Hadrons Nucl. **48**, 73 (1928)
19. V. Fock, Z. Phys. **61**, 126 (1930)
20. D.R. Hartree, W. Hartree, Proc. R. Soc. Lond. Ser. A – Math. Phys. Sci. **150**(869), 9 (1935)
21. W. Kohn, L.J. Sham, Phys. Rev. **140**(4A), A1133 (1965)
22. K. Burke. The ABC of DFT. http://dft.uci.edu/doc/g1.pdf (created April 10, 2007)
23. K. Burke, L.O. Wagner, Int. J. Quant. Chem. **113**, 96 (2013)
24. F. Schwabl, *Quantum Mechanics* (Springer, Berlin/Heidelberg/New York, 2007)
25. J.J. Sakurai, *Modern Quantum Mechanics*, Rev. Edn. (Addison Wesley, Reading, 1993)
26. E. Engel, R.M. Dreizler, *Density Functional Theory: An Advanced Course* (Springer, Heidelberg/Dordrecht/London/New York, 2011)
27. C.D. Sherrill, J. Chem. Phys. **132**(11), 110902 (2010)
28. R.M. Dreizler, E.K.U. Gross, *Density Functional Theory: An Approach to the Quantum Many-Body Problem* (Springer, Berlin/New York, 1990)
29. W. Kohn, in *Highlight of Condensed Matter Theory*, ed. by F. Bassani, F. Fumi, M.P. Tosi (North-Holland, Amsterdam, 1985), p. 1

30. G.F. Giuliani, G. Vignale (eds.), *Quantum Theory of the Electron Liquid* (Cambridge University Press, Cambridge, 2008)
31. R. Dreizler, J. da Providência, N.A.T.O.S.A. Division (eds.), *Density Functional Methods in Physics*. NATO ASI B Series (Springer, Dordrecht, 1985)
32. H. Englisch, R. Englisch, Phys. A Stat. Mech. Appl. **121**(1–2), 253 (1983)
33. F.W. Averill, G.S. Painter, Phys. Rev. B **15**, 2498 (1992)
34. S.G. Wang, W.H.E. Scharz, J. Chem. Phys. **105**, 4641 (1996)
35. P.R.T. Schipper, O.V. Gritsenko, E.J. Baerends, Theor. Chem. Acc. **99**, 4056 (1998)
36. P.R.T. Schipper, O.V. Gritsenko, E.J. Baerends, J. Chem. Phys. **111**, 4056 (1999)
37. C.A. Ullrich, W. Kohn, Phys. Rev. Lett. **87**, 093001 (2001)
38. M. Reed, B. Simon, *I: Functional Analysis (Methods of Modern Mathematical Physics)* (Academic, San Diego, 1981)
39. J. Harriman, Phys. Rev. A **24**, 680 (1981)
40. R.G. Parr, W. Yang, *Density Functional Theory of Atoms and Molecules* (Oxford University Press, New York, 1989)
41. R. van Leeuwen, Adv. Q. Chem. **43**, 24 (2003)
42. M. Levy, Phys. Rev. A **26**, 1200 (1982)
43. S.V. Valone, J. Chem. Phys. **73**, 1344 (1980)
44. H. Englisch, R. Englisch, Phys. Stat. Solidi B **123**, 711 (1984)
45. H. Englisch, R. Englisch, Phys. Stat. Solidi B **124**, 373 (1984)
46. J.T. Chayes, L. Chayes, M.B. Ruskai, J. Stat. Phys **38**, 497 (1985)
47. J. Perdew, Phys. Rev. B **33**, 8822 (1986)
48. A.D. Becke, Phys. Rev. A **38**(6), 3098 (1988)
49. A.D. Becke, J. Chem. Phys. **98**(7), 5648 (1993)
50. C. Lee, W. Yang, R.G. Parr, Phys. Rev. B **37**(2), 785 (1988)
51. J.P. Perdew, K. Schmidt, in *Density Functional Theory and Its Applications to Materials*, ed. by V.E.V. Doren, K.V. Alsenoy, P. Geerlings (American Institute of Physics, Melville, 2001)
52. J.P. Perdew, K. Burke, M. Ernzerhof, Phys. Rev. Lett. **77**(18), 3865 (1996); Ibid. **78**, 1396(E) (1997)
53. J. Perdew, K. Burke, M. Ernzerhof, Phys. Rev. Lett. **80**, 891 (1998)
54. V.V. Karasiev, R.S. Jones, S.B. Trickey, F.E. Harris, Phys. Rev. B **80**, 245120 (2009)
55. V.V. Karasiev, R.S. Jones, S.B. Trickey, F.E. Harris, Phys. Rev. B **87**, 239903 (2013)
56. V. Karasiev, S. Trickey, Comput. Phys. Commun. **183**(12), 2519 (2012)
57. Y.A. Wang, E.A. Carter, in *Theoretical Methods in Condensed Phase Chemistry*, ed. by S.D. Schwartz (Kluwer, Dordrecht, 2000), chap. 5, p. 117
58. J.C. Snyder, M. Rupp, K. Hansen, K.R. Mueller, K. Burke, Phys. Rev. Lett. **108**, 253002 (2012)
59. M. Levy, J. Perdew, Phys. Rev. A **32**, 2010 (1985)
60. J.P. Perdew, S. Kurth, in *A Primer in Density Functional Theory*, ed. by C. Fiolhais, F. Nogueira, M.A.L. Marques (Springer, Berlin/Heidelberg, 2003), pp. 1–55
61. O. Gunnarsson, B. Lundqvist, Phys. Rev. B **13**, 4274 (1976)
62. D. Langreth, J. Perdew, Solid State Commun. **17**, 1425 (1975)
63. M. Ernzerhof, K. Burke, J.P. Perdew, in *Recent Developments and Applications in Density Functional Theory*, ed. by J.M. Seminario (Elsevier, Amsterdam, 1996)
64. R. Jones, O. Gunnarsson, Rev. Mod. Phys. **61**, 689 (1989)
65. K. Burke, J. Perdew, D. Langreth, Phys. Rev. Lett. **73**, 1283 (1994)
66. A.C. Cancio, C.Y. Fong, J.S. Nelson, Phys. Rev. A **62**, 062507 (2000)
67. H. Eschrig, Phys. Rev. B **82**, 205120 (2010)
68. A. Theophilou, J. Phys. C **12**, 5419 (1979)
69. E. Gross, L. Oliveira, W. Kohn, Phys. Rev. A **37**, 2809 (1988)
70. L. Oliveira, E. Gross, W. Kohn, Phys. Rev. A **37**, 2821 (1988)
71. A. Nagy, Phys. Rev. A **57**, 1672 (1998)
72. N.I. Gidopoulos, P.G. Papaconstantinou, E.K.U. Gross, Phys. Rev. Lett. **88**, 033003 (2002)
73. F. Perrot, Phys. Rev. A **20**, 586 (1979)
74. F. Perrot, M.W.C. Dharma-wardana, Phys. Rev. A **30**, 2619 (1984)

75. F. Perrot, M.W.C. Dharma-wardana, Phys. Rev. B **62**(24), 16536 (2000)
76. R.G. Dandrea, N.W. Ashcroft, A.E. Carlsson, Phys. Rev. B **34**(4), 2097 (1986)
77. J. Perdew, in *Density Functional Method in Physics*, ed. by R. Dreizler, J. da Providencia. NATO Advanced Study Institute, Series B: Physics, vol. 123 (Plenum Press, New York, 1985)
78. S. Kurth, J.P. Perdew, in *Strongly Coupled Coulomb Systems*, ed. by G.J. Kalman, J.M. Rommel, K. Blagoev (Plenum Press, New York, 1998)
79. J.P. Perdew, Int. J. Quantum Chem. **28**(Suppl. 19), 497 (1985)
80. M. Greiner, P. Carrier, A. Görling, Phys. Rev. B **81**, 155119 (2010)
81. V.V. Karasiev, T. Sjostrom, S.B. Trickey, Phys. Rev. B **86**, 115101 (2012)
82. K.U. Plagemann, P. Sperling, R. Thiele, M.P. Desjarlais, C. Fortmann, T. Döppner, H.J. Lee, S.H. Glenzer, R. Redmer, New J. Phys. **14**(5), 055020 (2012)

Innovations in Finite-Temperature Density Functionals

Valentin V. Karasiev, Travis Sjostrom, Debajit Chakraborty, James W. Dufty, Keith Runge, Frank E. Harris, and S.B. Trickey

Abstract Reliable, tractable computational characterization of warm dense matter is a challenging task because of the wide range of important aggregation states and effective interactions involved. Contemporary best practice is to do ab initio molecular dynamics on the ion constituents with the forces from the electronic population provided by density functional calculations. Issues with that approach include the lack of reliable approximate density functionals and the computational bottleneck intrinsic to Kohn-Sham calculations. Our research is aimed at both problems, via the so-called orbital-free approach to density functional theory. After a sketch of the relevant properties of warm dense matter to motivate our research, we give a survey of our results for constraint-based non-interacting free energy functionals and exchange-correlation free-energy functionals. That survey includes comparisons with novel finite-temperature Hartree-Fock calculations and also presents progress on both pertinent exact results and matters of computational technique.

V.V. Karasiev • T. Sjostrom • D. Chakraborty • K. Runge • F.E. Harris • S.B. Trickey (✉)
Quantum Theory Project, Departments of Physics and of Chemistry, University of Florida, P.O. Box 118435, Gainesville, FL 32611-8435, USA
e-mail: trickey@qtp.ufl.edu; trickey@UFL.EDU
http://www.qtp.ufl.edu/ofdft

J.W. Dufty
Department of Physics, University of Florida, P.O. Box 118435, Gainesville, FL 32611-8435, USA

F. Graziani et al. (eds.), *Frontiers and Challenges in Warm Dense Matter*, Lecture Notes in Computational Science and Engineering 96,
DOI 10.1007/978-3-319-04912-0_3,

1 Setting and Perspective

Materials under extreme conditions have been a major and rewarding focus of condensed matter physics for well over a century. Low-temperature physics is a familiar example, with its modern form dating to the liquefaction of Hydrogen by Dewar in 1898 and of Helium by Kamerlingh Onnes in 1908 [1]. High energy-density physics (HEDP), the focus of the 2012 IPAM Long Program "Computational Methods in High Energy Density Plasmas", is less well-known, even in the scientific community. But from the perspective of shock and detonation phenomena at least, HEDP has as long and deep a scientific heritage as does low temperature physics. A history of early work on detonation waves [2], for example, notes the first detonation velocity measurements in condensed explosives by Abel in 1869, the beginnings of detonation theory (antecedent to the Chapman-Jouguet equation) by Michelson in 1890, and the Chapman-Jouguet relation itself by Chapman in 1899 and Jouguet in 1905 and 1906. Other examples could be given from plasma physics.

A sign of the richness of HEDP is that only much more recently has it been recognized that HEDP includes a complicated condensed matter regime now called warm dense matter (WDM). We discuss WDM traits briefly below. The challenge and opportunity to develop computationally tractable, predictive methods suitable for a comparatively unexplored condensed matter regime drew us together as a team. That same combination of challenge and opportunity led to Workshop IV "Computational Challenges in Warm Dense Matter" of the IPAM HEDP Long Program. The Workshop was a *de facto* review of theory, modeling, and simulation for WDM, with conversations among those already involved as well as with workers whose research clearly could be of relevance. In that context, we offer here a survey of our contributions, confident that we do not have to survey the entire area and, hence, the entire Workshop!

It is fitting to begin with a system perspective. WDM comprises a condensed matter regime characterized roughly by electron temperatures $T \approx 1$–$15\,\text{eV}$ and pressures to 1 Mbar or greater. (Aside: Theoretical and computational treatment of WDM involves diverse topics with many abbreviations. Most of the more common ones are listed in the Glossary; see the Appendix.) Recently WDM has attracted attention because of its importance in diverse physical systems, including exo-planet interiors, the path to inertial confinement fusion, and neutron stellar atmospheres [3–13]. The relationship between the WDM regime and other states of matter is illustrated schematically in Fig. 1. The regime is inherently challenging because its thermodynamics cannot be framed in terms of small perturbations from ideal, solvable models. WDM temperature and pressure ranges correspond to values of order unity in the two relevant expansion parameters: the Coulomb coupling parameter $\Gamma = Q^2/r_s k_B T$ and the electron degeneracy parameter $\Theta = k_B T/\epsilon_F$. (Here Q = the relevant charge, r_s = Wigner radius, ϵ_F = electron Fermi energy, T = temperature, k_B = Boltzmann constant.) WDM thus does not fall neatly within the parameter space typical of either ordinary condensed-matter physics or plasma physics. One result is that plasma physics methods which originate in the classical

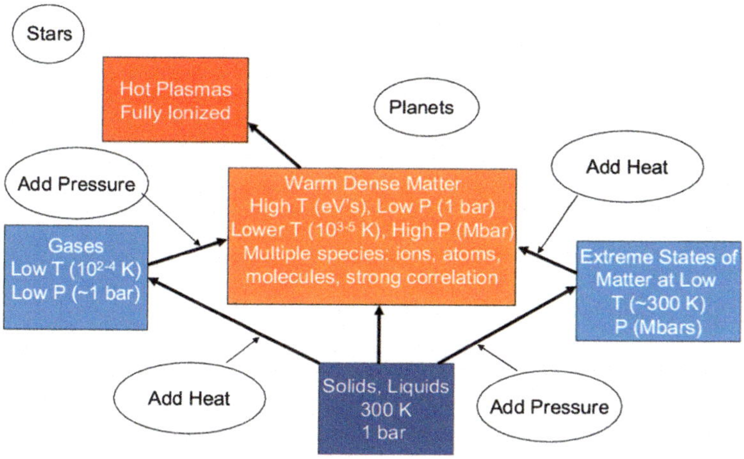

Fig. 1 Qualitative positioning of WDM relative to other physical states and systems

limit do not extend well into the WDM regime. At a fundamental level, such approaches cannot be entirely successful; the quantum limit cannot be recovered from classical physics.

Conversely, condensed-matter physics methods must be extended well beyond their normal realms of application. The logic is inexorable. A foundational element for understanding and manipulating material behavior is the equation of state (EOS). For materials under near-ambient conditions, best practice has evolved to be a combination of electronic structure calculations using density functional theory (DFT) and molecular dynamics (MD). In many cases, the zero-temperature EOS (or cold curve), including crystalline phase transitions, can be predicted quite accurately with DFT alone [14,15]. For the nuclear (or ionic) contribution, so-called ab initio MD (AIMD; see Refs. [16–20]) is quite successful. In its simplest form (Born-Oppenheimer MD), AIMD gets the electronic forces on the nuclei (or ions) from a DFT calculation at each nuclear step. AIMD thereby combines chemical realism from explicit quantum mechanical treatment of the electrons with the more-or-less classical contributions of the nuclear species.

There are both conceptual and technical problems associated with this rosy picture, however, and they are worsened by extending into the WDM regime. A brief sketch of the customary DFT approach is needed for context. The variational minimization of the density functional customarily is via the well-known Kohn-Sham (KS) procedure [21]. It introduces a model (sometimes called “fictitious”) non-interacting many-fermion system of the same density, n, as the physical system of interest. The density is expanded in the one-body states (orbitals) of the KS Hamiltonian

$$h_{\mathrm{KS}}[n]\varphi_i = \varepsilon_i \varphi_i \,, \tag{1}$$

$$n(\mathbf{r}) = \sum_i f_i |\varphi_i(\mathbf{r})|^2 \,, \tag{2}$$

with f_i the occupation numbers. For ambient T, it is appropriate to use $f_i = 0, 1$ (for the spin-polarized case); the T-dependence of the Fermi-Dirac distribution is negligible. The KS equation follows from decomposition of the fundamental Hohenberg-Kohn, Levy-Lieb functional [22–24] into non-interacting kinetic energy $\mathscr{T}_s$, Hartree, exchange, and correlation energies [25,26]. Observe that exchange and correlation are defined in terms of the KS decomposition, hence differ subtly from the quantities with the same names defined in other parts of many-electron theory. A free-energy (i.e., non-zero T) version of the theory exists; we give relevant details below. The essential point for now is that the KS equation continues to be valid but with full Fermi-Dirac occupation number T-dependence:

$$f_j \equiv f(\varepsilon_j - \mu) = [1 + \exp(\beta(\varepsilon_j - \mu))]^{-1} , \tag{3}$$

with $\beta = 1/k_B\mathrm{T}$ and μ the chemical potential. In the WDM regime, the many small, but non-negligible values of f_j for levels far above μ constrict drastically a computational bottleneck that already is evident at $\mathrm{T} = 0\,\mathrm{K}$. Simply, at $\mathrm{T} = 0\,\mathrm{K}$, diagonalization of the KS Hamiltonian means that the KS calculational cost scales as N_e^3 at best, with N_e = total number of electrons. If a conventional KS DFT calculation drives the AIMD, each MD step therefore also scales as N_e^3 or worse. Matters worsen as T increases. Then the computational burden increases proportionally to N_b^3, where N_b is the number of thermally occupied levels, with $N_b \gg N_e$ for WDM. This scaling of computational cost is a severe limitation for thorough exploration of WDM, as noted in recent practitioner remarks (Sept. 2012) [27,28]. Even with the massive computational power of the DOE labs, realistic KS-driven AIMD simulations of WDM remain at the level of heroic.

Our research program aims to eliminate the KS bottleneck in WDM simulations by making orbital-free DFT (OF-DFT) a viable alternative. Orbitals enter the KS procedure primarily as a device for computing the major part of the KE and, for nonzero T, also for computing the non-interacting entropy from the occupation numbers. Putting aside for now the present-day emphasis on developing explicitly orbital-dependent exchange-correlation functionals, the other free-energy contributions in conventional DFT are all explicit density functionals.

Schematically, the OF-DFT simulation is compared with a conventional KS-DFT simulation in Fig. 2. The crucial difference is that the N_b-dimensional Kohn-Sham eigenvalue problem is replaced with a single 3-dimensional Euler-Lagrange equation. As already stated, the challenge addressed in this project is to generate the functionals which are ingredients of that equation. To illustrate the power of orbital-free methods, Fig. 3 shows our comparison for Hydrogen at $0.983\,\mathrm{g/cm^3}$, about a two-fold compression from ambient density. Discussion of these computations is below. Here, the essential point is that the OF-DFT simulation replaces the power-law scaling of the conventional KS-DFT-driven MD with a nearly invariant scaling.

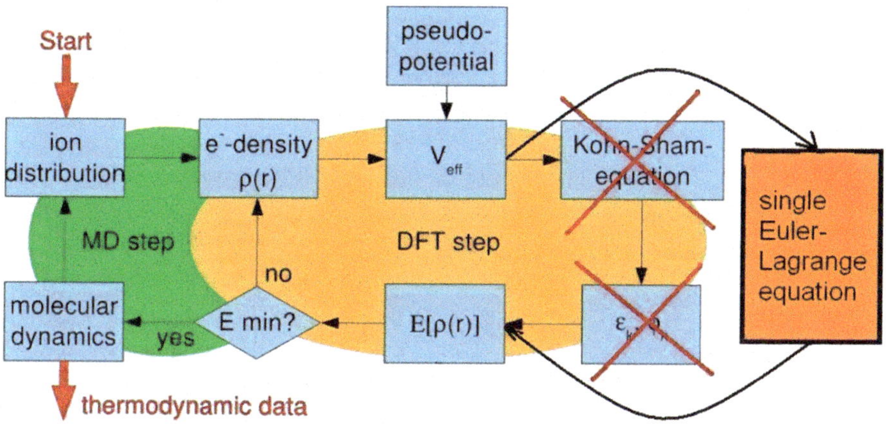

Fig. 2 Comparison of Kohn-Sham and orbital-free MD computational cycles (Credit for original: W. Lorenzen)

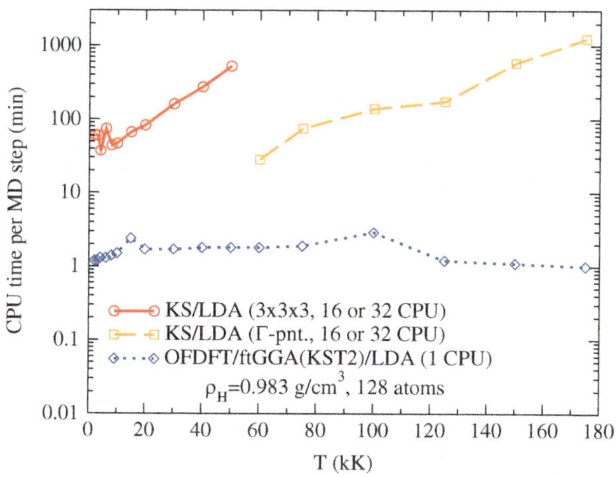

Fig. 3 Illustration of the dramatic advantage of OF-DFT: time per AIMD step as a function of T for KS-DFT vs. OF-DFT for Hydrogen

2 Primary Challenges

Identification of the primary challenges to use of OF-DFT for predictive WDM studies requires that we set out a few details of free-energy DFT (FE-DFT). At the level of the Born-Oppenheimer approximation, FE-DFT follows from a constrained search approach within the grand canonical ensemble [25, 29–32]. It should be stressed that this approach provides a firm theoretical foundation for FE-DFT, so the resulting equations cannot be dismissed as ad hoc or heuristic. The analysis yields the key result that, in parallel with ground-state DFT [22–24], there exists a universal functional $\mathscr{F}[n, \mathrm{T}]$ which gives the equilibrium free energy of a many-electron system with Coulomb interactions in terms of n and T. Once one has $\mathscr{F}$ for a given configuration of nuclei or ions $\{\mathbf{R}\}$, the electronic forces on those ions or nuclei can be computed from $\nabla_{\mathbf{R}}(\mathscr{F} + E_{\mathrm{ext}})$ with $E_{\mathrm{ext}} = \int d\mathbf{r} n(\mathbf{r}) \mathrm{v}_{\mathrm{ext}}(\mathbf{r})$ and $\mathrm{v}_{\mathrm{ext}}$ the external potential. (The ion-ion Coulomb repulsion term is omitted here for simplicity.).

To compute $\mathscr{F}$, one introduces, just as in zero-temperature DFT [21], a non-interacting system with the same density as the physical (interacting) system. The universal functional $\mathscr{F}[n, \mathrm{T}]$ thus partitions into a non-interacting part $\mathscr{F}_{\mathrm{s}}[n, \mathrm{T}]$ plus the classical Coulomb repulsion energy (Hartree energy) $\mathscr{F}_{\mathrm{H}}[n]$, and the exchange-correlation remainder $\mathscr{F}_{\mathrm{xc}}[n, \mathrm{T}]$

$$\mathscr{F}[n, \mathrm{T}] = \mathscr{F}_{\mathrm{s}}[n, \mathrm{T}] + \mathscr{F}_{\mathrm{H}}[n, \mathrm{T}] + \mathscr{F}_{\mathrm{xc}}[n, \mathrm{T}] , \tag{4}$$

where

$$\mathscr{F}_{\mathrm{s}}[n, \mathrm{T}] = \mathscr{T}_{\mathrm{s}}[n, \mathrm{T}] - \mathrm{T}\mathscr{S}_{\mathrm{s}}[n, \mathrm{T}] . \tag{5}$$

Here $\mathscr{T}_{\mathrm{s}}$ and $\mathscr{S}_{\mathrm{s}}$ are the non-interacting kinetic energy (KE) and entropy. The motivation is that $\mathscr{F}_{\mathrm{s}}$ and $\mathscr{F}_{\mathrm{H}}$ constitute the largest contributions to $\mathscr{F}$, yet can be calculated explicitly, while the smaller contribution $\mathscr{F}_{\mathrm{xc}}[n, \mathrm{T}]$ can be approximated. Analogously with the zero-T formulation, $\mathscr{F}_{\mathrm{xc}}$ contains not only Coulombic correlation and exchange effects, but also the difference between the kinetic energies and entropies of the actual and non-interacting systems.The expressions for the non-interacting KE and entropy are:

$$\mathscr{T}_{\mathrm{s}}[n, \mathrm{T}] := -\frac{1}{2} \sum_{j=1}^{\infty} \int d\mathbf{r} \, f_j \, \varphi_j^*(\mathbf{r}) \nabla^2 \varphi_j(\mathbf{r}) := \int d\mathbf{r} \, \tau_{\mathrm{orb}}(\mathbf{r}, \mathrm{T}) , \tag{6}$$

$$\mathscr{S}_{\mathrm{s}}[n, \mathrm{T}] = -k_{\mathrm{B}} \sum_j \{ f_j \ln f_j + (1 - f_j) \ln(1 - f_j) \} . \tag{7}$$

Variation with respect to the density then gives the KS Hamiltonian as

$$h_{\mathrm{KS}} = -\frac{1}{2}\nabla^2 + \mathrm{v}_{\mathrm{KS}}([n]; \mathbf{r}) \equiv -\frac{1}{2}\nabla^2 + \mathrm{v}_{\mathrm{ext}} + \mathrm{v}_{\mathrm{H}} + \mathrm{v}_{\mathrm{xc}} . \tag{8}$$

Here $\mathrm{v}_{\mathrm{H}} = \delta\mathscr{F}_{\mathrm{H}}/\delta n$, $\mathrm{v}_{\mathrm{xc}} = \delta\mathscr{F}_{\mathrm{xc}}/\delta n$, and the density is given in terms of the Fermi-Dirac occupation numbers f_j and the KS orbitals φ_j as in Eqs. (2) and (3).

The expressions for the non-interacting KE and entropy just given illustrate some of the challenges to OF-FE-DFT. In place of Eqs. (6) and (7), we must find adequately accurate, general, orbital-free representations for both, namely

$$\mathscr{T}_{\mathrm{s}}[n, \mathrm{T}] = \int d\mathbf{r} \, \tau_{\mathrm{s}}[n(\mathbf{r}), \mathrm{T}] \quad , \quad \mathscr{S}_{\mathrm{s}}[n, \mathrm{T}] = \int d\mathbf{r} \, \sigma_{\mathrm{s}}[n(\mathbf{r}), \mathrm{T}] . \tag{9}$$

Further, we must restrict consideration to XC functionals that are not explicitly orbital-dependent, i.e., functionals which correspond to rungs one or two on the Perdew-Schmidt Jacob's ladder [33] of complexity. This restriction motivates continued effort (including contributions from our group) to improve lower-rung zero-T XC functionals, despite their de-emphasis in the quantum chemistry community (which does not focus on orbital-free DFT). We therefore work with XC functionals of the form

$$\mathscr{F}_{\mathrm{xc}} = \int d\mathbf{r}\, f_{\mathrm{xc}}[n(\mathbf{r}), \mathrm{T}] \ . \tag{10}$$

For simplicity of notation we have omitted possible dependence on density gradients and higher derivatives of the density in several of the foregoing expressions. With all contributions in orbital-free form, the Euler equation becomes

$$\frac{\delta \mathscr{T}_{\mathrm{s}}}{\delta n} - \mathrm{T}\frac{\delta \mathscr{S}_{\mathrm{s}}}{\delta n} + \frac{\delta \mathscr{F}_{\mathrm{H}}}{\delta n} + \frac{\delta \mathscr{F}_{\mathrm{xc}}}{\delta n} = \mu - \mathrm{v}_{\mathrm{ext}} \ . \tag{11}$$

The primary challenges to the use of OF-DFT for WDM studies now can be identified succinctly. First, there is no constructive route from the exact DFT theorems [21–24] to viable approximations for the XC free energy [25, 29–32], an observation that is hardly surprising in view of the fact that no such route has been found for zero-temperature XC functionals. Secondly, the abandonment of an orbital description means that there also is no obvious constructive path for approximate functionals for either the non-interacting kinetic energy or entropy solely in terms of the density. Third, our work has shown that the functionals $\mathscr{T}_{\mathrm{s}}$, $\mathscr{S}_{\mathrm{s}}$, and $\mathscr{F}_{\mathrm{xc}}$ have a significant dependence on temperature that thus far has received insufficient attention.

Our research can be conceptualized in terms of these three challenges. The discussion which follows surveys the main points of our work to

- Uncover the T-dependence of $\mathscr{T}_{\mathrm{s}}$, $\mathscr{S}_{\mathrm{s}}$ for ranges of material densities and pressures appropriate to WDM by finite-T KS calculations
- Develop new, T-dependent generalized gradient approximations for $\mathscr{T}_{\mathrm{s}}$ and $\mathscr{S}_{\mathrm{s}}$
- Characterize the T-dependence of the exchange functional $\mathscr{F}_{\mathrm{x}}$ by comparison of both T-dependent local density approximation (LDA) [34] and ordinary T-independent LDA with thermal (finite-T) Hartree-Fock (tHF) calculations [35–37]
- Derive exact constraints and properties of the various contributions to the free energy density functional
- Improve $\mathrm{T} = 0\,\mathrm{K}$ orbital-free $\mathscr{T}_{\mathrm{s}}$ and $\mathscr{F}_{\mathrm{xc}}$ functionals (the obvious low-T limits of our target functionals)
- Develop the requisite technical means (algorithms, local pseudopotentials, projector augmented wave data sets, etc.).

A brief discussion of the context of our work is useful. Zero-temperature OF-KE functional development today has two streams. One uses two-point functionals, of the generic form

$$\mathscr{T}_s[n] = \int d\mathbf{r}_1 d\mathbf{r}_2 n^{\lambda}(\mathbf{r}_1) K[n(\mathbf{r}_1), n(\mathbf{r}_2), \mathbf{r}_1, \mathbf{r}_2] n^{\gamma}(\mathbf{r}_2) \tag{12}$$

with $\lambda + \gamma = 8/3$ and the kernel K referenced to the response of the homogeneous electron gas (HEG). See Ref. [38] for a didactic discussion. The motivation is to describe Friedel oscillations in solids and corresponding atomic and molecular electronic shell structure. While progress continues, the inherent non-locality of this approach causes computational burden (for example, see remarks in Ref. [39]). The two-point strategy also has conceptual limitations. The linear response formulation implicitly requires a reference density (think of the homogeneous electron gas). But a reference density is ill-defined for a truly unenclosed system, e.g., an atom or molecule, a problematic limitation for proper low-density limits such as in WDM (including the T = 0 K gas phase). So-called density-independent two-point kernels also are intrinsically unstable [40]. More complicated, density-dependent kernels thus are inescapable.

Another form of two-point functional development starts with modeling of the first-order reduced density matrix [41]. Recently, one of us worked on algorithms for two-point optimization of density matrices with respect to various idempotency constraints [42] (D. Chakraborty, R. Cuevas-Saavedra, P.W. Ayers, J. Chem. Phys., Unpublished) for atoms and molecules. Though this approach requires ingenuity in accurate modeling of a density matrix, it could be an alternative route to a more accurate two-point OFKE functional. With emphasis on computational speed, however, our main focus for $\mathscr{T}_s$ approximations is on one-point functionals. We return to that emphasis at various points below.

One other matter of context also is important, namely that we are not working on the embedding form of OF-DFT [43, 44]. The distinction is important because the requirements on OF-KE embedding functionals differ substantially from those on approximations for $\mathscr{T}_s$.

3 Advances

3.1 T-Dependence of $\mathscr{T}_s$, $\mathscr{S}_s$, and New GGA Non-interacting Functionals

Despite skepticism about obtaining adequate approximations for τ_s and σ_s, in fact recent times have seen significant progress. Suffice it to say here that in roughly the last decade a modest-sized but growing community of researchers is making headway on effective zero-temperature OF-DFT functionals. Several of us were involved [45–48]. A bibliographic list we keep lists 115 papers in that decade (and a total of 340). Though thorough, the list surely is not complete.

To have predictive capability, the approximate functionals should not be parameterized to voluminous data. That requirement leads to what are called constraint-based functionals. For the non-interacting functionals, our constraint-based development starts with a rigorous decomposition of $\mathcal{T}_s$ that yields important positivity constraints. Rearrange Eq. (6) into the form with a positive-definite integrand,

$$\mathcal{T}_s = \frac{1}{2}\sum_j f_j \int d\mathbf{r} \mid \nabla\varphi_j \mid^2 . \tag{13}$$

It follows that

$$\mathcal{T}_s[n] := \mathcal{T}_W[n] + \mathcal{T}_\theta[n],$$

$$\text{with} \quad \mathcal{T}_W[n] = \frac{1}{8}\int \frac{|\nabla n(\mathbf{r})|^2}{n(\mathbf{r})}\, d\mathbf{r} \ \text{ and } \ \mathcal{T}_\theta[n] \geq 0 . \tag{14}$$

Note the absence of a Thomas-Fermi term and the importance of $\mathcal{T}_W$, the von Weizsäcker functional (vW) [49]. At T $=$ 0 K, the non-negativity of $\mathcal{T}_\theta$ ("Pauli term") is a rigorous requirement [50–53]. (Related discussions are in Refs. [48] and [54].) Recently, two of us proved that the non-negativity requirement also holds at T $>$ 0 K [55]. In consequence, the "Pauli potential", $v_\theta \equiv \delta\mathcal{T}_\theta[n]/\delta n(\mathbf{r})$, also must be non-negative for all $\mathbf{r}$. This constraint is extremely strict; see our earlier work at zero temperature for examples [47]. (As occasionally happened during the IPAM Long Program, the vocabulary "Pauli potential" can cause confusion. Clearly the quantity discussed here is not the same as the effective potential used to mimic quantum effects in classical potential models.)

The only nonzero-T non-interacting free-energy functionals known prior to the work we are about to discuss are the Thomas-Fermi (ftTF) [56] and the gradient expansion. It usually is truncated at second order, leading to the nonzero-T second-order gradient approximation functional (ftSGA) [57–59]. Both functionals are exact in the high-T limit, but are inadequate to serve as a $\mathcal{T}_s$ approximation in the WDM temperature range.

To address this lack of an adequate nonzero-T non-interacting free-energy functional, we have developed a generalized gradient approximation (GGA) framework [60]. Analysis of the gradient expansion shows clearly that the non-interacting kinetic energy and entropy contributions to the free energy differ in their dependence on T. From this, we were able to define appropriate, T-dependent reduced density gradients for parameterizing the kinetic and entropic contributions separately. By analogy with zero-T GGA, the nonzero-T GGAs are formed as the LDA energy density multiplied by a kinetic or entropic enhancement factor. Those factors are

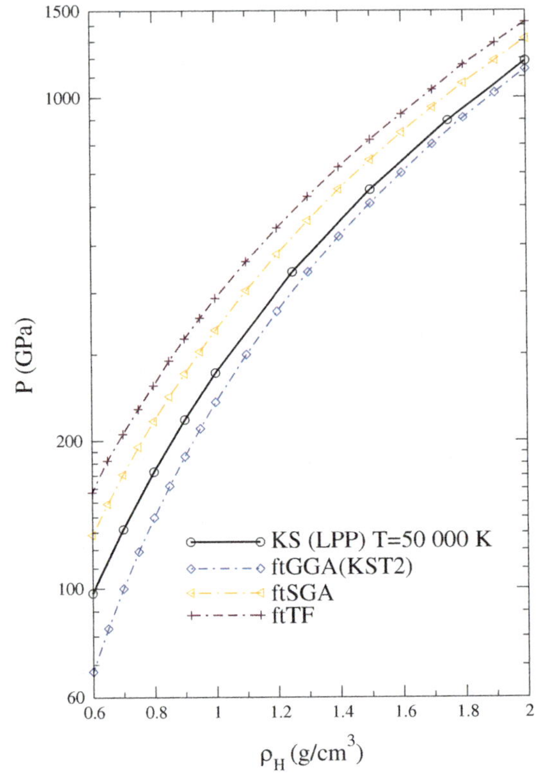

Fig. 4 Pressure vs. material density of simple cubic H at T = 50,000 K for various non-interacting free-energy functionals $\mathscr{F}_s = \mathscr{T}_s - \mathrm{T}\mathscr{S}_s$ compared to Kohn-Sham calculation. Note that the pressure axis is logarithmic

functions of the corresponding kinetic and entropic reduced density gradients. By construction, these nonzero-T GGA functionals satisfy exact scaling properties [55].

A simple, straightforward choice for the form of the nonzero-T GGA KE enhancement factor is a zero-T GGA enhancement factor (at least that choice provides a correct zero-T limit) but with the new kinetic reduced density variable instead of the zero-T reduced density gradient $s(\mathbf{r}) = (24\pi^2)^{-1/3}|\nabla n|/n^{4/3}$. There is no analogous choice for the entropic contribution. Instead, from a basic thermodynamic identity, we derived a differential equation which relates the two enhancement factors such that, once the KE enhancement factor is specified, its entropic counterpart can be calculated [60]. As a first example, we proposed a simple approximate solution to that rather complicated relationship and used it and a kinetic energy enhancement factor based on the zero-T "modified conjoint GGA" kinetic energy of Ref. [47]. Compared to either ftTF or ftSGA, the new Karasiev-Sjostrom-Trickey (KST2) functional is substantially superior, though still in need of improvement. An example of the effects on the EOS is in Fig. 4.

3.2 T-Dependence of $\mathscr{F}_{\mathrm{xc}}$

Contemporary nonzero-T DFT calculations for AIMD typically use [61, 62] an XC functional derived for zero temperature, i.e.,

$$\mathscr{F}_{\mathrm{xc}}[n(\mathbf{r},\mathrm{T}),\mathrm{T}] \approx E_{\mathrm{xc}}[n(\mathbf{r},\mathrm{T})]\,. \tag{15}$$

A few workers have used nonzero-T LDA functionals that are based on interpolation between high and low-T limits, or on calculations in the intermediate temperature region obtained via approximate methods such as RPA or local-field corrections [34,63–65]. Beyond LDA, only the gradient expansion term for nonzero-T exchange has been given [66–68]. In short, there is not a well-established set of XC functionals suitable for DFT computations at nonzero temperature, in sharp contrast with the enormous diversity of approximate ground-state functionals.

This situation makes it a priority item to generate XC free-energy functionals of documented reliability at WDM temperatures. An important component of this task is the establishment of reference data for calibration and/or validation of candidate functionals. Somewhat surprisingly (and contrary to a referee comment we once received), there is little reference data for nonzero temperature of anywhere near the quality of the QMC data on which zero-T XC LDA is based [69]. This state of affairs is in contrast to that for zero temperature, for which there are QMC data, jellium surface data, and more-or-less standard test sets of molecular data and well-established solid-state data for lattice constants and bulk moduli that can be used to assess the quality of a new functional. Such benchmark data are not easy to generate from first-principles calculations for $\mathrm{T} > 0\,\mathrm{K}$.

In response, we have studied the effects of explicit T-dependence in the exchange free energy functional $\mathscr{F}_{\mathrm{x}}$, by comparing nonzero-T Hartree-Fock (thermal HF) calculations with nonzero-T DFT KS calculations for both extended (periodically bounded) and confined systems. The DFT calculations used exchange-only functionals, of both the nonzero-T [34] and ordinary zero-T types. Remark: Thermal HF is rigorously established in the grand ensemble; see Refs. [35–37]. In particular, the full T-dependence of the exchange free energy is obtained through thermally occupied orbitals which appear in the familiar HF exchange term.

The periodically bounded case that we studied [70] was bcc Li, compressed at one to three times ambient density and at $0 \le \mathrm{T} \le 100{,}000\,\mathrm{K}$. A careful study (see below) eliminated any pseudopotential effects. The resulting tHF and T-dependent LDA-exchange free energies agree semi-quantitatively. Both increase (decrease in magnitude) with increasing T. In marked contrast, the zero-T E_{x} functional with T-dependent density, $n(\mathbf{r},\mathrm{T})$, is essentially constant with increasing T. The electronic pressure for the nonzero-T exchange functional tracks the thermal HF results much more closely than the pressure from the zero-T functional. The pressure difference from either of the former two with respect to the zero-T functional is roughly 10 % at 100,000 K. See Fig. 5 for an example.

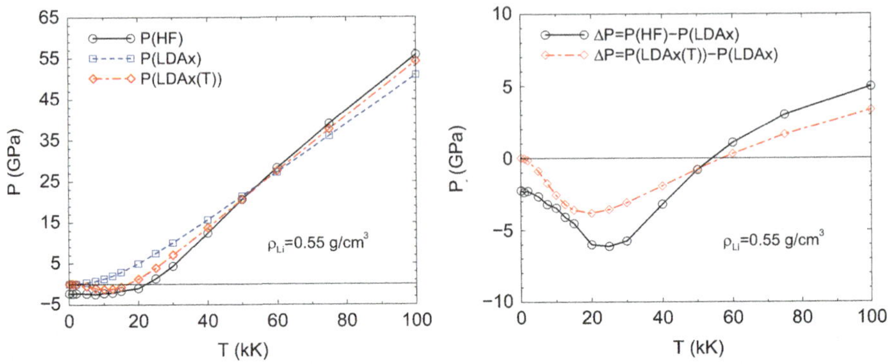

Fig. 5 Total pressure (*left*) and pressure difference (*right*) between two T-dependent $\mathscr{F}_x$ and a T-independent E_x used in a T-dependent calculation; bcc Li, $\rho = 0.55\,\mathrm{g/cm^3}$

Secondly, we devised and programmed an entirely new all-electron methodology (discussed below) for treatment of multi-atom systems confined in a hard-walled box [71]. For eight H atoms in a cubically symmetric array and $0 \leq \mathrm{T} \leq 300{,}000\,\mathrm{K}$, we find semi-quantitative agreement between the exchange free energy from tHF and that from the nonzero-T LDA X-functional F_{x}. As before, there is virtually no T-dependence from the zero-T $E_{\mathrm{xc}}[n(\mathbf{r}, \mathrm{T})]$. The energy difference between the N_e and $N_e + 1$ levels (the highest occupied and lowest unoccupied or virtual levels at $\mathrm{T} = 0\,\mathrm{K}$) decreases with increasing T for tHF but increases for DFT (for both zero-T and T-dependent functionals). This differing shift of initial virtuals as they become populated comes from their different self-interaction behavior. For DFT with an approximate XC functional, the initially occupied KS orbitals suffer spurious self-repulsion, whereas the original virtuals do not. In tHF, however, the virtuals arise from a full N_e-electron potential. Thus, as T increases, the initially occupied KS orbitals are partially depopulated and their self-interaction error decreases. In tHF, the initial virtuals become fractionally populated and thereby fractionally corrected (lowered).

For the correlation free energy, we compared several proposed nonzero-T XC functionals [34, 64, 65] with zero-T functionals for warm dense Li and H [72]. We find a non-trivial reduction in the total pressure. For simple cubic H (scH) with densities 1–4 $\mathrm{g/cm^3}$ and temperatures 100–300 kK, we find a pressure lowering of 2–4 %, relative to values predicted using zero-T XC functionals.

By use of the QUANTUM ESPRESSO @ PROFESS code capacity discussed below (see Methodological and Technical Advances) , we have done orbital-free AIMD simulations for Hydrogen and Deuterium at material densities from near-equilibrium at $\mathrm{T} = 0\,\mathrm{K}$ to about a four-fold compression in the range $2{,}000\,\mathrm{K} \leq \mathrm{T} \leq 4{,}000{,}000\,\mathrm{K}$. We compared our newly developed KST2 non-interacting GGA functional (discussed above) to the reference KS AIMD for $\mathrm{T} < 200{,}000\,\mathrm{K}$ and to

Fig. 6 *Top*: Instantaneous pressures for OF-DFT-AIMD compared to KS-AIMD for three OF-DFT non-interacting functionals: tTF, TW, and KST2. All with ground-state LDA XC. The system is Deuterium at material density $\rho = 1.9636\,\mathrm{g/cm^3}$. *Bottom*: percentage differences relative to KS-AIMD. Note the logarithmic temperature scale for both panels

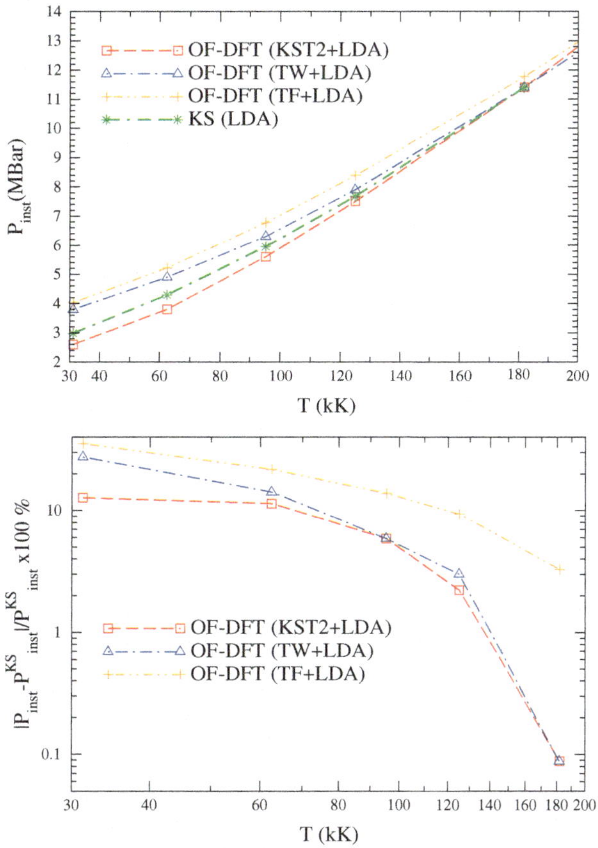

both tTF and our finite-T generalization of the Tran-Wesołowski KE functional [73] and the corresponding approximate entropy functional obtained by the procedure in Ref. [60]. That functional is labeled ftTW. (We remark, in fairness to Tran and Wesołowski, that their functional never was intended for this purpose. We simply use it as an example of a zero-T functional from the literature which might, naïvely perhaps, be generalized to non-zero T.) To isolate the effects of the non-interacting functionals, all the simulations were done with simple ground-state LDA XC. Note that for T > 200 kK, the KS procedure becomes an intractable task. Figure 6 shows both the actual EOSs (with "instantaneous pressure", that is, without the ideal gas kinetic contribution) and the percentage error relative to the KS calculation for Deuterium at $\rho_D = 1.96361\,\mathrm{g/cm^3}$ ($r_s = 1.40$ bohr). The EOS from the KST2 functional clearly is in better agreement with the reference KS results than the EOS from either the ftTF or ftTW (which actually behaves rather like ftSGA) functionals, particularly in the lower part of the temperature range. Note added in proof: Subsequent development of the non-empirical VT84F functional provides essentially the same improvement [74].

3.3 Exact Properties

Success for the constraint-based strategy obviously depends on having a sufficient number of meaningful constraints. The many formal relationships known for zero-temperature DFT have played that role in the development of constraint-based ground state XC functionals. Corresponding relationships at nonzero temperature would be useful, hence it is important to determine the extent to which the zero-T results remain valid or can be generalized for non-zero T. We have established a number of formal results for free energy DFT.

Homogeneous scaling [75–78] has been particularly important in developing zero-T XC functionals. We have extended this scaling to finite-T non-interacting functionals with a thorough analysis [55] and showed that (i) exact representation in terms of the one-particle Fermi density matrix is extremal among all single-particle density matrices; (ii) the von Weizsäcker functional is a lower bound to the non-interacting kinetic energy for all T; (iii) a single-particle density matrix corresponding to the TF approximation can be identified, thus giving an upper bound for the free energy; and (iv) exact scaling and associated upper and lower bounds can be established for the free energy, energy, and entropy functionals. Similar scaling results for the interacting system were published shortly thereafter by Pittalis et al. [79] via an analysis analogous with ground-state DFT scaling. Our scaling uses dimensional analysis and a demonstration of the relationship of DFT functionals to those of inhomogeneous-system thermodynamics. We have extended the treatment to the interacting case [80]. For it, both T and charge scaling are involved.

We also have investigated the phenomenological nonzero-T Thomas-Fermi functional (ftTF) for the non-interacting free energy [81–83]. From a functional expansion of the local free-energy density about its value at an arbitrary point, the LDA emerges as the leading term. This is the TF approximation for very large N_e, but not for finite N_e, as demonstrated by an exact evaluation for $N_e = 4$ in three dimensions at $\mathrm{T} = 0\,\mathrm{K}$. The non-interacting HEG free energy was calculated at finite T for arbitrary N_e, which gives the basis for a generalized TF approximation. Carried out to second order at nonzero T, the procedure generalizes the Perrot result [59] to arbitrary gradients. The expansion coefficients are non-linear response functions for the ideal Fermi gas as functions of the local density. The result is an exact limit for a class of non-local, linear response functionals introduced phenomenologically by others, for example, Refs. [38, 84–86].

Further, we have addressed the failure of the TF approximation for singular external potentials (e.g., Coulomb potentials) by a formal map of the exact functional for a given external potential onto a fictitious TF functional for an effective external potential [87], which we find to be a thermal TF approximation without the singularity.

The classical-map procedure [88] is an alternative route to functionals. It represents an equilibrium non-uniform quantum system by a corresponding classical system chosen to reproduce thermodynamic and structural properties. The connection allows application of classical strong-coupling methods. To obtain a better formal basis, we analyzed the grand canonical ensemble correspondence for the two systems [89] and showed that it enables definition of the classical-system effective T, local chemical potential, and pair potential. Approximate inversion of these definitions follows from the integral equations of classical liquid state theory. The structure and thermodynamics of the HEG (jellium) were calculated this way across a wide range of T and densities, including various high-T, weak-coupling, and low-density limits. We find good agreement with zero-T diffusion Monte Carlo results. The shell structure of harmonically confined charges (nonzero-T "harmonium") also was studied, with the result that both diffraction and exchange degeneracy can produce shell structure even in the absence of Coulomb correlations.

WDM transport properties are beyond the scope of strict DFT (which describes thermodynamic properties). Nevertheless, the KS orbitals and eigenvalues often are assumed to provide an effective medium, or mean field theory for the dynamical properties as well (e.g., transport properties, opacities, scattering cross sections). These uncontrolled approximations appear to have some empirical basis for validity. To clarify this context and provide improved approximations, the following initial results [90] have been obtained: (i) The exact Green-Kubo expression for the frequency-dependent conductivity has been evaluated in an arbitrary single-particle basis which is assumed to approximately diagonalize the Hamiltonian. The Kubo-Greenwood form follows, providing the basis for a variational formulation and associated improvements. (ii) A formal quantum kinetic theory for the Green-Kubo time-correlation functions provides a second approach to the context for the Kubo-Greenwood approximation. In particular, a natural mean-field approximation based on exact initial correlations goes beyond Hartree-Fock and is closely connected to the choice of a Kohn-Sham basis.

*3.4 Better Zero-*T *Limit Functionals*

To have at least reasonable levels of chemical realism in AIMD simulations of WDM, the zero-T KE functional $\mathcal{T}_s[n, \mathrm{T} = 0]$ is an important limit. In Ref. [91], we showed that an intriguing and novel information-theoretic orbital-free form for the zero-T KE functional is seriously flawed by violation of various positivity constraints. The analysis led us to propose a modified form which is positive definite. It remains to explore and develop that concept.

The zero-T XC functional $E_{xc}[n]$ is a similarly important limiting case for WDM simulations. In Refs. [92–95], one of us, with colleagues in Mexico, showed ways to make significant improvements in GGA E_x energetics. In addition, we gave both an improved (relative to B3LYP [96] and revTPSS [97]) meta-GGA and an improved hybrid functional. If a proposed rung-reduction procedure (analogous with the construction of the LYP functional [98]) works, these will give us better GGA X functionals, hence also contribute to the OF-DFT agenda.

3.5 Methodological and Technical Advances

With Eq. (14) and the substitution $\delta\mathcal{T}_{\rm W}/\delta n = -n^{-1/2}\nabla^2(n^{1/2})/2$, the Euler equation, Eq. (11), takes the form of a one-orbital eigenvalue equation,

$$\left\{-\frac{1}{2}\nabla^2 + \frac{\delta\mathcal{T}_\theta}{\delta n} - \mathrm{T}\frac{\delta\mathcal{S}_{\rm s}}{\delta n} + \mathrm{v}_{\rm KS}\right\} n^{1/2} = \mu\, n^{1/2} \,. \tag{16}$$

The form of this equation led to the seemingly obvious notion that numerical solution would be straightforward with a standard KS code. However, one of our studies [99] confirmed the contradictory prior claim [100] that, in general, such codes cannot handle the peculiar potential which appears in Eq. (16). The difficulty is the repulsive nature of any KE functional. The problem is worsened by the singular nature of approximate GGA KE functionals. Direct minimization algorithms are essential.

Most KS codes used to study WDM employ non-local pseudopotentials (PPs). The PPs are fitted to free ground-state atoms, hence their transferability to WDM conditions is not guaranteed but must be verified. We developed a new procedure to test PP transferability to high compression [70] by comparison to all-electron results for clusters of the same local symmetry and near-neighbor distances. For Li, we found that standard norm-conserving PPs and projector-augmented wave (PAW) data sets used in packages such as ABINIT [101, 102], QUANTUM ESPRESSO [103] and VASP [104–107] have relatively small transferability compression ranges, typically up to 1.5-, 4.5- or 7-fold compression depending on details. The compensation charge density (CCD) sometimes included in PAW data sets was found to cause problems at high compressions. Without CCD and with cutoff radius reduced by about a factor of two, we were able to build all-electron PAW LDA and GGA data sets for Li which are transferable to at least 140-fold compression.

Though OF-DFT in principle does not need PPs to exclude chemically inactive core states, regularization of the nuclear-electron interaction singularity is required for efficient implementation, e.g., in a plane-wave basis. PPs developed for KS calculations generally are non-local, i.e., orbital-dependent, thus are inapplicable in OF-DFT calculations. Local pseudopotentials (LPP) must be developed for OF-DFT. But there is no standard or even dominant method to do that. We developed a new method [99] aimed specifically at OF-DFT calculations. The LPP is constructed as a normalized linear combination of l-components of a norm-conserving non-local PP (NLPP). The coefficients are constrained to reproduce some bulk property. The method was tested for bcc Li. An alternative procedure, used for scH, is to fit the parameters of the simple Heine-Abarenkov model [108, 109] such that a bulk property (in our case, the lattice parameter) from a more sophisticated PP is reproduced. We showed that bulk properties are reproduced for material densities to at least $15\,\mathrm{g/cm^3}$ and for $\mathrm{T} \leq 100{,}000\,\mathrm{K}$.

Most of our OF-DFT calculations were done with a locally modified version of the PROFESS code [110, 111]. Originally set up with two-point KE functionals, it performs periodic OF-DFT calculations. Our modifications included adding four nonzero-T non-interacting free-energy functionals: (i) standard Thomas-Fermi (ftTF); (ii) gradient-corrected ftTF, i.e., the nonzero-T second-order gradient approximation (ftSGA)); and (iii) our new two-parameter Karasiev-Sjostrom-Trickey (KST2) form and the nonzero-T extension of the Tran-Wesołowksi zero-T KE (see details in Ref. [60]). For orbital-free AIMD, we coded an interface between PROFESS and QUANTUM-ESPRESSO [103] to replace KS forces with those from OF-DFT. This modified package enables comparison between KS AIMD and OF-DFT AIMD on the same footing [112].

The thermal Hartree-Fock, hard-walled rectangular box calculations mentioned in Sect. 3.2 [71] obviously require a basis set which is compatible with the box boundary conditions and also yields a computationally tractable problem. Sine functions, our original choice, turned out to have drawbacks for rapid matrix-element evaluation. Thus, we exploited familiarity with Gaussian type orbitals (GTOs) to provide efficient four-center integral calculation. The new basis is Cartesian GTOs truncated to match the boundary conditions and scaled to retain continuity at the atomic sites.

A Cartesian GTO of order ν has factors of the form

$$g^{\nu}(x) = (x - x_c)^{\nu} e^{-\alpha(x-x_c)^2} \tag{17}$$

The function can be made into one which is zero at the box boundaries $x = 0$ and $x = L_x$ by subtracting a shift equal to the original function value at each end. When x_c is not at the box center, that shift differs for the two ends. Thus, we split the original function into two pieces, then shift by the two end-values, and scale the two shifted pieces such that the resulting function is continuous. The outcome is an un-normalized Cartesian factor

$$\begin{aligned} g^{\nu}_{box}(x) &= a_0 \left(g^{\nu}(x) - \Delta_0\right) \quad 0 \le x \le x_c \\ &= a_L \left(g^{\nu}(x) - \Delta_{L_x}\right) \quad x_c \le x \le L_x \end{aligned} \tag{18}$$

with $\Delta_0 = g^{\nu}(0)$, $\Delta_{L_x} = g^{\nu}(L_x)$. These are the Cartesian factors of our truncated GTO (tGTO) basis. Because the tGTOs may not have continuous derivatives, kinetic energy matrix elements require attention. In fact, nothing unusual happens. The kinetic energy is the expected sum of piecewise contributions, except for the case of p-type functions, which have a simple correction term. Appropriate matrix element expressions were rederived to take account of these modifications from normal molecular GTO-basis techniques. The procedure was implemented in an entirely new code which supports both thermal HF and KS DFT calculations for any ion arrangement and T. So far we have used 1–32 H atoms and T up to 300 kK. The highly optimized code is written in C++ utilizing MPI. Future development includes nonzero-T post-tHF calculations for correlation and AIMD with analytically calculated forces.

Faced with a dearth of nonzero-T HEG data for calibrating and validating new OF-DFT functionals, we have undertaken a focused study using the restricted path-integral MD (PIMD) technique [113, 114]. This choice is deliberately complementary to the PIMC work underway elsewhere [69]. The original PIMD code had been applied to a rather low density HEG ($r_s = 5$ bohr) and T < 2,500 K, which is low for WDM. We have modified the code to extend the density and temperature ranges as well as remedy some instabilities. Initial results at $r_s = 4$ and 5 bohr and T < 7,000 K are encouraging. For example, the average KE of these HEGs can be fit with a Sommerfeld expansion

$$\begin{aligned} \bar{\mathcal{T}}(\mathrm{T}) &= \mathcal{T}_0[1 + ax^2 + bx^4 + \cdots], \\ x &:= \frac{1}{\beta \mathcal{T}_0}, \end{aligned} \tag{19}$$

where $\mathcal{T}_0 := \mathcal{T}(\mathrm{T} = 0)$. Through fourth order, the data are fit well by $a = 5.48$, $b = 78.0$ for both densities. Additional useful data for functional development is being calculated (V. Kapila, K. Runge, P.A. Deymier, unpublished).

4 Outlook and Challenges

While authors' perspectives on their own work may merit some skepticism, we think the spirit of the IPAM Long Program is exemplified by the word "Challenges" in the program name. So we venture comment on three questions.

1. What does this all mean, especially for application to the WDM problem?
2. What do we recommend as preferred practice at this point?
3. What are the shortcomings of the approaches and approximations we have used thus far?

About all these, prudence requires keeping clearly in mind that there is much still to be learned about free-energy DFT, about its orbital-free form, and about dynamics. That in mind, we plunge ahead.

The finding relative to question (1) that is perhaps most intriguing is that there is significant temperature-dependence in the XC free energy. Equation (15) does not hold in general. But how, then, are the apparent successes, e.g. Ref. [115], in using Eq. (15) to be understood? We suspect that the underlying distinction is between calculating the EOS or the Hugoniot for a system versus calculating a Kubo-Greenwood conductivity. Such a conductivity is given by

$$\sigma(\omega) = \frac{1}{N_{\rm config}} \sum_{I=1}^{N_{\rm config}} \sum_{\mathbf{k}} w_{\mathbf{k}} \sigma_{\mathbf{k}}(\omega; \{\mathbf{R}\}_I),$$

$$\sigma_{\mathbf{k}}(\omega; \{\mathbf{R}\}_I) = \frac{2\pi}{3\omega\Omega} \sum_{i,j}^{N_b} \sum_{\nu=1}^{3} (f_{i,\mathbf{k},I} - f_{j,\mathbf{k},I}) |\langle \varphi_{j,\mathbf{k},I} | \nabla_\nu | \varphi_{i,\mathbf{k},I} \rangle|^2 \times \delta(\epsilon_{j,\mathbf{k},I} - \epsilon_{i,\mathbf{k},I} - \omega) \,. \quad (20)$$

Here Ω is the system volume, $w_{\mathbf{k}}$ is the weight of point $\mathbf{k}$ in the Brillouin zone and the KS quantities have multiple indices in consequence of configurational sampling ("snapshotting") over the AIMD trajectories. The obvious distinction with an EOS or Hugoniot calculation is that transport coefficients depend on matrix elements with respect to KS orbitals and upon the associated KS eigenvalues. It has been known since the earliest days of ground-state DFT that one can have an $E_{\rm xc}$ functional that gives good bulk properties (e.g. lattice parameters) but a poor band structure. However, another simple functional, call it $\tilde{E}_{\rm xc}$, usually can be found such that $\tilde{\rm v}_{\rm xc} = \delta \tilde{E}_{\rm xc}/\delta n$ gives a good band structure (but poor bulk properties) [116]. We suspect that a similar distinction may be relevant to the T-dependent case. What we have found is that the T-dependence of $\mathscr{F}_{\rm xc}$ is important for the EOS. It does not necessarily follow that the T-dependence of the associated ${\rm v}_{\rm xc}$ is important. If the T-dependence is important in principle, it is not necessarily true that an approximate ${\rm v}_{\rm xc}$ has the *correct* T-dependence for Eq. (20) even if its precursor $\mathscr{F}_{\rm xc}$ does well for the EOS or Hugoniot.

For OF-DFT, question (1) invites a discussion of the obviously related issue of the T-dependence of the non-interacting functionals $\mathscr{T}_{\rm s}$ and $\mathscr{S}_{\rm s}$. For these, we think the evidence is clear. The only functionals that come reasonably close to reproducing the KS results are our KST2 and the combination of full vW with full TF, i.e., ftVWTF. We have not discussed ftVWTF here, but refer to Ref. [60]. What is relevant here is that the two functionals behave quite similarly. They also share a drawback, in that both have empirical elements. But ftVWTF is, we think, irretrievably empirical in the sense that it is an ad hoc combination of two functionals, one for the high-T, high density limit, the other for the 1- and 2-electron localized limit. Our KST2 is empirical only in the sense of having two constants chosen to match KS calculations on a few reference silicate systems. (Silicates were chosen for historical reasons having nothing to do with WDM.) We have formulated a procedure by which we believe we can eliminate that empirical dependence. That procedure is under study at this writing.

Question (1) also implicates the issue of approximate $\mathscr{F}_{\rm xc}$ functionals that are orbital free. Were we to have the exact $\mathscr{T}_{\rm s}$ and $\mathscr{S}_{\rm s}$, we still would be stuck with XC approximations which, in the T $\rightarrow$ 0 K limit would be no better than today's XC GGAs. The problem is that no current $E_{\rm xc}^{\rm GGA}$ does equally well on molecular and extended systems. For a given GGA functional form, constraint-based arguments can be used (and have been) to give parameterizations that are well-suited to molecules, but do poorly on solids. A different choice of constraints leads to the opposite result. (Discussion and references are in Ref. [94].) Since WDM can comprise both extended phase and localized regimes concurrently, this is a problem.

So what do we recommend? For now, use KST2 for $\mathscr{T}_s$ and $\mathscr{S}_s$, but be aware that better functionals will be forthcoming from our group as well as from others. For $\mathscr{F}_x$, what we have designated above as LDAx(T) is the choice for the moment if one wants to include T-dependence at least semi-quantitatively. We do not know what to say about $\mathscr{F}_c$ at the moment. Whatever the choice of OF-functionals, for those computing transport coefficients, the need to do KS calculations at snapshotted ion configurations provides an internal consistency test on the OF calculation. For a given ion configuration $\{\mathbf{R}\}$, simply compare the KS and OF-DFT densities and energies. This also leads to recommending, once again, the graded-sequence-of-approximations (GSA) procedure [117]. In the present context, the simplest GSA simulation is to run a few AIMD-MD steps with approximate OF-DFT functionals, then do a single KS step to get force corrections, then do another set of OF-DFT steps, etc.

Some of the shortcomings of the present state of affairs already have been mentioned, e.g. the lack of uniformly good predictive behavior from any E_{xc}^{GGA} available today. It is worth emphasizing that the problem is more tangled (and may be worse) in WDM than in ordinary materials because of the thermal occupation of KS levels which, at $T = 0\,K$, are virtuals (unoccupied). Self-interaction error, which is known in E_{xc}^{GGA} and $\mathscr{F}_{xc}^{GGA}$, may also play an unexpected role in $\mathscr{T}_s$ and $\mathscr{S}_s$. This problem is easy to state. Improper self-repulsion allowed by approximate E_{xc} and $\mathscr{F}_{xc}$ functionals (zero T, non-zero T respectively) delocalizes the density n incorrectly. Presumably that delocalization has the effect of reducing the numerical values of $\mathscr{T}_s[n]$ (zero-T) or $\mathscr{T}_s[n, T]$ (non-zero T) from what they would be were one to use the exact (non-self-interacting) XC functionals. We do not know what such delocalization does to $\mathscr{S}_s$. A second shortcoming of all current OF-DFT functionals at $T = 0\,K$ is that none of them is built to satisfy the differential virial relationship between the KE and XC functional [75]. Presumably this problem carries over to non-zero temperature.

Acknowledgements We acknowledge, with thanks, many informative conversations with participants in the IPAM Long Program, including Kieron Burke, Jerome Daligault, Mike Desjarlais, Andreas Görling, Frank Graziani, Leslie Greengard, Hardy Gross, Stephanie Hansen, Walter Johnson, David Levermore, Winfried Lorenzen, Patrick Ludwig, Andreas Markmann, Michael Murillo, Aurora Pribram-Jones, Ronald Redmer, Luke Shulenberger, and Brian Wilson. We thank Russell Caflisch, Christian Ratsch, and Roland McFarland for the arrangements for those of us who visited IPAM. This work was supported under U.S. Dept. of Energy BES (TCMP, TMS) grant DE-SC 0002139.

Appendix

Acronyms

AIMD	Ab initio molecular dynamics
B3LYP	Empirical XC functional [96] which combines Becke 3-parameter X with Lee, Yang, Parr parameterization of Colle-Salvetti correlation [118]
BO; BOMD	Born-Oppenheimer; Born-Oppenheimer molecular dynamics
DFT	Density functional theory
ftSGA	Finite-T second-order gradient approximation
ftTF	Finite-T Thomas-Fermi
FE-DFT	Free-energy density functional theory, i.e. nonzero-T DFT
GGA	Generalized gradient approximation; most commonly an XC functional but also used here for KS (non-interacting) kinetic energy and entropy functionals
GTO	Gaussian-type orbital
HEDP	High energy-density physics
HEG	Homogeneous electron gas
meta-GGA	An approximate functional (most commonly XC) dependent on higher derivatives of the density than a GGA. Most use the KS KE orbital density τ_{orb}
KE	Kinetic energy
KS	Kohn-Sham
LDA; LSDA	Local-density approximation; local spin-density approximation
OF-DFT	Orbital-free density functional theory
OF-KE	Orbital-free kinetic energy (usually, orbital-free KS kinetic energy)
PAW	Projector Augmented Wave
PIMD	Restricted Path Integral Molecular Dynamics
PP	Pseudopotential
QMC	Quantum Monte-Carlo
RPA	Random phase approximation
tHF	thermal Hartree-Fock
TF	Thomas-Fermi
VASP	Vienna Ab-initio Simulation Program
vW	von Weizsäcker functional
WDM	Warm dense matter
X	Exchange
XC	Exchange-correlation

Symbols

Not infrequently the same symbol is used for different quantities in the T > 0 K and T = 0 K DFT literature. An example is temperature and kinetic energy. In our papers we have worked toward a simple standard, as close as reasonably possible to the T = 0 K DFT literature. In its present form, that usage is as follows.

$$\beta = 1/k_B\mathrm{T}$$
$$E_x,\ E_{xc} = \text{exchange, exchange-correlation functional}$$
$$\mathcal{F} = \text{Helmholtz free energy}$$
$$\mathcal{F}_{xc},\ \mathcal{F}_s = \text{XC and KS free energies}$$
$$k_B = \text{Boltzmann's constant}$$
$$\mathcal{S},\ \mathcal{S}_s = \text{total, KS entropy}$$
$$n(\mathbf{r}) = \text{electron density at point } \mathbf{r}$$
$$\mathrm{T} = \text{temperature}$$
$$\mathcal{T},\ \mathcal{T}_s = \text{total, KS kinetic energy}$$

References

1. K. Gavroglu, Eur. J. Phys. **15**, 9 (1994) and references therein
2. P.A. Bauer, E.K. Dabora, N. Manson, in *Dynamics of Detonations and Explosions: Detonations*, ed. by A.L. Kuhl, J.-C. Leyer, A.A. Borisov, W.A. Siriganano (American Institute of Aeronautics and Astronautics, Washington, DC, 1991), pp. 3ff and references therein
3. *Basic Research Needs for High Energy Density Laboratory Physics* (Report of the Workshop on Research Needs, Nov. 2009). U.S. Department of Energy, Office of Science and National Nuclear Security Administration (2010); see Chapter 6 and references therein
4. H.F. Wilson, B. Militzer, Phys. Rev. Lett. **108**, 111101 (2012)
5. R.P. Drake, Phys. Today **63**, 28 (2010) and references therein
6. B.A. Remington, R.P. Drake, D.D. Ryutov, Rev. Mod. Phys. **78**, 755 (2006) and references therein
7. R.P. Drake, *High-Energy-Density Physics* (Springer, Berlin, 2006) and references therein
8. J. Li, C.B. Agee, Nature **381**, 686 (1996)
9. M.J. Drake, K. Righter, Nature **416**, 39 (2002)
10. E. Jagoutz, H. Palme, H. Baddenhausen, K. Blum, M. Cendales, G. Dreibus, B. Spettel, V. Lorenz, H. Wanke, Proc. Lunar Planet. Sci. Conf. **10**, 2031 (1979)
11. H. Wanke, Phil. Trans. Roy. Soc. Lond. A **303**, 287 (1981)
12. C.J. Allegre, J.-P. Poirer, E. Humler, A.W. Hofmann, Earth Planet. Sci. Lett. **134**, 515 (1995)
13. K. Righter, M.J. Drake, Earth Planet. Sci. Lett. **146**, 541 (1997)
14. J.C. Boettger, S.B. Trickey, Phys. Rev. B **51**, R15623 (1995)
15. J.C. Boettger, S.B. Trickey, Phys. Rev. B **53**, 3007 (1996)
16. R.N. Barnett, U. Landman, Phys. Rev. B **48**, 2081 (1993)

17. D. Marx, J. Hutter, Ab initio molecular dynamics: theory and implementation, in *Modern Methods and Algorithms of Quantum Chemistry*, ed. by J. Grotendorst, NIC Series, vol. 1 (John von Neumann Institute for Computing, J ulich, 2000), pp. 301ff and references therein
18. J.S. Tse, Annu. Rev. Phys. Chem. **53**, 249 (2002)
19. D. Marx, J. Hutter, *Ab Initio Molecular Dynamics: Basic Theory and Advanced Methods* (Cambridge University Press, Cambridge, 2009) and references therein
20. T.D. Kühne, Ab-initio molecular dynamics arXiv:1201.5945 [physics.chem-ph]
21. W. Kohn, L.J. Sham, Phys. Rev. **140**, A1133 (1965)
22. P. Hohenberg, W. Kohn, Phys. Rev. **136**, B864 (1964)
23. M. Levy, Proc. Natl. Acad. Sci. USA **76**, 6062 (1979)
24. E.H. Lieb, Int. J. Quantum Chem. **24**, 243 (1983)
25. R.G. Parr, W. Yang, *Density Functional Theory of Atoms and Molecules* (Oxford, New York, 1989). For T-dependent theory see Sections 3.6 and 9.1
26. R.M. Dreizler, E.K.U. Gross, *Density Functional Theory* (Springer, Berlin, 1990)
27. L.A. Collins (Los Alamos National Lab), Equilibrium and non-equilibrium orbital-free molecular dynamics simulations at extreme conditions, CECAM Workshop, Paris, 05 Sept 2012
28. D.O. Gericke (Univ. Warwick), Effective interactions and ion dynamics in warm dense matter, Invited Talk I8, Physics of Non-ideal Plasmas 14, Rostock Germany, 11 Sept 2012
29. N.D. Mermin, Phys. Rev. **137**, A1441 (1965)
30. A.K. Rajagopal, A density functional formalism for condensed matter systems, in *Density Functional Methods in Physics*, ed. by J. da Providencia, R.M. Dreizler, NATO ASI B, vol. 123 (Plenum, New York, 1985), pp. 159ff
31. M.V. Stoitsov, I.Zh. Petkov, Ann. Phys. **184**, 121 (1988)
32. R.M. Dreizler, Density functional theory at finite temperatures, in *The Nuclear Equation of State, Part A*, ed by W. Greiner, H. Stöcker, NATO ASI B, vol. 216 (Plenum, NY, 1989), pp. 521ff
33. J.P. Perdew, K. Schmidt, A.I.P. Conf. Proc. **577**, 1 (2001)
34. F. Perrot, M.W.C. Dharma-wardana, Phys. Rev. A **30**, 2619 (1984)
35. N.D. Mermin, Ann. Phys. (NY) **21**, 99 (1963)
36. J. Sokoloff, Ann. Phys. (NY) **45**, 186 (1967)
37. J. Dolbeault, P. Felmer, M. Lewin, Math. Model. Methods Appl. Sci. **19**, 347 (2009)
38. Y.A. Wang, E.A. Carter, Orbital-free kinetic-energy density functional theory, Chap. 5 in *Theoretical Methods in Condensed Phase Chemistry*, ed. by S.D. Schwartz (Kluwer, New York, 2000), pp. 117ff and references therein
39. M. Hodak, W. Lu, J. Bernholc, J. Chem. Phys. **128**, 014101 (2008)
40. X. Blanc, E. Cances, J. Chem. Phys. **122**, 214106 (2005)
41. E.V. Ludeña, J. Chem. Phys. **79**, 6174 (1983)
42. R. Cuevas-Saavedra, D. Chakraborty, P.W. Ayers, Phys. Rev. A **85**, 042519 (2012)
43. T.A. Wesołowski, Phys. Rev. A **77**, 012504 (2008) and references therein
44. J.W. Kaminski, S. Gusarov, T.A. Wesołowski, A. Kovalenko, J. Phys. Chem. A, **114**, 6082 (2010) and references therein
45. V.V. Karasiev, S.B. Trickey, F.E. Harris, J. Comput. Aided Mat. Des. **13**, 111 (2006)
46. V.V. Karasiev, R.S. Jones, S.B. Trickey, F.E. Harris, Recent advances in developing orbital-free kinetic energy functionals, in *New Developments in Quantum Chemistry*, ed. by J.L. Paz, A.J. Hernández (Transworld Research Network, Kerala, 2009), pp. 25ff
47. V.V. Karasiev, R.S. Jones, S.B. Trickey, F.E. Harris, Phys. Rev. B **80**, 245120 (2009)
48. S.B. Trickey, V.V. Karasiev, R.S. Jones, Int. J. Quantum Chem. **109**, 2943 (2009)
49. C.F. von Weizsäcker, Z. Phys. **96**, 431 (1935)
50. Y. Tal, R.F.W. Bader, Int. J. Quantum Chem. **S12**, 153 (1978)
51. L.J. Bartolotti, P.K. Acharya, J. Chem. Phys. **77**, 4576 (1982)
52. J.E. Harriman, The interface between reduced density matrices and density functional theory, in *Density Matrices and Density Functionals*, ed. by R. Erdahl, V.H. Smith Jr. (D. Reidel, Dordrecht, 1987), pp. 359ff

53. M. Levy, H. Ou-Yang, Phys. Rev. A **38**, 625 (1988)
54. R. Baltin, J. Chem. Phys. **86**, 947 (1987)
55. J.W. Dufty, S.B. Trickey, Phys. Rev. B **84**, 125118 (2011)
56. R.P. Feynman, N. Metropolis, E. Teller, Phys. Rev. **75**, 1561 (1949)
57. J. Bartel, M. Brack, M. Durand, Nucl. Phys. A **445**, 263 (1985)
58. M. Brack, C. Guet, H.-B. Hakansson, Phys. Rep. **123**, 275 (1985)
59. F. Perrot, Phys. Rev. A **20**, 586 (1979)
60. V.V. Karasiev, T. Sjostrom, S.B. Trickey, Phys. Rev. B **86**, 115101 (2012)
61. B. Holst, R. Redmer, M. Desjarlais, Phys. Rev. B **77**, 184201 (2008)
62. D.A. Horner, F. Lambert, J.D. Kress, L.A. Collins, Phys. Rev. B **80**, 024305 (2009)
63. W. Ebeling, W. Richert, W.D. Kraeft, W. Stolzmann, Phys. Stat. Solidi (b) **104**, 193 (1981)
64. S. Tanaka, S. Mitake, S. Ichimaru, Phys. Rev. A **32**, 1896 (1985)
65. F. Perrot, M.W.C. Dharma-wardana, Phys. Rev. B **62** 16536 (2000); Ibid. **67**, 079901(E) (2003)
66. D.J.W. Geldart, E. Dunlap, M.L. Glasser, M.R.A. Shegelski, Solid State Commun. **88**, 81 (1993)
67. E. Dunlap, D.J.W. Geldart, Can. J. Phys. **72**, 1 (1994)
68. M.L. Glasser, D.J.W. Geldart, E. Dunlap, Can. J. Phys. **72**, 7 (1994)
69. E. Brown. D. Ceperley, J. Dubois, Simulation of the warm, dense homogeneous electron gas. Paper X24-2, American Physical Society, March meeting, 1 Mar 2012
70. V.V. Karasiev, T. Sjostrom, S.B. Trickey, Phys. Rev. E **86**, 056704 (2012)
71. T. Sjostrom, F.E. Harris, S.B. Trickey, Phys. Rev. B **85**, 045125 (2012)
72. T. Sjostrom, Finite-temperature hartree-fock exchange and exchange-correlation free energy functionals, *IPAM Workshop IV, Computational Challenges in Warm Dense Matter*, UCLA, Los Angeles, May 2012, https://www.ipam.ucla.edu/publications/plws4/plws410624.pdf and to be published
73. F. Tran, T.A. Wesołowski, Int. J. Quantum Chem. **89**, 441 (2002)
74. V.V. Karasiev, D. Chakraborty, O.A. Shukruto, S.B. Trickey, Phys. Rev. B **88**, 161108(R) (2013)
75. M. Levy, J.P. Perdew, Phys. Rev. A **32**, 2010 (1985)
76. M. Levy, The coordinate scaling requirements in density functional theory, in *Single-Particle Density in Physics and Chemistry*, ed. by N.H. March (Academic, London/San Diego, 1987), pp. 45ff
77. M. Levy, Phys. Rev. A **43**, 4637 (1991)
78. M. Levy, Coordinate scaling requirements for approximating exchange and correlation, in *Density Functional Theory*, ed. by E.K.U. Gross, R.M. Dreizler, NATO ASI B, vol. 337 (Plenum, New York, 1995), pp. 11ff
79. S. Pittalis, C.R. Proetto, A. Floris, A. Sanna, C. Bersier, K. Burke, E.K.U. Gross, Phys. Rev. Lett. **107**, 163001 (2011)
80. J.W. Dufty, S.B. Trickey, Finite temperature scaling in density functional theory (in preparation)
81. T. Gál, J.W. Dufty, S.B. Trickey, Local density approximation to the kinetic energy functional for finite N_e, in preparation
82. J. Dufty, S. Dutta, Phys. Rev. E **87**, 032101 (2013)
83. J. Dufty, S. Dutta, Phys. Rev. E **87**, 032102 (2013)
84. L.W. Wang, M.P. Teter, Phys. Rev. B **45**, 13196 (1992)
85. B.J. Zhou, V.L. Ligneres, E.A. Carter, J. Chem. Phys. **122**, 044103 (2005)
86. C. Huang, E.A. Carter, Phys. Rev. B **81**, 045206 (2010)
87. J.W. Dufty, S.B. Trickey, Revised Thomas-Fermi functional for singular potentials, American Physical Society, March 2013 meeting abstract
88. F. Perrot, M.W.C. Dharma-wardana, Phys. Rev. B **62**, 16536 (2000)
89. J.W. Dufty, S. Dutta, Contrib. Plasma Phys. **52**, 100 (2012)
90. J.W. Dufty, Derivation and context of Kubo-Greenwood transport coefficients (in preparation)
91. S.B. Trickey, V.V. Karasiev, A. Vela, Phys. Rev. B **84**, 075146 (2011)

92. A. Vela, J.C. Pacheco-Kato, J.L. Gázquez, J.M. del Campo, S.B. Trickey, J. Chem. Phys. **136**, 144115 (2012)
93. J. Martín del Campo, J.L. Gázquez, R.J. Alvarez-Mendez, S.B. Trickey, A. Vela, Analysis of the generalized gradient approximation for the exchange energy, in *Concepts and Methods in Modern Theoretical Chemistry*, vol. 1, ed. by volume honoring Professor B.M. Deb, S.K. Ghosh and P.K. Chattaraj (Taylor and Francis/CRC, Boca Raton, 2012) in press
94. J.M. del Campo, J.L. Gázquez, S.B. Trickey, A. Vela, J. Chem. Phys. **136**, 104108 (2012)
95. J. Martín del Campo, A. Vela, J.L. Gázquez, S.B. Trickey, Chem. Phys. Lett. **543**, 179 (2012)
96. P.J. Stephens, F.J. Devlin, C.F. Chabalowski, M.J. Frisch, J. Phys. Chem. **98**, 11623 (1994)
97. J.P. Perdew, A. Ruzsinszky, G.I. Csonka, L.A. Constantin, J.W. Sun, Phys. Rev. Lett. **103**, 026403 (2009). Erratum, Ibid. **106**, 179902 (2011)
98. C. Lee, W. Yang, R.G. Parr, Phys. Rev. B **37**, 785 (1984)
99. V.V. Karasiev, S.B. Trickey, Comput. Phys. Commun. **183**, 2519 (2012)
100. G.K-L. Chan, A.J. Cohen, N.C. Handy, J. Chem. Phys. **114**, 631 (2001)
101. X. Gonze, B. Amadon, P.-M. Anglade, J.-M. Beuken, F. Bottin, P. Boulanger, F. Bruneval, D. Caliste, R. Caracas, M. Cote, T. Deutsch, L. Genovese, Ph. Ghosez, M. Giantomassi, S.G̃oedecker, D.R. Hamann, P. Hermet, F. Jollet, G. Jomard, S. Leroux, M. Mancini, S.M̃azevet, M.J.T. Oliveira, G. Onida, Y. Pouillon, T. Rangel, G.-M. Rignanese, D. Sangalli, R. Shaltaf, M. Torrent, M.J. Verstraete, G. Zerah, J.W. Zwanziger, ABINIT: first-principles approach to material and nanosystem properties, Comput. Phys. Commun. **180**, 2582 (2009)
102. X. Gonze, G.-M. Rignanese, M. Verstraete, J.-M. Beuken, Y. Pouillon, R. Caracas, F. Jollet, M. Torrent, G. Zerah, M. Mikami, Ph. Ghosez, M. Veithen, J.-Y. Raty, V. Olevano, F.B̃runeval, L. Reining, R. Godby, G. Onida, D.R. Hamann, D.C. Allan, A brief introduction to the ABINIT software package, Zeit. Kristallogr. **220**, 558 (2005)
103. P. Giannozzi, S. Baroni, N. Bonini, M. Calandra, R. Car, C. Cavazzoni, D. Ceresoli, G.L.C̃hiarotti, M. Cococcioni, I. Dabo, A.D. Corso, S. de Gironcoli, S. Fabris, G. Fratesi, R. Gebauer, U. Gerstmann, C. Gougoussis, A. Kokalj, M. Lazzeri, L. Martin-Samos, N. Marzari, F. Mauri, R. Mazzarello, S. Paolini, A. Pasquarello, L. Paulatto, C. Sbraccia, S. Scandolo, G. Sclauzero, A.P. Seitsonen, A. Smogunov, P. Umari, R.M. Wentzcovitch, Quantum espresso: a modular and open-source software project for quantum simulations of materials, J. Phys. Condens. Matter **21**, 395502 (2009)
104. G. Kresse, J. Hafner, Phys. Rev. B **47**, 558 (1993)
105. G. Kresse, J. Hafner, Phys. Rev. B **49**, 14251 (1994)
106. G. Kresse, J. Furthmüller, Comput. Mat. Sci. **6**, 15 (1996)
107. G. Kresse, J. Furthmüller, Phys. Rev. B **54**, 11169 (1996)
108. V. Heine, I. Abarenkov, Phil. Mag. **9**, 451 (1964)
109. L. Goodwin, R.J. Needs, V. Heine, J. Phys. Condens. Matter **2**, 351 (1990)
110. G.S. Ho, V.L. Lignères, E.A. Carter, Comput. Phys. Commun. **179**, 839 (2008)
111. L. Hung, C. Huang, I. Shin, G.S. Ho, V.L. Lignères, E.A. Carter, Comput. Phys. Commun. **181**, 2208 (2010)
112. V.V. Karasiev, T. Sjostrom, S.B. Trickey, Finite-temperature orbital-free DFT molecular dynamics: coupling PROFESS and QUANTUM ESPRESSO (in preparation)
113. Ki-dong Oh, P.A. Deymier, Phys. Rev. B **58**, 7577 (1998) and references therein
114. Ki-dong Oh, P.A. Deymier, Phys. Rev. B **59**, 11276 (1999) and references therein
115. M.P. Desjarlais, J.D. Kress, L.A. Collins, Phys. Rev. E **66**, 025401 (2002)
116. D.A. Liberman, This distinction was recognized in the different values of α in the Xα exchange functional of the late 1960s and early 1970s, Phys. Rev. **171**, 1 (1968) as well as references in J.W.D. Connolly, in *Semiempirical Methods of Electronic Structure Calculation, Part A: Techniques*, ed. by G.A. Segal (Plenum, NY 1977), pp. 105ff
117. D.E. Taylor, V.V. Karasiev, K. Runge, S.B. Trickey, F.E. Harris, Computat. Mat. Sci. **39**, 705 (2007)
118. R. Colle, O. Salvetti, Theor. Chim. Acta **37**, 329 (1975)

Toward a Comprehensive Treatment of Temperature in Electronic Structure Calculations: Non-zero-Temperature Hartree-Fock and Exact-Exchange Kohn-Sham Methods

Hannes Schulz and Andreas Görling

Abstract Non-zero-temperature exact-exchange Kohn-Sham and Hartree-Fock methods are discussed and compared to standard Kohn-Sham methods using exchange-correlation functionals within the generalized gradient approximation (GGA). It is shown that the PBE exchange functional, chosen as representative example of a GGA functional, cannot capture the strong temperature dependence of the exchange energy found in the exact-exchange Kohn-Sham and Hartree-Fock methods. This has the drastic consequences of leading to a qualitatively wrong behavior of the PBE free energy with temperature as compared to the exact-exchange Kohn-Sham and Hartree-Fock methods. This indicates that conventional GGA functionals are unsuitable for the treatment of the electronic structure of matter at high temperatures. The exact evaluation of the exchange energy, the construction of the local multiplicative KS exchange potential as well as the handling of the nonlocal Hartree-Fock exchange potential require integrations in **k** space for integrands with integrable singularities. It is shown how these singularities can be treated numerically in the case of non-zero temperatures.

1 Introduction

The treatment of matter by first-principles methods, i.e., methods based exclusively on quantum mechanics without recourse to empirical information, is distinctively more challenging at high temperatures than at ambient conditions for a number of reasons. At ambient temperatures usually the Born-Oppenheimer approximation is invoked and, at least as a first step, the electronic ground state is considered

H. Schulz • A. Görling (✉)
Lehrstuhl für Theoretische Chemie, Universität Erlangen-Nürnberg, Egerlandstraße 3, 91058 Erlangen, Germany
e-mail: hannes.schulz@fau.de; andreas.goerling@fau.de; goerling@chemie.uni-erlangen.de

F. Graziani et al. (eds.), *Frontiers and Challenges in Warm Dense Matter*, Lecture Notes in Computational Science and Engineering 96, DOI 10.1007/978-3-319-04912-0_4,

for a fixed geometrical arrangement of the nuclei, typically at their equilibrium ground-state geometry. If nuclear motion is considered then in most cases it is via Born-Oppenheimer molecular dynamics. That is, in each geometry step of the molecular dynamics simulation the electronic ground-state energy and its derivatives with respect to the nuclear coordinates are calculated for the current geometry within the Born-Oppenheimer approximation.

For ambient temperatures, these procedures rely on the fact that thermal electronic excitations play no role. The reason is that in finite systems and insulating solids kT (product of Boltzmann constant k and temperature T) is small compared to typical electronic excitation energies or the band gap at ambient temperatures and that in metals, for which thermal excitations occur at all non-zero temperatures to some degree, the latter affect many materials properties very little. For high temperatures this is different. Indeed we consider high temperatures as those with kT being at least of the order 1 eV or roughly 10,000 K and above. In this case the following points need to be addressed: (i) Instead of staying exclusively in a ground state formalism it is necessary to switch to an ensemble formalism that takes into account thermal electronic excitations. This implies use (implicit or explicit) of density matrices build, in principle, from all the electronic states of the system instead of those exclusively from the ground state. (ii) Concentration on the ground-state equilibrium geometry no longer makes sense, instead, a sampling of geometries, e.g., by molecular dynamics or Monte Carlo methods, is required. (iii) Because the potential energy surfaces of excited electronic states typically intersect, the Born-Oppenheimer approximation becomes questionable.

In this work we will concentrate exclusively on the first point and discuss first-principles electronic structure methods for high temperatures. Calculations with these methods will be carried out for clamped nuclei, despite the fact that this is not realistic for high temperatures. However, in this way, we can consider the implication of high temperatures exclusively for the treatment of the electronic structure. Being able to take into account high temperatures in electronic structure calculations for clamped nuclei obviously is a prerequisite for a more nearly complete treatment that also considers nuclear motions.

First-principles electronic structure calculations in materials science nowadays in most cases are carried out within density-functional theory (DFT), more precisely within the Kohn-Sham (KS) formalism of DFT [1–3]. In the simplest cases KS methods invoke the local density approximation (LDA), or generalized gradient approximations (GGAs) for the density functionals giving the exchange-correlation energy and potential [1–3]. The KS exchange-correlation potential is defined as the functional derivative of the exchange-correlation energy with respect to the electron density. The approximation of the exchange-correlation energy therefore is the central quantity which also determines the exchange-correlation potential. Within the LDA the exchange-correlation energy is a functional that exclusively depends directly on the electron density, within the GGA exchange-correlation functionals depend on the electron density and its gradients. Within an LDA or GGA scheme temperature can be taken into account, in a first step, by simply introducing fractional occupation numbers for the KS orbitals (KS one-electron functions) according to a Fermi distribution and evaluating zero-temperature LDA or GGA exchange-correlation functionals with the electron density depending on temperature via the occupation numbers. This procedure can be rationalized in terms of free energy density functional theory [4,5]. Such a scheme, however, ignores that the exchange-correlation functional has a genuine temperature-dependence. That is the exchange-correlation energy for a given density differs with temperature. For high temperatures neglect of this temperature-dependence might not be justified. Evidence to support that concern at least at the level of the LDA recently has been given in Ref. [6]. Because GGA functionals typically are developed for and applied to the zero-temperature case knowledge on the genuine temperature-dependence of GGA functionals is limited.

One type of DFT methods going beyond the LDA and GGA are hybrid methods that combine elements of the Hartree-Fock (HF) method and of KS methods [7–10]. In these hybrid methods exchange is treated to a certain fraction, which is typically determined by parameters, in the HF way. The one-electron equations for the orbitals of hybrid methods are obtained by treating the energy expression as a modified HF expression, i.e., executing the variational principle with respect to the orbitals. The result is, exactly as expected, a modified HF-like one-electron equation with a *nonlocal*, i.e., orbital-dependent, potential. In other words a fraction of the exchange energy is calculated exactly from the orbitals and the exchange potential contains a corresponding fraction of the nonlocal exchange operator known from the HF method. Because of the occurrence of a nonlocal exchange operator, hybrid methods no longer fall in the realm of the KS formalism. They can, however, be rigorously justified within a generalized KS framework [9, 10] and thus still are density-functional methods. Because in hybrid methods the correlation energy and the fraction of exchange not taken into account exactly in the HF way are treated via approximate density functionals the problem of treating the temperature dependence of approximate density functionals is also present for hybrid methods.

In order to learn more about the temperature-dependence of density functionals but also as alternative electronic structure methods in material science, approaches are desirable that treat at least part of the KS exchange-correlation energy and potential exactly including their temperature-dependence. Indeed this is possible for the exchange part. Such exact-exchange (EXX) methods [11–17] are the central topic of this work. For comparison we will also consider HF and GGA methods. It is important to note that EXX methods treat exchange exactly within the KS formalism and that this is different from the way exchange is treated in the HF approach. The exchange energy in terms of the orbitals has the same form in the HF and the KS formalism. However, the exchange potential is completely different in HF and EXX methods. The HF exchange potential is a nonlocal operator known explicitly in terms of the orbitals while the exact KS exchange potential by definition is a local multiplicative potential defined as the functional derivative of the exchange energy with respect to the electron density. Because KS and HF orbitals are different, the values of the KS and HF exchange energy also differ, despite the fact that they both have the same form in terms of the orbitals. It is important to realize that the differences between the KS and the HF formalism concerning the value of the exchange energy and the form of the exchange potentials do not imply that one formalism is treating exchange approximately. The two formalisms define exchange differently but neither definition, per se, is better or worse or exact or approximate. Therefore questions asking, e.g., "whether the true exchange potential can be local or not" are ill-posed. It depends on the definition whether the exchange potential is a local multiplicative potential or not.

Despite the fact that the calculation of the exact KS exchange energy is straightforward and technically identical to the well-established calculation of the HF exchange energy, EXX methods for molecules or solids [13–16] were developed decades after LDA or GGA methods and still are not in common use. The reasons are (i) the increased computational cost of an exact treatment of exchange, (ii) the loss of error cancellation between exchange and correlation energies, a cancellation which is crucial for the performance of LDA and GGA functionals, and (iii) the difficulty of constructing the corresponding exact KS exchange potential [15, 17–23]. All three argument have become questionable nowadays. The computational costs of EXX methods are not higher than for the hybrid methods which are frequently employed these days. Concerning the argument of error cancellation, it has to be kept in mind that relying on error cancellation, besides being unsatisfying from a formal point, is dangerous because the extent of such error cancellations in general will be system-dependent. In particular, for nonequilibrium geometries that occur in transition states of chemical reaction but also are relevant at high temperatures, error cancellation between GGA exchange and correlation energies often is much less substantial than for equilibrium geometries and LDA and GGA methods therefore cannot correctly describe such cases. On the other hand, recently developed methods [24–55] combining orbital-dependent correlation functionals based on the adiabatic-connection fluctuation-dissipation theorem [56, 57] (often termed random phase approximation functionals despite the fact that some of these approaches go beyond the random phase approximation) with an exact treatment of exchange yield accuracies that are unprecedented for DFT methods and are applicable to cases which LDA and GGA methods fail to describe [47, 55]. Finally, the construction of exact KS exchange potentials is possible via the optimized effective potential (OEP) procedure [11, 12, 17]. OEP procedures and thus EXX methods can be implemented with acceptable programming effort in plane-wave [13, 14] as well fully-linearized augmented plane wave (FLAPW) [58, 59] codes for periodic solids in a numerical stable way. For EXX methods for finite systems employing Gaussian basis sets, numerical stability is a more severe problem [15, 17–23, 60] but progress also has been made there [61, 62]. In summary, EXX methods have developed into an interesting option in recent years.

In the context of high temperature electronic structure methods, a KS approach that treats exchange exactly has the advantage that it can be rigorously generalized within the framework of ensemble density functional theory [4] to include temperature without making any approximations [63, 64]. In this way, at least for exchange, the effect of high temperature in KS approaches can be studied without making assumptions or approximations. In this work we will consider the effect of high temperatures in an exact-exchange-only KS method, i.e., a method treating exchange exactly and neglecting correlation completely. The approach is implemented in a plane-wave code for solids [64]. The EXX method is compared with a non-zero-temperature HF method and use of a ground state GGA functional. The HF method also takes into account temperature exactly while the GGA method includes temperature only via fractional occupations due to the Fermi distribution but apart from that employs the standard zero-temperature exchange-correlation functional.

When discussing temperature effects we place special emphasis on the band structures because the latter not only are crucial for characterizing the electronic properties but also exhibit distinctive differences among EXX, HF, and ground-state GGAs. An advantage of EXX compared to LDA and GGA methods is that EXX methods are free of Coulomb self-interaction. Within the KS formalism, the exchange energy cancels the unphysical interaction of electrons with itself which is contained in the Hartree energy, the classical electrostatic energy of the electron density. In EXX methods, unphysical Coulomb self-interaction is removed completely, while the removal is incomplete in LDA and GGA functionals because the exchange energy is approximated. As a results, LDA and GGA exchange-correlation potentials are too repulsive and LDA and GGA orbital and eigenvalue spectra are qualitatively wrong. For example, in finite systems, LDA and GGA spectra do not contain Rydberg series and the eigenvalue of the highest occupied molecular orbital, which should equal the negative of the ionization energy [65], typically is found at much too high energies due to the overly repulsive KS potential. In semiconductors, LDA and GGA band gaps are much smaller than experimental band gaps and it can happen, e.g., in Germanium, that semiconductors are obtained as metals. EXX methods cure all these problems. Regarding semiconductors, one might argue that even the exact KS band gap differs from true quasi-particle band gaps by the derivative discontinuity [66–68] and in that extent it is not clear whether EXX band gaps are closer to the true KS band gaps than LDA or GGA band gaps. This is tantamount to arguing that inclusion of the exact correlation potential which, of course, is unknown, would bring back EXX band gaps to LDA or GGA band gaps. The effect the exact correlation potential would have, indeed, is subject of controversial discussions [68–73]. However, in any case a method like the EXX method that yields qualitatively correct orbital and eigenvalues spectra and does not treat semiconductors as metals seems to be superior compared to methods like LDA or GGA methods that fail in both regards.

Compared to the HF method, an exact-exchange-only KS method yields quite similar total electronic energies. In the zero temperature case, HF total energies are always slightly lower than EXX total energies of necessity because the HF determinant by construction yields the lowest total electronic energy of all Slater determinants. Thus, the EXX total energy, the energy of the EXX determinant has to be higher. The differences, however, are remarkably small for zero temperature [74].

The spectra of the unoccupied orbitals and their eigenvalue, however, are completely different in the HF and EXX case. Unoccupied HF orbitals of finite systems at zero temperature typically have positive eigenvalues and are unphysical. If calculated with finite basis sets unoccupied HF eigenvalues with positive values always approach zero with increasing basis sets. In semiconductors or insulators, HF band gaps are much larger than LDA, GGA, and EXX band gaps. Compared to experimental band gaps HF band gaps are way too large. The EXX approach thus is the only approach among HF, LDA, GGA, and EXX methods that yields physical orbital and eigenvalue spectra.

A physical orbital and eigenvalue spectrum not only is important for a correct description and interpretation of electronic structures but also has important implications for the calculations of total electronic energies and, consequently, geometric structures and phase diagrams, in particular in the high temperature regime. If one-electron wave functions, i.e., orbitals, are occupied according to a Fermi distribution, then the occupation of orbitals that are unoccupied in the ground state at zero temperature depends crucially on the band gap. In the HF case with band gaps that are typically 10 or more eV larger than EXX band gaps, unoccupied one-particle states will acquire significant occupation only at much higher temperatures than in the EXX case, while in LDA or GGA approaches which yield smaller band gaps than EXX approaches, unoccupied states will become significantly occupied at lower temperatures. Indeed this behavior has been demonstrated in the case of HF versus density-functional methods for a confined test system of eight one-electron atoms in Ref. [75]. Such behaviour could have an influence on total energies and geometries. Consider, for example, a typical diatomic molecule with a closed shell electronic structure. The lowest unoccupied molecular orbital (LUMO) then typically is of antibonding character. An occupation of this orbital at high temperatures will lead to a weakening of the chemical bond between the two atoms of the molecule and an increase in the bond length. This means temperature would affect bond strength and geometry of the molecule directly via the electronic structure. This effect, which adds to the effect of temperature on the nuclear motion, can be assumed to be quite different for LDA/GGA, EXX, and HF methods. A comparison of band structures and occupations of one-electron states in the different cases therefore seems to be interesting and, indeed, is one of the goals of this work.

Another point to be discussed in this work is of more technical nature but a prerequisite for doing EXX or HF calculations correctly at high temperatures. In a periodic system the evaluation of the exchange energy requires an integration in reciprocal space, i.e., an integration in **k**-space. The integrand in this integration, however, is singular. A simple integration on a mesh of **k**-points as it is performed for other contributions of the total energy would not treat this singularity correctly. Different approaches to deal with the singularity have been proposed in the literature [76–79]. We here consider their high-temperature generalizations and compare their performance at high temperatures.

2 Exchange Energies and Potentials

In this section we discuss the exchange energy and potential in the HF and the exact-exchange KS method in the case of non-zero temperatures.

2.1 Exchange Energy and Potential in the HF Method for Non-zero Temperatures

In a non-zero-temperature HF approach within the grand-canonical-ensemble formalism [80,81], the kinetic energy, T^{HF}, the Coulomb energy, U, and the interaction energy with the nuclei are simply given by

$$T^{\text{HF}} = \sum_i f_i \, \langle\phi_i| - \frac{1}{2}\nabla^2|\phi_i\rangle \,, \tag{1}$$

$$U = \frac{1}{2}\int d\mathbf{r}d\mathbf{r}' \, \frac{\rho(\mathbf{r})\rho(\mathbf{r}')}{|\mathbf{r}-\mathbf{r}'|} \,, \tag{2}$$

and

$$\int d\mathbf{r} \; v(\mathbf{r}) \, \rho(\mathbf{r}) \tag{3}$$

with the electron density

$$\rho(\mathbf{r}) \;=\; \sum_i f_i \;\, \phi_i^\dagger(\mathbf{r}) \, \phi_i(\mathbf{r}) \tag{4}$$

and fractional occupation numbers

$$f_i \;=\; \frac{1}{1 \,+\, e^{(1/kT)(\varepsilon_i - \mu)}} \tag{5}$$

determined by Fermi statistics. In Eqs. (1)–(4) ϕ_i denotes orbitals, i.e., one electron functions, which, throughout this work, shall be two-dimensional spinors containing the spin degree of freedom. By $v(\mathbf{r})$ the electrostatic potential of the nuclei is designated. The parameters k, T, μ, and ε_i are the Boltzmann constant, the temperature, the chemical potential, and the eigenvalue of orbital ϕ_i, respectively. Instead of the chemical potential μ the electron number N is chosen to characterize the underlying grand canonical ensemble in addition to the parameters volume and temperature. The reason is that in a treatment of periodic systems charge neutrality is required which means that the electron number is determined by the nuclear charges. (Except if positive background charges are invoked.) For a given number of electrons N the chemical potential is determined by requiring that

$$\sum_i f_i = \sum_i \frac{1}{1 + e^{(1/kT)(\varepsilon_i - \mu)}} = N \,. \tag{6}$$

The exchange energy is given by

$$E_x = -\frac{1}{2} \sum_i \sum_j f_i \, f_j \, \langle ij | ji \rangle \,, \tag{7}$$

with

$$\langle ij |\, k\ell \rangle = \int d\mathbf{r} d\mathbf{r}' \, \frac{\phi_i^\dagger(\mathbf{r})\phi_k(\mathbf{r})\phi_j^\dagger(\mathbf{r}')\phi_\ell(\mathbf{r}')}{|\mathbf{r} - \mathbf{r}'|} \,. \tag{8}$$

The weight of the contributions $\langle ij|ji\rangle$ to the exchange energy, Eq. (7), is given simply by the product of the occupation numbers f_i and f_j of the corresponding orbitals [5, 64, 81]. The occupation numbers f_i describe the weight of contributions from the corresponding orbitals ϕ_i in the evaluation of one-electron operators like the kinetic energy or the operator emerging from the electrostatic potentials of the nuclei, see Eqs. (1) or (3). That their products f_i and f_j occur in the evaluation of the exchange energy, which emerges from the operator of the electron-electron interaction, a two electron-operator, is shown, e.g., in Refs. [5, 64, 81]. As in the zero-temperature case, also in the non-zero-temperature case the sum $U + E_x$ of Coulomb and exchange energy represents the energy of the electron-electron interaction within the HF approach.

The HF orbitals are obtained by self-consistently solving the HF equations

$$\left[-\frac{1}{2}\nabla^2 + \hat{v} + \hat{v}_H + \hat{v}_x^{\mathrm{NL}} \right] \phi_i = \varepsilon_i \, \phi_i \tag{9}$$

containing the kinetic energy operator, $-\frac{1}{2}\nabla^2$, the operator $\hat{v}$ corresponding to the potential of the nuclei, the operator $\hat{v}_H$ emerging from the Hartree potential

$$v_H(\mathbf{r}) = \int d\mathbf{r}' \, \frac{\rho(\mathbf{r}')}{|\mathbf{r} - \mathbf{r}'|} \,, \tag{10}$$

and the nonlocal HF exchange potential $\hat{v}_x^{\mathrm{NL}}$ defined by its kernel

$$v_x^{\mathrm{NL}}(\mathbf{r}, \mathbf{r}') = -\sum_j f_j \, \frac{\phi_j(\mathbf{r}) \, \phi_j^\dagger(\mathbf{r}')}{|\mathbf{r} - \mathbf{r}'|} \,. \tag{11}$$

2.2 Exchange Energy and Potential in the KS Formalism for Non-zero Temperatures

An exact-exchange non-zero-temperature KS formalism was derived in Ref. [64] based on the ensemble DFT formalism of Mermin [4]. Implementations of non-zero-temperature exact-exchange KS methods were presented in Refs. [63] and [64] in the framework of plane-wave solid state codes for periodic systems.

In an exact-exchange-only KS approach for non-zero temperatures, the kinetic energy, the interaction energy of the electrons with the potential of the nuclei, the Coulomb energy, and the exchange energy are given by the same expressions in terms of orbitals as in the HF case, i.e., by Eqs. (1)–(3) and (7). We note at this point that in KS methods in general, the total energy contains a correlation energy with a correlation contribution to the kinetic energy and the electron-electron interaction energy. Only in the special case of an exact-exchange-only KS method does the kinetic energy given in Eq. (1) become the complete kinetic energy. In general the kinetic energy according to Eq. (1) is denoted as the noninteracting kinetic energy, typically abbreviated by the symbol T_s, and supplemented by a correlation contribution, often abbreviated by T_c, to give the complete kinetic energy. Similarly the sum of Coulomb and exchange energy given in Eq. (2) and (7) has to be supplemented by the electron-electron contribution of the correlation energy to give the complete energy of the electron-electron interaction. For the correlation energy and its parts, of course, no exact expressions are known, in contrast to the other contributions of the total energy given by the expressions (1)–(3) and (7). Therefore, the correlation energy has to be treated by approximate functionals in the general case. In the exchange-only case considered here, however, the total energy is given by the sum of the contributions (1)–(3) and (7) and thus has the same form in terms of the orbitals as in the HF case. The expressions for the contributions to the total energy, however, are evaluated with KS instead of HF orbitals and therefore the values for the various energy contributions as well as the total energy differ in the exact-exchange-only KS case from those in the HF case.

The exact-exchange-only KS orbitals are obtained from the KS equations

$$\left[-\frac{1}{2}\nabla^2 + \hat{v} + \hat{v}_H + \hat{v}_x\right] \phi_i = \varepsilon_i \, \phi_i \tag{12}$$

which differ from the HF equations only by the operator $\hat{v}_x$ of the KS exchange potential $v_x(\mathbf{r})$. The KS exchange potential $v_x(\mathbf{r})$, in contrast to the nonlocal HF exchange potential is a local multiplicative potential, which, as mentioned in the introduction, is defined as functional derivative of the exchange energy with respect to the electron density. Because we do not know how the orbitals depend on the electron density, we cannot simply take this functional derivative via the chain rule by first taking the derivative of expression (7) with respect to the orbitals and by then taking the derivative of the orbitals with respect to the electron density. Instead the local multiplicative KS exchange potential is accessible via the optimized effective potential (OEP) method [11, 12] by taking the functional derivative of the exchange energy with respect to the effective KS potential v_s in two different ways, see e.g. Ref. [17] for a discussion of OEP methods. For the non-zero temperature case we follow Ref. [64] and obtain

$$\int d\mathbf{r}' \left[\frac{\delta E_x}{\delta \rho(\mathbf{r}')} \frac{\delta \rho(\mathbf{r}')}{\delta v_s(\mathbf{r})} \right] = \sum_i \int d\mathbf{r}' \left[\frac{\delta E_x}{\delta \phi_i(\mathbf{r}')} \frac{\delta \phi_i(\mathbf{r}')}{\delta v_s(\mathbf{r})} + \text{c.c.} \right] + \sum_i \left[\frac{\delta E_x}{\delta f_i} \frac{\delta f_i}{\delta v_s(\mathbf{r})} \right] \tag{13}$$

as derivatives of expression (7) for the exchange energy with respect to v_s. The derivative $\delta\rho(\mathbf{r}')/\delta v_s(\mathbf{r})$ of the electron density with respect to the KS potential is the KS response function $X_s(\mathbf{r}', \mathbf{r})$, i.e.,

$$X_s(\mathbf{r}, \mathbf{r}') = \frac{\delta \rho(\mathbf{r})}{\delta v_s(\mathbf{r}')} . \tag{14}$$

The functional derivative $\delta E_x/\delta\rho(\mathbf{r}')$ is the exchange potential $v_x(\mathbf{r}')$, the quantity we are interested in. The right hand side of Eq. (13) shall be abbreviated by $t(\mathbf{r})$. We furthermore exploit that the KS response function X_s is symmetric in its arguments, see Eqs. (18)–(20) below, and then obtain the OEP equation

$$\int d\mathbf{r}' \; X_s(\mathbf{r}, \mathbf{r}') \; v_x(\mathbf{r}') = t_x(\mathbf{r}) . \tag{15}$$

for the KS exchange potential in the non-zero temperature case.

Next we consider in more detail the response function X_s of Eq. (14). The electron density, according to Eq. (4) is given in terms of the orbitals ϕ_i and, via the occupation numbers f_i, see Eq. (5), in terms of eigenvalues ε_i. The functional derivatives of the orbitals and the eigenvalues with respect to the effective potential $v_s(\mathbf{r})$ are known from perturbation theory and given by

$$\frac{\delta \phi_i(\mathbf{r})}{\delta v_s(\mathbf{r}')} = \sum_{\ell \neq i} \phi_\ell(\mathbf{r}) \; \frac{\phi_\ell^\dagger(\mathbf{r}')\phi_i(\mathbf{r}')}{\varepsilon_i - \varepsilon_\ell} \tag{16}$$

and

$$\frac{\delta \varepsilon_i}{\delta v_s(\mathbf{r})} = \phi_i^\dagger(\mathbf{r})\phi_i(\mathbf{r}) \tag{17}$$

respectively. Note that the use of perturbation theory usually implies invocation of an approximation. This is not the case here. We are interested in derivatives, that is first order quantities which are exactly given by first order perturbation theory. With Eqs. (16) and (17) the functional derivative of the electron density with respect to $v_s(\mathbf{r})$, that is the KS response function $X_s(\mathbf{r}, \mathbf{r}')$, is obtained [64] as

$$X_s(\mathbf{r},\mathbf{r}')=\sum_i f_i \sum_{\ell\neq i}\left[\frac{\phi_i^\dagger(\mathbf{r})\,\phi_\ell(\mathbf{r})\,\phi_\ell^\dagger(\mathbf{r}')\,\phi_i(\mathbf{r}')}{\varepsilon_i-\varepsilon_\ell}+\text{c.c.}\right]$$
$$+\sum_i \phi_i^\dagger(\mathbf{r})\,\phi_i(\mathbf{r})\,\frac{\delta f_i}{\delta v_s(\mathbf{r})} \tag{18}$$

with

$$\frac{\delta f_i}{\delta v_s(\mathbf{r})}=\frac{df_i}{d\varepsilon_i}\frac{\delta\varepsilon_i}{\delta v_s(\mathbf{r})}+\frac{df_i}{d\mu}\frac{\delta\mu}{\delta v_s(\mathbf{r})}=-\frac{f_i(1-f_i)}{kT}\left[\phi_i^\dagger(\mathbf{r})\phi_i(\mathbf{r})-\frac{\delta\mu}{\delta v_s(\mathbf{r})}\right] \tag{19}$$

and with

$$\frac{\delta\mu}{\delta v_s(\mathbf{r})}=\frac{\sum_i f_i(1-f_i)\phi_i^\dagger(\mathbf{r})\phi_i(\mathbf{r})}{\sum_\ell f_\ell(1-f_\ell)}. \tag{20}$$

Similarly the right hand side t_x of the OEP equation (15), i.e., the right hand side of Eq. (13), is obtained as

$$t_x(\mathbf{r})=\sum_i f_i \sum_{\ell\neq i}\left[\frac{\langle\phi_i|\hat{v}_x^{\mathrm{NL}}|\phi_\ell\rangle\phi_\ell^\dagger(\mathbf{r})\,\phi_i(\mathbf{r})}{\varepsilon_i-\varepsilon_\ell}+\text{c.c.}\right]$$
$$+\sum_i \langle\phi_i|\hat{v}_x^{\mathrm{NL}}|\phi_i\rangle\,\frac{\delta f_i}{\delta v_s(\mathbf{r})} \tag{21}$$

3 Singularity Corrections

In periodic systems summations over orbitals turn into summations over bands and integrations over **k** points in the first Brillouin zone. In the evaluation of the exchange energy, integrable singularities are present in integrations in **k** space. While this is no problem from a formal point of view, it requires special care in practical implementations of HF or exact-exchange KS methods because in such implementations integrations in **k** space have to be carried out on integration grids. This can lead to numerical problems if singularities are present. There are two basic strategies to handle these problems. The first, due to Gygi and Baldereschi [76] subtracts and adds to the singular integrand an auxiliary function which has the same singularity as the integrand but is analytically integrable. The difference between the original integrand and the auxiliary function is free of singularities and thus can be integrated numerically while the integral of the subtracted auxiliary function can be carried out analytically and added to the numerically integrated difference.

Gygi and Baldereschi presented an auxiliary function for systems with fcc lattice [76] while Wenzien et al. [77] presented auxiliary functions for bcc, sc, hexagonal, and orthorhombic systems. In Ref. [78] an auxiliary function for arbitrary crystal structures was introduced that is not analytically integrable but can be integrated with negligible computational effort by an adaptive grid procedure.

The second approach, due to Spencer and Alavi [79], modifies the Coulomb interaction $1/|\mathbf{r} - \mathbf{r}'|$ occurring in the exchange integral (7) by an attenuated Coulomb interaction $v_{atten}(|\mathbf{r} - \mathbf{r}'|)$ which equals the Coulomb interaction for $|\mathbf{r} - \mathbf{r}'| \leq R_c$ and is set to zero for $|\mathbf{r} - \mathbf{r}'| > R_c$. The cutoff radius R_c is chosen such that the volume of a sphere of radius R_c equals to the product of the unit cell volume times the number of **k** points in the numerical integration in **k** space.

Both approaches converge towards the exact integral in the limit of an infinite **k** point mesh. Furthermore both approaches do not require significant computational effort. The second approach is somewhat easier to implement, but does, however, manipulate the basic interaction, the Coulomb interaction. We will present a generalization of the first approach to non-zero temperature, i.e., fractional occupation numbers and show results obtained with this approach. However, we have implemented also the second approach and in all cases got identical results within the limits of convergence in **k** space except for HF band structures of aluminum. Indeed, the treatment of exchange singularities turned out to be problematic for both approaches in the case of HF band structures of metals, in particular at low temperatures, i.e. further work is required in this case, see later on for details.

3.1 Singularity Correction for Exchange Energy

The singularity correction for the exchange energy in the non-zero-temperature case can be obtained by generalizing the derivation for the zero-temperature case given in Ref. [78]. In a plane wave code for periodic systems, expression (7) for the exchange energy, initially has the form

$$E_x = -\frac{1}{2N_k} \sum_{i\mathbf{k}} \sum_{j\mathbf{q}} f_{i\mathbf{k}}\, f_{j\mathbf{q}} \int d\mathbf{r} \int d\mathbf{r}' \; \frac{\phi_{i\mathbf{k}}^{\dagger}(\mathbf{r}')\phi_{j\mathbf{q}}(\mathbf{r}')\phi_{j\mathbf{q}}^{\dagger}(\mathbf{r})\phi_{i\mathbf{k}}(\mathbf{r})}{|\mathbf{r} - \mathbf{r}'|} \tag{22}$$

if we take into account that in periodic systems energies including the exchange energy are calculated per unit cell, i.e., if the total exchange energy is divided by the number N_k of **k**-points. The summations $i, \mathbf{k}$ and $j, \mathbf{q}$ in Eq. (22) run over all bands i and j and all integration points **k** and **q** in **k** space. In practice the number of bands is limited to those which exhibit nonnegligible occupation numbers $f_{i\mathbf{k}}$ or $f_{j\mathbf{q}}$. The integrations over $\mathbf{r}$ and $\mathbf{r}'$ run over the crystal volume Ω, taking, however, Born-von-Karman periodic boundary conditions into account, see below Eq. (26).

We represent products of orbitals by plane waves $e^{i(\mathbf{G}+\mathbf{k}-\mathbf{q})\cdot\mathbf{r}}$ according to

$$\phi^{\dagger}_{j\mathbf{q}}(\mathbf{r})\phi_{i\mathbf{k}}(\mathbf{r}) = \frac{1}{\Omega}\sum_{\mathbf{G}} Y_{j\mathbf{q},i\mathbf{k}}(\mathbf{G})\; e^{i(\mathbf{G}+\mathbf{k}-\mathbf{q})\cdot\mathbf{r}}\,, \tag{23}$$

with

$$Y_{j\mathbf{q},i\mathbf{k}}(\mathbf{G}) = \int_{\Omega} d\mathbf{r}\; e^{-i(\mathbf{G}+\mathbf{k}-\mathbf{q})\cdot\mathbf{r}}\; \phi^{\dagger}_{j\mathbf{q}}(\mathbf{r})\,\phi_{i\mathbf{k}}(\mathbf{r})\,. \tag{24}$$

The exchange energy then assumes the form

$$E_x = -\frac{2\pi}{N_k\Omega}\sum_{i\mathbf{k}}\sum_{j\mathbf{q}} f_{i\mathbf{k}}\, f_{j\mathbf{q}} \sum_{\mathbf{G}} \frac{Y^*_{j\mathbf{q},i\mathbf{k}}(\mathbf{G})\, Y_{j\mathbf{q},i\mathbf{k}}(\mathbf{G})}{|\mathbf{G}+\mathbf{k}-\mathbf{q}|^2} \tag{25}$$

if we use the fact that

$$\int_{\Omega} d\mathbf{r} \int d\mathbf{r}'\; \frac{e^{i\mathbf{G}\cdot\mathbf{r}}\, e^{-i\mathbf{G}'\cdot\mathbf{r}'}}{|\mathbf{r}-\mathbf{r}'|} = \int_{\Omega} d\mathbf{r}\int d\mathbf{z}\; \frac{e^{-i\mathbf{G}\cdot\mathbf{z}}}{|\mathbf{z}|} = \frac{4\pi\Omega}{|G|^2}\;\delta_{\mathbf{G}\mathbf{G}'}\,. \tag{26}$$

In Eq. (26) Born-von-Karman periodic boundary conditions are taken into account by carrying out a coordinate transformation from the original coordinates $\mathbf{r}$ and $\mathbf{r}'$ to the coordinates $\mathbf{r}$ and $\mathbf{z} = \mathbf{r}' - \mathbf{r}$. In the second integral of Eq. (26) the integration over $\mathbf{r}$ runs over the crystal volume while the integration over $\mathbf{z}$ runs over all space due to the Born-von-Karman periodic boundary conditions.

Expression (25) for the exchange energy contains singular terms, namely those with $\mathbf{G} = \mathbf{0}$, $\mathbf{k} = \mathbf{q}$, and $j = i$. For $j \neq i$ no singularities occur for any value of $\mathbf{G}$, $\mathbf{k}$, or $\mathbf{q}$ due to the relation

$$Y_{j\mathbf{k},i\mathbf{k}}(\mathbf{0}) = \delta_{ji} \tag{27}$$

which holds because Eq. (24) defining $Y_{j\mathbf{q},i\mathbf{k}}(\mathbf{0})$ for $\mathbf{G} = \mathbf{0}$ just turns into the orthonormality condition of the one-particle functions for $\mathbf{G} = \mathbf{0}$ and $\mathbf{k} = \mathbf{q}$. Thus, contributions with $\mathbf{G} = \mathbf{0}$, $\mathbf{k} = \mathbf{q}$, and $j \neq i$ vanish because the plane wave representations of the products $\phi^{\dagger}_{j\mathbf{k}}(\mathbf{r})\phi_{i\mathbf{k}}(\mathbf{r})$ with $j \neq i$ do not contain contributions from a plane wave with $\mathbf{G} = \mathbf{0}$.

Because of the presence of the singularities described above the exchange energy is well-defined only in the limit of an infinite number N_k of unit cells, i.e., for $N_k \to \infty$. In this case, the summations over $\mathbf{k}$ and $\mathbf{q}$ turn into integrals over the Brillouin zone and Eq. (25) for the exchange energy assumes the form

$$E_x = -\frac{2\pi}{N_k\Omega}\frac{\Omega^2}{(2\pi)^6}\sum_{i}\int_{BZ} d\mathbf{k}\; f_{i\mathbf{k}} \sum_{j}\int_{BZ} d\mathbf{q}\; f_{j\mathbf{q}} \sum_{\mathbf{G}} \frac{Y^*_{j\mathbf{q},i\mathbf{k}}(\mathbf{G})\, Y_{j\mathbf{q},i\mathbf{k}}(\mathbf{G})}{|\mathbf{G}+\mathbf{k}-\mathbf{q}|^2}\,. \tag{28}$$

Because the singularities in integral (28) are integrable the exchange energy is now well-defined.

Following Ref. [78] we now adopt an idea of Gygi and Baldereschi[76] and manipulate those contributions on the right hand side of Eq. (28) for the case $\mathbf{G} = \mathbf{0}$ and $j = i$, i.e., those contributions which contain the integrable singularities, by adding and subtracting a function $s(\mathbf{q})$ which shall obey the three following conditions: (i) $s(\mathbf{q})$ is periodic within the reciprocal lattice, (ii) $s(\mathbf{q})$ diverges as $1/q^2$ for $\mathbf{q} \to \mathbf{0}$ and is smooth elsewhere, and (iii) $s(\mathbf{q}) = s(-\mathbf{q})$. These requirements lead to

$$
\begin{aligned}
&-\frac{2\pi}{N_k \Omega}\frac{\Omega^2}{(2\pi)^6}\sum_i \int_{BZ} d\mathbf{k}\; f_{i\mathbf{k}} \int_{BZ} d\mathbf{q}\; f_{i\mathbf{q}}\, \frac{Y^*_{i\mathbf{q},i\mathbf{k}}(\mathbf{0})\, Y_{i\mathbf{q},i\mathbf{k}}(\mathbf{0})}{|\mathbf{k}-\mathbf{q}|^2} \\
&\quad = -\frac{2\pi}{N_k \Omega}\frac{\Omega^2}{(2\pi)^6}\sum_i \int_{BZ} d\mathbf{k}\; f_{i\mathbf{k}} \int_{BZ} d\mathbf{q} \left[f_{i\mathbf{q}}\, \frac{Y^*_{i\mathbf{q},i\mathbf{k}}(\mathbf{0})\, Y_{i\mathbf{q},i\mathbf{k}}(\mathbf{0})}{|\mathbf{k}-\mathbf{q}|^2} - f_{i\mathbf{k}}\, s(\mathbf{k}-\mathbf{q}) \right] \\
&\qquad -\frac{2\pi}{N_k \Omega}\frac{\Omega^2}{(2\pi)^6}\sum_i \int_{BZ} d\mathbf{k}\; f_{i\mathbf{k}} \int_{BZ} d\mathbf{q}\; f_{i\mathbf{k}}\, s(\mathbf{k}-\mathbf{q}) \\
&\approx -\frac{2\pi}{N_k \Omega}\sum_i \sum_{\mathbf{k}} f_{i\mathbf{k}} \sum_{\mathbf{q}\neq\mathbf{k}} \left[f_{i\mathbf{q}}\, \frac{Y^*_{i\mathbf{q},i\mathbf{k}}(\mathbf{0})\, Y_{i\mathbf{q},i\mathbf{k}}(\mathbf{0})}{|\mathbf{k}-\mathbf{q}|^2} - f_{i\mathbf{k}}\, s(\mathbf{k}-\mathbf{q}) \right] \\
&\qquad -\frac{2\pi}{N_k \Omega}\frac{\Omega^2}{(2\pi)^6}\sum_i \int_{BZ} d\mathbf{k}\; f^2_{i\mathbf{k}} \int_{BZ} d\mathbf{q}\, s(\mathbf{q}) \\
&= -\frac{2\pi}{N_k \Omega}\sum_i \sum_{\mathbf{k}} f_{i\mathbf{k}} \sum_{\mathbf{q}\neq\mathbf{k}} f_{i\mathbf{q}}\, \frac{Y^*_{i\mathbf{q},i\mathbf{k}}(\mathbf{0})\, Y_{i\mathbf{q},v\mathbf{k}}(\mathbf{0})}{|\mathbf{k}-\mathbf{q}|^2} \\
&\qquad +\frac{1}{2N_k}\sum_i \sum_{\mathbf{k}} f^2_{i\mathbf{k}}\, \frac{4\pi}{\Omega} \sum_{\mathbf{q}\neq\mathbf{k}} s(\mathbf{k}-\mathbf{q}) \\
&\qquad -\frac{1}{2N_k}\sum_i \sum_{\mathbf{k}} f^2_{i\mathbf{k}}\, \frac{4\pi}{(2\pi)^3} \int_{BZ} d\mathbf{q}\, s(\mathbf{q}) \\
&= -\frac{2\pi}{N_k \Omega}\sum_i \sum_{\mathbf{k}} f_{i\mathbf{k}} \sum_{\mathbf{q}\neq\mathbf{k}} f_{i\mathbf{q}}\, \frac{Y^*_{i\mathbf{q},i\mathbf{k}}(\mathbf{0})\, Y_{i\mathbf{q},i\mathbf{k}}(\mathbf{0})}{|\mathbf{k}-\mathbf{q}|^2} + \frac{1}{2N_k}\sum_i \sum_{\mathbf{k}} f^2_{i\mathbf{k}}\, [\tilde{S}_{\mathbf{k}} - S] \\
&= -\frac{2\pi}{N_k \Omega}\sum_i \sum_{\mathbf{k}} f_{i\mathbf{k}} \sum_{\mathbf{q}\neq\mathbf{k}} f_{i\mathbf{q}}\, \frac{Y^*_{i\mathbf{q},i\mathbf{k}}(\mathbf{0})\, Y_{i\mathbf{q},i\mathbf{k}}(\mathbf{0})}{|\mathbf{k}-\mathbf{q}|^2} \\
&\qquad + \left[\frac{1}{2N_k}\sum_i \sum_{\mathbf{k}} f^2_{i\mathbf{k}} \right] [\tilde{S} - S]
\end{aligned}
\tag{29}
$$

with

$$\tilde{S}_{\mathbf{k}} = \frac{4\pi}{\Omega} \sum_{\mathbf{q}\neq\mathbf{k}} s(\mathbf{k}-\mathbf{q})\,, \tag{30}$$

$$S = \frac{4\pi}{(2\pi)^3} \int_{BZ} d\mathbf{q}\; s(\mathbf{q})\,, \tag{31}$$

and

$$\tilde{S} = \frac{4\pi}{\Omega} \sum_{\mathbf{q}\neq\mathbf{0}} s(\mathbf{q})\,. \tag{32}$$

Condition (ii) for the function s and Eq. (27) guarantee that $Y^*_{i\mathbf{q},i\mathbf{k}}(\mathbf{0})\, Y_{i\mathbf{q},i\mathbf{k}}(\mathbf{0})/|\mathbf{k}-\mathbf{q}|^2 - s(\mathbf{k}-\mathbf{q})$ for any given $\mathbf{k}$ is a function of $\mathbf{q}$ that equals zero at $\mathbf{q}=\mathbf{k}$, i.e., no longer contains the $1/q^2$ singularity. Therefore, the first integral over $\mathbf{q}$ and $\mathbf{k}$ after the first equality sign of Eq. (29) can be evaluated by summations over the finite grid of $\mathbf{k}$ points omitting the terms with $\mathbf{q}=\mathbf{k}$. Due to the periodicity and inversion symmetry of $s(\mathbf{q})$ (condition (i) and (iii) for s) the integrals of $s(\mathbf{k}-\mathbf{q})$ over the Brillouin zone can be replaced by Brillouin zone integrals of $s(\mathbf{q})$. Furthermore, for the case of a uniform grid of $\mathbf{k}$ points the functions $\tilde{S}_{\mathbf{k}}$ of Eq. (30) are all equal to $\tilde{S}$ of Eq. (32), i.e. $\tilde{S}_{\mathbf{k}} = \tilde{S}$.

The exchange energy can now be evaluated as

$$\begin{aligned} E_x = &-\frac{2\pi}{N_k \Omega} \sum_{i,\mathbf{k}} f_{i\mathbf{k}} \sum_{j,\mathbf{q}\neq\mathbf{k}} f_{j\mathbf{q}} \sum_{\mathbf{G}} \frac{Y^*_{j\mathbf{q},i\mathbf{k}}(\mathbf{G})\, Y_{j\mathbf{q},i\mathbf{k}}(\mathbf{G})}{|\mathbf{G}+\mathbf{k}-\mathbf{q}|^2} \\ &-\frac{2\pi}{N_k \Omega} \sum_{i,\mathbf{k}} f_{i\mathbf{k}} \sum_{j} f_{j\mathbf{k}} \sum_{\mathbf{G}\neq\mathbf{0}} \frac{Y^*_{j\mathbf{k},i\mathbf{k}}(\mathbf{G})\, Y_{j\mathbf{k},i\mathbf{k}}(\mathbf{G})}{|\mathbf{G}|^2} \\ &+\left[\frac{1}{2N_k} \sum_{i,\mathbf{k}} f^2_{i\mathbf{k}}\right] [\tilde{S} - S]\,. \end{aligned} \tag{33}$$

Thus in the evaluation of the exchange energy, the singular terms in the original expression (25) first can simply be omitted and then be taken into account by the correction $\left[\frac{1}{2N_k} \sum_{i,\mathbf{k}} f^2_{i\mathbf{k}}\right] [\tilde{S} - S]$, i.e., by adding the correction to the exchange energy obtained if the singular terms are simply omitted. The factor $[\tilde{S} - S]$ contained in the correction is calculated only once before the self-consistency procedure. In fact, the factor $[\tilde{S} - S]$ depends only on the lattice vectors defining the unit cell and the positions of the $\mathbf{k}$ points but not on the number, type, or positions of the atoms within the unit cell.

3.2 Singularity Correction for Local Kohn-Sham and Nonlocal Hartree-Fock Exchange Potentials

The calculation of the exchange energy including the involved treatment of singularities is identical in HF and exact-exchange KS methods. The treatment of the exchange potential, on the other hand, is fundamentally different in the two methods; recall the discussion in Sects. 1 and 2.

The exact local multiplicative KS exchange potential is calculated by solving the OEP equation (15). The right hand side t_x of the OEP equation is given in Eq. (21) which in the case of a plane-wave code for periodic systems, initially is

$$t_x(\mathbf{r}) = \frac{1}{N_k} \sum_{i\mathbf{k}} f_{i\mathbf{k}} \sum_{\ell \neq i} \left[\frac{\langle \phi_{i\mathbf{k}} | \hat{v}_x^{\mathrm{NL}} | \phi_{\ell\mathbf{k}} \rangle \phi_{\ell\mathbf{k}}^{\dagger}(\mathbf{r})\, \phi_{i\mathbf{k}}(\mathbf{r})}{\varepsilon_{i\mathbf{k}} - \varepsilon_{\ell\mathbf{k}}} + \text{c.c.} \right] + \frac{1}{N_k} \sum_{i\mathbf{k}} \langle \phi_{i\mathbf{k}} | \hat{v}_x^{\mathrm{NL}} | \phi_{i\mathbf{k}} \rangle \frac{\delta f_{i\mathbf{k}}}{\delta v_s(\mathbf{r})} \tag{34}$$

With the definition (11) of the nonlocal exchange potential $\hat{v}_x^{\mathrm{NL}}$ the matrix elements $\langle \phi_{i\mathbf{k}} | \hat{v}_x^{\mathrm{NL}} | \phi_{\ell\mathbf{k}} \rangle$ can be expressed as

$$\langle \phi_{i\mathbf{k}} | \hat{v}_x^{\mathrm{NL}} | \phi_{\ell\mathbf{k}} \rangle = -\sum_{j\mathbf{q}} f_{j\mathbf{q}} \int d\mathbf{r} \int d\mathbf{r}' \, \frac{\phi_{i\mathbf{k}}^{\dagger}(\mathbf{r}) \phi_{j\mathbf{q}}(\mathbf{r}) \phi_{j\mathbf{q}}^{\dagger}(\mathbf{r}') \phi_{\ell\mathbf{k}}(\mathbf{r}')}{|\mathbf{r} - \mathbf{r}'|} \tag{35}$$

and calculated from

$$\langle \phi_{i\mathbf{k}} | \hat{v}_x^{\mathrm{NL}} | \phi_{\ell\mathbf{k}} \rangle = -\frac{4\pi}{\Omega} \sum_{j\mathbf{q}} f_{j\mathbf{q}} \sum_{\mathbf{G}} \frac{Y_{j\mathbf{q},i\mathbf{k}}^{*}(\mathbf{G})\, Y_{j\mathbf{q},\ell\mathbf{k}}(\mathbf{G})}{|\mathbf{G} + \mathbf{k} - \mathbf{q}|^2} \,. \tag{36}$$

For $\ell = i$ the right hand side of Eq. (36) contains a singular term namely the one with $\mathbf{G} = \mathbf{0}$, $\mathbf{q} = \mathbf{k}$, and $j = i = \ell$. For $\ell \neq i$ no singular terms occurs because in this case for $\mathbf{G} = \mathbf{0}$ and $\mathbf{q} = \mathbf{k}$ at least one factor in the product $Y_{j\mathbf{q},i\mathbf{k}}^{*}(\mathbf{0})\, Y_{j\mathbf{q},\ell\mathbf{k}}(\mathbf{0})$ has to be zero due to the orthonormality condition (27).

Because of the presence of the singularity in Eq. (36) for $\ell = i$ the expectation value $\langle \phi_{i\mathbf{k}} | \hat{v}_x^{\mathrm{NL}} | \phi_{\ell\mathbf{k}} \rangle$ is well-defined only in the limit of an infinite number N_k of $\mathbf{k}$ points. In this case Eq. (36) turns into the integral

$$\langle \phi_{i\mathbf{k}} | \hat{v}_x^{\mathrm{NL}} | \phi_{i\mathbf{k}} \rangle = -\frac{4\pi}{\Omega} \frac{\Omega}{(2\pi)^3} \sum_{j} \int_{BZ} d\mathbf{q} \; f_{j\mathbf{q}} \sum_{\mathbf{G}} \frac{Y_{j\mathbf{q},i\mathbf{k}}^{*}(\mathbf{G})\, Y_{j\mathbf{q},i\mathbf{k}}(\mathbf{G})}{|\mathbf{G} + \mathbf{k} - \mathbf{q}|^2} \tag{37}$$

that is well-defined because the singularity in the integrand is integrable. In the evaluation of the expectation value $\langle \phi_{i\mathbf{k}} | \hat{v}_x^{\mathrm{NL}} | \phi_{\ell\mathbf{k}} \rangle$ in practice we proceed as in the evaluation of the exchange energy by adding and subtracting the auxiliary function

$s(\mathbf{q})$ that exhibits the same singularity as the integrand in Eq. (37). This procedure leads to

$$
\begin{aligned}
&-\frac{4\pi}{\Omega}\frac{\Omega}{(2\pi)^3}\int_{BZ} d\mathbf{q}\ f_{i\mathbf{q}}\ \frac{Y^*_{i\mathbf{q},i\mathbf{k}}(\mathbf{0})\ Y_{i\mathbf{q},i\mathbf{k}}(\mathbf{0})}{|\mathbf{k}-\mathbf{q}|^2} \\
&\quad = -\frac{4\pi}{\Omega}\frac{\Omega}{(2\pi)^3}\int_{BZ} d\mathbf{q}\left[\ f_{i\mathbf{q}}\ \frac{Y^*_{i\mathbf{q},i\mathbf{k}}(\mathbf{0})\ Y_{i\mathbf{q},i\mathbf{k}}(\mathbf{0})}{|\mathbf{k}-\mathbf{q}|^2} - f_{i\mathbf{k}}\ s(\mathbf{k}-\mathbf{q})\right] \\
&\qquad\quad -\frac{4\pi}{\Omega}\frac{\Omega}{(2\pi)^3}\int_{BZ} d\mathbf{q}\ f_{i\mathbf{k}}\ s(\mathbf{k}-\mathbf{q}) \\
&\quad \approx -\frac{4\pi}{\Omega}\sum_{\mathbf{q}\neq\mathbf{k}}\left[\ f_{i\mathbf{q}}\ \frac{Y^*_{i\mathbf{q},i\mathbf{k}}(\mathbf{0})\ Y_{i\mathbf{q},i\mathbf{k}}(\mathbf{0})}{|\mathbf{k}-\mathbf{q}|^2} - f_{i\mathbf{k}}\ s(\mathbf{k}-\mathbf{q})\right] \\
&\qquad\quad -\frac{4\pi}{\Omega}\frac{\Omega}{(2\pi)^3}\ f_{i\mathbf{k}}\int_{BZ} d\mathbf{q}\ s(\mathbf{q}) \\
&\quad = -\frac{4\pi}{\Omega}\sum_{\mathbf{q}\neq\mathbf{k}} f_{i\mathbf{q}}\ \frac{Y^*_{i\mathbf{q},i\mathbf{k}}(\mathbf{0})\ Y_{i\mathbf{q},v\mathbf{k}}(\mathbf{0})}{|\mathbf{k}-\mathbf{q}|^2} \\
&\qquad\quad + f_{i\mathbf{k}}\ \frac{4\pi}{\Omega}\sum_{\mathbf{q}\neq\mathbf{k}} s(\mathbf{k}-\mathbf{q}) - f_{i\mathbf{k}}\ \frac{4\pi}{(2\pi)^3}\int_{BZ} d\mathbf{q}\ s(\mathbf{q}) \\
&\quad = -\frac{4\pi}{\Omega}\sum_{\mathbf{q}\neq\mathbf{k}} f_{i\mathbf{q}}\ \frac{Y^*_{i\mathbf{q},i\mathbf{k}}(\mathbf{0})\ Y_{i\mathbf{q},i\mathbf{k}}(\mathbf{0})}{|\mathbf{k}-\mathbf{q}|^2} + f_{i\mathbf{k}}\ [\tilde{S}_{\mathbf{k}} - S] \\
&\quad = -\frac{4\pi}{\Omega}\sum_{\mathbf{q}\neq\mathbf{k}} f_{i\mathbf{q}}\ \frac{Y^*_{i\mathbf{q},i\mathbf{k}}(\mathbf{0})\ Y_{i\mathbf{q},i\mathbf{k}}(\mathbf{0})}{|\mathbf{k}-\mathbf{q}|^2} + f_{i\mathbf{k}}\ [\tilde{S} - S]\,.
\end{aligned}
\tag{38}
$$

Expectation values $\langle\phi_{i\mathbf{k}}|\hat{v}_x^{\mathrm{NL}}|\phi_{i\mathbf{k}}\rangle$ then are calculated according to

$$
\begin{aligned}
\langle\phi_{i\mathbf{k}}|\hat{v}_x^{\mathrm{NL}}|\phi_{i\mathbf{k}}\rangle = &-\frac{4\pi}{\Omega}\sum_{j,\mathbf{q}\neq\mathbf{k}} f_{j\mathbf{q}}\sum_{\mathbf{G}}\frac{Y^*_{j\mathbf{q},i\mathbf{k}}(\mathbf{G})\ Y_{j\mathbf{q},i\mathbf{k}}(\mathbf{G})}{|\mathbf{G}+\mathbf{k}-\mathbf{q}|^2} \\
&-\frac{4\pi}{\Omega}\sum_{j} f_{j\mathbf{k}}\sum_{\mathbf{G}\neq\mathbf{0}}\frac{Y^*_{j\mathbf{k},i\mathbf{k}}(\mathbf{G})\ Y_{j\mathbf{k},i\mathbf{k}}(\mathbf{G})}{|\mathbf{G}|^2} \\
&+ f_{i\mathbf{k}}\,[\tilde{S} - S]\,,
\end{aligned}
\tag{39}
$$

i.e., the singular term in the original expression (36) is omitted in the summations over j, $\mathbf{q}$, and $\mathbf{G}$ and then taken into account by adding the correction $f_{i\mathbf{k}}\ [\tilde{S} - S]$.

In the second line on the right hand side of the Eq. (34) for the right hand side t_x of the OEP equation for the exchange potential, the matrix elements $\langle\phi_{i\mathbf{k}}|\hat{v}_x^{\mathrm{NL}}|\phi_{i\mathbf{k}}\rangle$

occur in a sum over i and $\mathbf{k}$. Applying the singularity correction $f_{i\mathbf{k}}\,[\tilde{S}-S]$ to each of the matrix elements leads to a total singularity correction of

$$\frac{1}{N_k}\sum_{i\mathbf{k}} f_{i\mathbf{k}}\,[\tilde{S}-S]\,\frac{\delta f_{i\mathbf{k}}}{\delta v_s(\mathbf{r})} \tag{40}$$

to be considered in the evaluation of Eq. (34) for t_x.

Once the OEP equation for the KS exchange potential is solved, the calculation of the KS orbitals from the corresponding KS equations (12) is straightforward and not impaired by singularities. Indeed, because the KS exchange potential is a local multiplicative potential, its representation in real space is diagonal and it can be handled highly efficiently in plane wave codes via fast Fourier transformations. This is completely different from the HF case. In it, the exchange potential is a nonlocal operator constructed directly from the HF orbitals according to Eq. (11). This means a solution of an OEP equation is not required. However, when solving the HF equations, the action of the HF exchange operator on the HF orbitals has to be treated. This means the quantities $\hat{v}_x^{\mathrm{NL}}\,\phi_{i\mathbf{k}}$ have to be handled. To analyse the action of the HF exchange operator on the HF orbitals, we expand the quantities $\hat{v}_x^{\mathrm{NL}}\,\phi_{i\mathbf{k}}$ in terms of HF orbitals. Because the HF exchange operator, like the HF equations, is diagonal in $\mathbf{k}$ space, only HF orbitals $\phi_{\ell\mathbf{q}}$ with $\mathbf{q}=\mathbf{k}$, i.e., orbitals $\phi_{\ell\mathbf{k}}$, have to be considered in the representation of $\hat{v}_x^{\mathrm{NL}}\,\phi_{i\mathbf{k}}$. The matrix elements $\langle\phi_{\ell\mathbf{k}}|\hat{v}_x^{\mathrm{NL}}|\phi_{i\mathbf{k}}\rangle$ representing the expansion coefficients of $\hat{v}_x^{\mathrm{NL}}\,\phi_{i\mathbf{k}}$ are not affected by singularities except for $\ell=i$. This means that when solving the HF equation, singularities have a direct effect only on the HF eigenvalues but not on the HF orbitals. Therefore in the solution of the HF equation in an iteration step of the HF self-consistency procedure singular terms can be neglected at first. Afterwards the singularity correction $f_{i\mathbf{k}}\;[\tilde{S}-S]$, see Eq. (39), has to be added to the HF eigenvalues.

While there is no direct effect of the singularity correction on the HF orbital, there exists an indirect effect. This is because the HF eigenvalues, which are affected by a singularity correction, determine the occupation numbers. The latter enter the HF exchange potential and therefore the Fock operator and thus subsequently affect the HF orbitals. In the handling of the singularity correction within the HF self-consistency process the 'hen-egg' problem arises: the singularity corrections $f_{i\mathbf{k}}\;[\tilde{S}-S]$ for the HF eigenvalues depend on the occupation numbers which, on the other hand, depend according to the Fermi distribution (5) on the HF eigenvalues. In practice the singularity corrections for the HF eigenvalues of a given iteration cycle are calculated with the occupation numbers of the previous cycle. Upon self-consistency the singularity correction also becomes consistent. This procedure worked well for the semiconductor silicon. For the metal aluminium it worked for temperatures of 10,000 K and above, for low temperatures, of e.g. 300 K, it converged to slightly different results depending on the starting conditions of the self-consistent field procedure. This means for metals at low temperatures further work on the treatment of the exchange singularity is required in the HF case. Finally, we note that for an insulator in the zero-temperature limit occupation numbers are always zero or one. In this special case the HF orbitals are not even indirectly affected by singularity corrections and the latter only energetically shift the eigenvalues of all occupied orbitals in the same way while unoccupied orbitals are not affected at all, i.e., the singularity corrections only modify the band gap.

3.3 Singularity Correction in Band Structure Calculations

Band structures typically are calculated based on the potentials determined in a self-consistent HF or KS calculation. In the case of exact-exchange KS band structures, this is straightforwardly possible without the need of any singularity correction because all potentials occurring in the KS Hamiltonian operator, including the exchange potential, are local multiplicative. In the HF case this is different. The Fock operator contains the nonlocal HF operator and, as discussed in the previous subsection, this leads to the necessity of a singularity correction of the HF eigenvalues, i.e., of the band structure. The singularity corrections of the HF eigenvalues building the HF band structure, in principle, are those introduced in the HF self-consistency procedure, see previous subsection. The only difference is that in the band structure eigenvalues are required at **k** points that are not contained in the **k** point grid used in the self-consistency procedure, i.e., **k** points that are not contained in the set of **k** points that determines the HF exchange operator. As a consequence expression (39) for the evaluation of matrix elements $\langle\phi_{i\mathbf{k}}|\hat{v}_x^{\mathrm{NL}}|\phi_{i\mathbf{k}}\rangle$ has to be slightly modified to

$$\langle\phi_{i\mathbf{k}}|\hat{v}_x^{\mathrm{NL}}|\phi_{i\mathbf{k}}\rangle = -\frac{4\pi}{\Omega}\sum_{j,\mathbf{q}} f_{j\mathbf{q}} \sum_{\mathbf{G}}{}' \frac{Y^*_{j\mathbf{q},i\mathbf{k}}(\mathbf{G})\, Y_{j\mathbf{q},i\mathbf{k}}(\mathbf{G})}{|\mathbf{G}+\mathbf{k}-\mathbf{q}|^2} + f_{i\mathbf{k}}\,[\tilde{S}'_{\mathbf{k}} - S] \tag{41}$$

with

$$\tilde{S}'_{\mathbf{k}} = \frac{4\pi}{\Omega}\sum_{\mathbf{q}}{}' s(\mathbf{k}-\mathbf{q})\,. \tag{42}$$

In Eqs. (41) and (42) in contrast to Eqs. (39) and (30), the case $\mathbf{q} = \mathbf{k}$, in general, is not excluded simply because, in general, the point **k** does not equal any of the points **q** building the **k** point mesh that was used in the self-consistency process and that underlies the HF exchange potential. The special case that accidentally a **k** point of the band structure is identical to one of the **k** points **q** of the **k** point mesh is taken into account by the primes in the summations in Eqs. (41) and (42) which indicate that terms with $|\mathbf{G}+\mathbf{k}-\mathbf{q}| = 0$ or $|\mathbf{k}-\mathbf{q}| = 0$, respectively are omitted. In practice terms with $|\mathbf{G}+\mathbf{k}-\mathbf{q}|^2 \leq c_{thres}$ or $|\mathbf{k}-\mathbf{q}|^2 =\leq c_{thres}$ are omitted with c_{thres} being a threshold, here chosen as 10^{-13} a.u.

4 Results

In this section non-zero-temperature EXX, HF, and GGA calculations for silicon and aluminum are discussed. Silicon and aluminum were chosen as examples for a semiconducting and a metallic material. As example for a ground-state GGA functional the exchange-correlation functional due to Perdew, Burke, Ernzerhof (PBE) [82] was chosen.

4.1 Temperature Dependency of Energy and Band Structures

In Fig. 1 EXX band structures of silicon are shown for different temperatures and different unit cell volumes. The band structures in all cases are based on self-consistent EXX calculations with a plane wave cutoff of 30 a.u. and a regular $5 \times 5 \times 5\,\mathbf{k}$ point mesh. EXX pseudopotentials, i.e., pseudopotentials based on EXX calculations for atoms [83, 84], were constructed with the pseudopotential generation code of Engel [84] and applied not only for the EXX calculations but also for the HF and GGA calculations discussed later on. The reason for using the EXX pseudopotential in all cases is that in this way comparisons between different methods are not affected by the use of different pseudopotentials. The chosen unit cell volumes are (i) the one at ambient conditions, i.e., the one at 300 K and ambient pressure resulting in a lattice constant of 10.2631 a.u. (5.4310 Å) and (ii) a compressed unit cell volume with a fifth of the unit cell volume at ambient conditions, and (iii) an expanded unit cell that is five times as large as the unit cell at ambient conditions.

The EXX band structure at the ambient unit cell volume changes remarkably little with temperature. Even at 50,000 K, the band structure has almost exactly the same form as at 300 K. The only clearly noticeable effect of a raising of the temperature is a shift of the bands to higher energies with respect to the Fermi level. For 50,000 K this shift has the effect that silicon becomes metallic. The shift can easily be explained from the fact that at higher temperatures conduction bands, i.e., energetically higher lying bands, are occupied. At the same time, the occupation of valence bands is decreased. The number of valence bands that is depopulated, however, is smaller than the number of conduction bands that is populated. Therefore, in order to keep the number of electrons constant, the Fermi energy has to decrease with temperature which is tantamount to an energetic shift of the bands to higher energies with respect to the Fermi level. See Fig. 2 for a sketch of the Fermi distribution functions determining the occupation numbers of bands for the considered temperatures.

For higher temperatures conduction bands are significantly occupied while valence bands are significantly depopulated. At 50,000 K, e.g., bands that lie 10 eV above the Fermi level acquire an occupation of about 0.1 electron while the occupation of bands lying 10 eV below the Fermi level is reduced by roughly

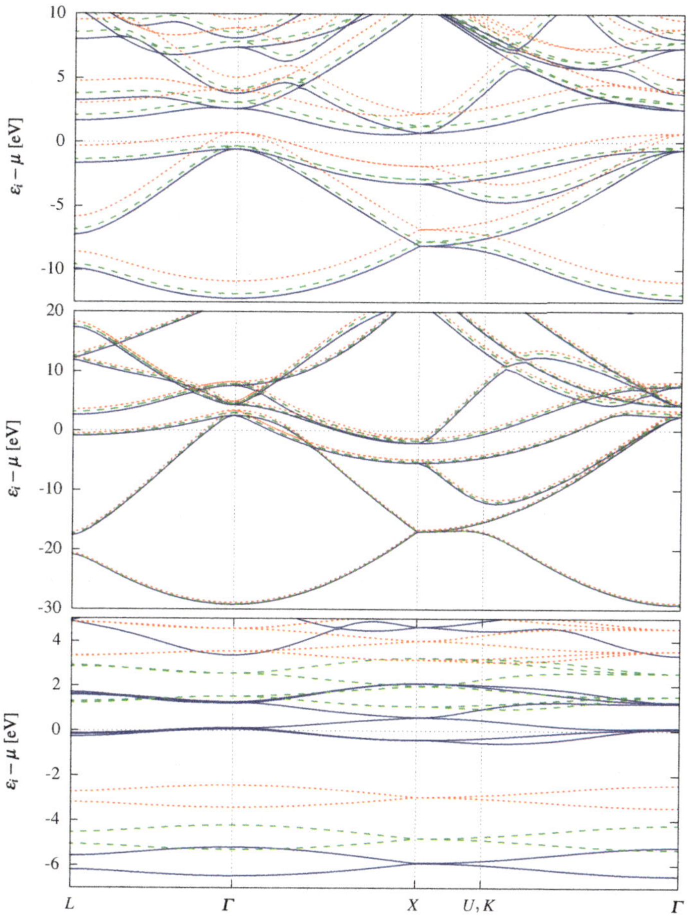

Fig. 1 EXX band structures of silicon at different temperatures and different unit cell volumes. *Upper panel*, unit cell volume at ambient conditions, i.e., lattice constant of 10.263 a.u. (5.431 Å); *middle panel* unit cell volume equalling one-fifth of the unit cell volume at ambient conditions; *lower panel*, unit cell volume equalling five times the unit cell volume at ambient conditions. *Solid blue lines* refer to 300 K, *dashed green lines* to 30,000 K, and *dotted red lines* to 50,000 K

0.1 electron. Despite these temperature induced changes in occupations the form of the bands is not changed and, except for the above-discussed energetic shift of the bands, temperature has no significant effect on the band structure. This means the changes of the Hartree potential and of the KS exchange potential due to the temperature induced modifications of the occupations remain so small that no significant effect on the shape of the bands is observed.

For silicon with a compressed unit cell, Fig. 1 shows that the effect of temperature on the band structure is even smaller than in the case of silicon with a unit cell at ambient conditions, i.e., the energetic shifts of the bands are distinctively smaller.

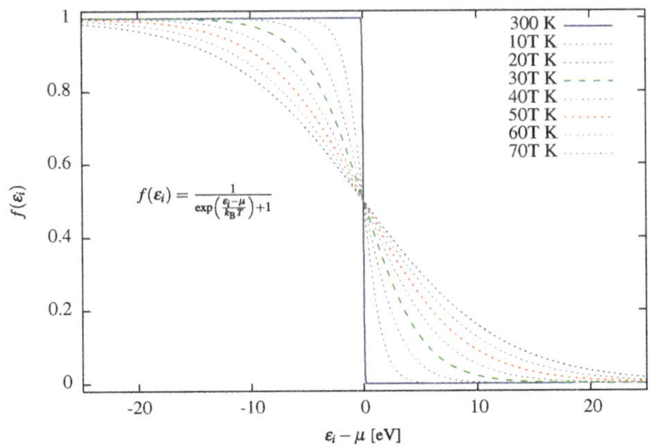

Fig. 2 Fermi function for different temperatures

This is not surprising. At smaller unit cell volume the width of all bands is much larger. Therefore a smaller part of the band structure is affected by significant changes of the occupations numbers. Note that silicon for the compressed unit cell is metallic for all temperatures considered. For silicon with an expanded unit cell the energetic shift of the band structure due to temperature, according to Fig. 1, is stronger than in the case of silicon with a unit cell corresponding to ambient conditions. This can be explained by the reduced width of the bands. Amazingly the form of the bands remains unaltered even for the expanded unit cell volume, for which the occupation numbers are strongly modified with temperature. Note that in the case of an expanded volume, silicon for all temperatures remains a semiconductor.

In Fig. 3 the HF and the PBE band structures of silicon for different temperatures are displayed. The calculation parameters (pseudopotentials, plane wave cutoff, **k** point grid) were chosen as in the EXX calculations. The unit cell is the one corresponding to ambient conditions. In the HF case, the temperature dependence of the band structure is more complicated than in the EXX case. Energetically low lying valence bands, like in the EXX case, only experience a shift to higher energies with temperature but remain unchanged as far as their shape is concerned. Energetically higher valence bands, however, are nonuniformly shifted to higher energies, that is their shape changes. The effect is most pronounced around the Γ point. In that region of the Brillouin zone, the energetically highest valence bands are shifted more than twice as much to higher energies than in other regions, e.g., around the X or K point. This difference can be explained as follows. While in the EXX case all bands are acted upon by the same local multiplicative KS exchange potential, the effect of the nonlocal HF exchange potential differs for the individual bands. Moreover, all the EXX bands, be they occupied or unoccupied, are free of Coulomb self-interactions in an average way, while in the HF case only the fully occupied bands have that property. If, due to temperature, the occupation of a valence band is decreased then this goes along with the occurrence of repulsive Coulomb self-interactions which increase the bands energy. In this case two effects

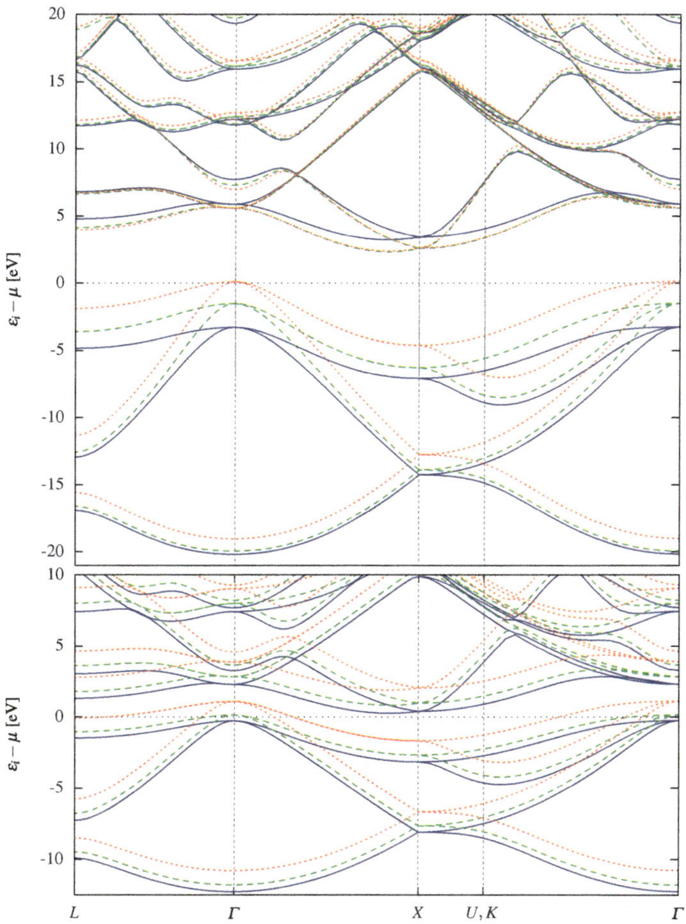

Fig. 3 HF and PBE band structures of silicon at different temperatures. *Upper panel*, HF band structure and *lower panel*, PBE band structure. *Solid blue lines* refer to 300 K, *dashed green lines* to 30,000 K, and *dotted red line* to 50,000 K

contribute to a temperature induced change of the band energy: (i) a raising of the band energy due to the requirement of a constant electron number, an effect also present in the EXX case, see above; (ii) a raising of the band energy due to the occurrence of Coulomb self-interactions. The two effects altering the energy of energetically high valence bands go in the same direction and therefore reinforce each other. For conduction bands a different situation arises. For low temperatures the conduction bands are unoccupied and thus are uncorrected with respect to Coulomb self-interaction, that is they describe one-electron states of the $N+1$-electron system instead of the N-electron system with N denoting the electron number. If, with increasing temperature, the energetically low conduction bands become occupied then an attractive correction of Coulomb self-interaction sets in.

This lowers the energy of these bands, while the requirement of a constant electron number leads to an upward shift of the bands energies. Figure 3 shows that for the energetically lowest conduction bands the former effect prevails, i.e., the bands are lowered in energy with temperature. The energetically higher conduction bands, on the other hand, remain almost unoccupied and therefore only experience the shift to higher energies due to the requirement of a constant electron number, i.e., they behave like the EXX bands. In summary, the HF band structure exhibits a distinctively different and more complicated behavior with temperature compared to the EXX case.

The influence of temperature on the PBE band structure of silicon is similar to that on the EXX band structure. This could be expected because in the PBE case, as in the EXX case, all bands are acted upon in the same way by a local multiplicative KS potential. In both the PBE and EXX cases, temperature-induced changes of the KS potential are small enough such that they do not affect the shape of the bands in a significant way. This is not surprising because the PBE functional for the exchange energy and thus also the PBE exchange potential does not exhibit any direct dependence on occupation numbers and therefore no direct temperature dependence. The PBE exchange potential only indirectly, through the electron density, depends on occupation numbers and thus on temperature. This indirect dependence, however, is weak.

In Figs. 4 and 5 the EXX, PBE, and HF bands for aluminum are shown. In the HF case we not only show band structures obtained with a treatment of the exchange singularity according to this work, but also band structures calculated with the singularity treatment of Ref. [79]. Again EXX pseudopotentials were applied in both cases [84]. A plane wave cutoff of 30 a.u. and a regular $6 \times 6 \times 6\,\mathbf{k}$ point mesh was used. A unit cell corresponding to ambient conditions, with a lattice constant of 7.6524 a.u. (4.0495 Å) was chosen. The effect of temperature on the band structures is similar as for silicon. The EXX band structure remains unaltered except for an almost uniform shift towards higher energies, whereas the HF band structure exhibits both the energetic shift and slight changes in the shape. The PBE and EXX band structures again exhibit similar behaviour with temperature. The HF band structures of aluminum at 300 K, in contrast to all other band structures considered here, depend on the way the exchange singularity is treated and exhibit small artefacts (bumps close to the Γ point and in the vicinity of the Fermi energy) originating from the treatment of the exchange singularity. This again indicates that further work is required concerning the handling of the exchange singularities in HF methods for metals.

In Fig. 6 the free energy and the exchange energy of aluminum are displayed as a function of the unit cell volume for different temperatures as obtained by finite-temperature EXX and HF and by a conventional KS calculation with the PBE exchange-correlation functional. The unit cell volume is given in units of the unit cell volume at ambient conditions which corresponds to a lattice constant of 7.6524 a.u. (4.0495 Å). Considering first the free energy, Fig. 6 shows that for 300 K the EXX and the HF method yield almost indistinguishable results, the corresponding free energy curves lie on top of each other with the EXX free energies being

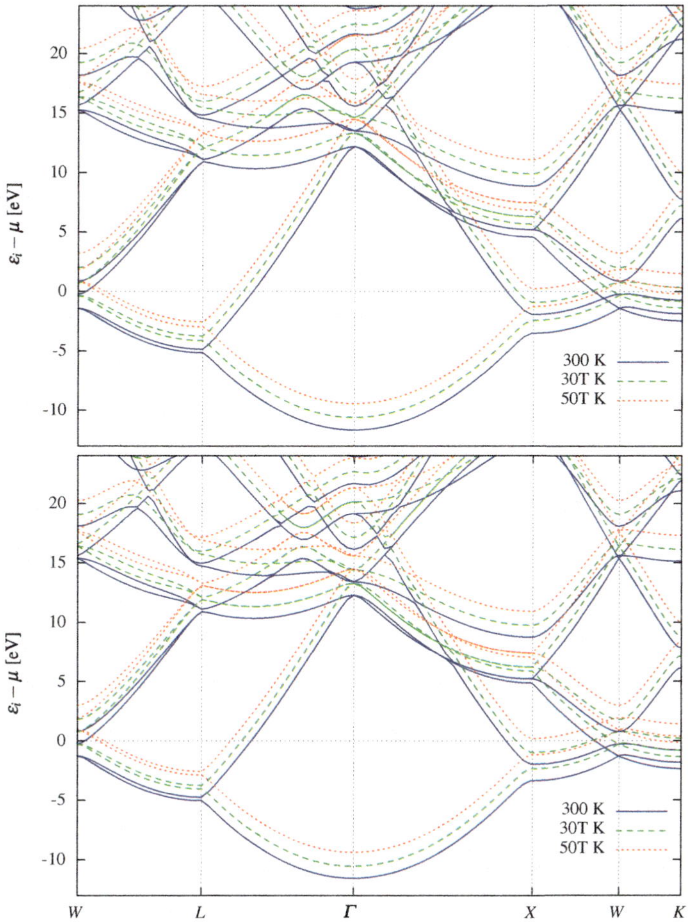

Fig. 4 EXX and PBE band structures of aluminum at different temperatures. *Solid blue lines* refer to 300 K, *dashed green lines* to 30,000 K, and *dotted red line* to 50,000 K

marginally higher. This could be expected. It is well-known [74] that in the zero-temperature case, i.e., if temperature is not considered, EXX and HF energies are almost identical with the HF energies always being a tiny bit lower. For higher temperatures, on the other hand, EXX and HF free energies differ distinctively. The EXX free energies exhibit a much stronger dependence on temperature than the HF ones. The reason is that the width of the HF band structure is much larger than that of the EXX band structure, see Figs. 4 and 5. As a result the temperature induced population or depopulations for most parts of the bands is much smaller than in the EXX case and with the smaller changes of the occupation number consequently also the change of the free energy with temperature is smaller. Note that the energy shift of the energetically low lying HF valence bands is larger than

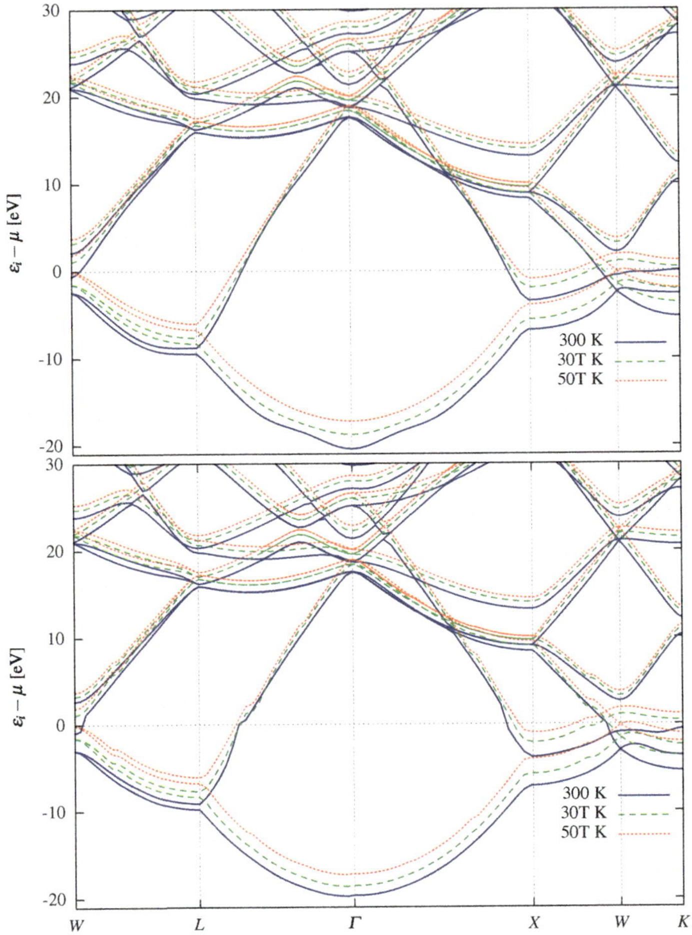

Fig. 5 HF band structure of aluminum at different temperatures. *Solid blue lines* refer to 300 K, *dashed green lines* to 30,000 K, and *dotted red line* to 50,000 K. *Upper panel* treatment of the exchange singularity according to Ref. [79], *lower panel* according to this work

in the EXX case despite the fact that changes in occupation numbers are smaller in the HF case. The reason is that, as discussed above, the band structure in the HF case is not only energetically shifted due to the requirement of keeping the particle number fixed but also due to an decreasing (increasing) degree of cancellation of Coulomb self-interaction for occupied (unoccupied) bands. The PBE free energies for aluminum also exhibit a temperature dependence. In this case, the free energy decreases with temperature while the EXX and HF free energies increase with increasing temperature. This means the PBE free energy shows a qualitatively different behavior with temperature compared to the EXX and HF case.

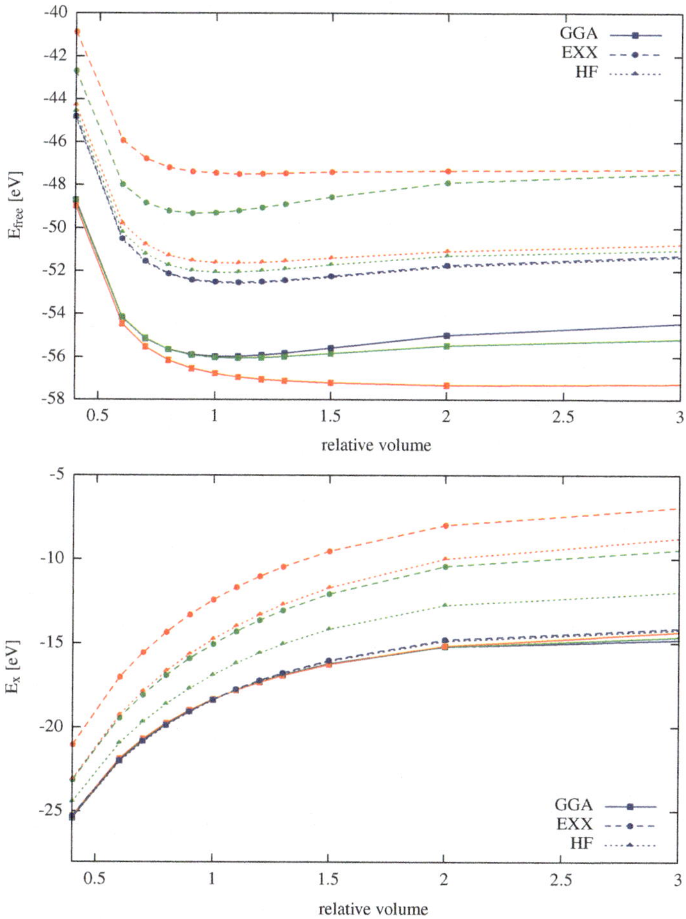

Fig. 6 Free energy (*upper panel*) and exchange energy (*lower panel*) for aluminum as function of the unit cell volume for different temperatures and different methods. *Solid lines* refer to PBE, *dashed* to EXX, and *dotted lines* to HF calculations. The color code is *blue* for 300 K, *green* for 30,000 K, and *red* for 50,000 K

In most cases the free energy exhibits a minimum at a unit cell volume roughly in the region of the unit cell volume at ambient condition. Only the EXX and PBE free energy for 50,000 K does not exhibit such a minimum. Note, however, that considering the free energy we have disregarded pressure, that is considered the case of zero pressure. For a non-zero pressure we would have to switch to the free enthalpy which always would exhibit a minimum at some unit cell volume.

The EXX KS and the HF exchange energies for 300 K are, as expected, almost identical. For higher temperatures both the EXX KS as well as the HF exchange energy strongly decrease in magnitude with the EXX case exhibiting a clearly stronger temperature dependence. The PBE exchange energy, on the other hand,

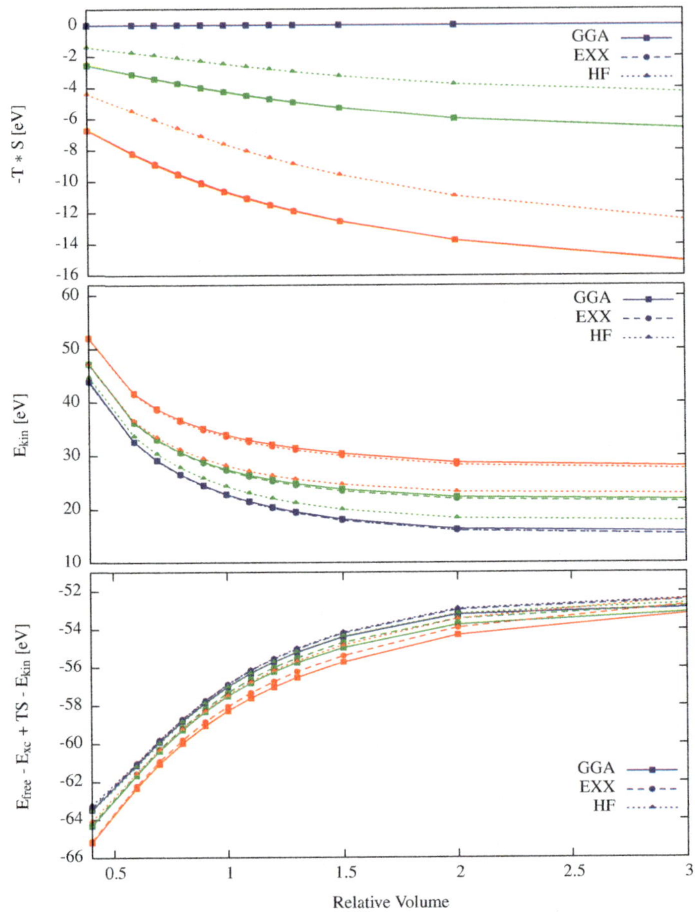

Fig. 7 Entropy term (*upper panel*), kinetic energy (*middle panel*), and Coulomb energy plus interaction energy with the nuclei (*lower panel*) for aluminum as function of the unit cell volume for different temperatures and different methods. *Solid lines* refer to PBE, *dashed* to EXX, and *dotted lines* to HF calculations. The color code is *blue* for 300 K, *green* for 30,000 K, and *red* for 50,000 K

does not have any noteworthy temperature dependence. This is not surprising because the PBE exchange functional itself does not exhibit a temperature dependence and the only way for the PBE exchange energy to depend on temperature is via the temperature dependence of the electron density which, however, is weak. The temperature dependence of the other contributions to the free energy that is the product $-TS$ of temperature times entropy, the kinetic energy, and the sum of the Coulomb energy and the interaction energy with the nuclei is found in Fig. 7. The product $-TS$ and the kinetic energy which depend explicitly on occupation numbers are strongly dependent on temperature, changes of 10 eV per unit cell or more are found for the unit cell corresponding to ambient conditions. In contrast, the sum of the Coulomb energy and the interaction energy with the nuclei which depend only implicitly via the electron density on temperature show a very weak dependence on the latter, only changes of less than 2 eV are observed for the unit cell corresponding to ambient conditions. For all these contributions to the free energy, i.e., all contributions except the exchange energy, the EXX and PBE method yields almost identical values for all temperatures. (Indeed the EXX curves sometimes cannot be seen in Fig. 7 because they lie underneath the corresponding PBE curves.) The reason is that these contributions to the free energy are calculated exactly in both the PBE and the EXX case and the band structures and thus the occupation numbers are very similar for all temperatures. In the HF calculations, on the other hand, values obtained for the product $-TS$ and the kinetic energy are close to the corresponding EXX and PBE values only at 300 K but differ distinctively for higher temperatures, exhibiting a less strong temperature-dependence.

In summary, all contributions to the free energy of aluminum except the exchange energy are obtained by the PBE method in good agreement with the EXX method for all considered temperatures and unit cell volumes. The fact that the PBE exchange functional does not capture the temperature dependence of the exchange energy however, has the severe consequence of leading to a qualitatively wrong behavior of the free energy with temperature in comparison with the EXX and HF results. These findings are in agreement with Ref. [6] which compares non-zero-temperature HF and LDA exchange and free energies for lithium. Because HF and EXX methods treat exchange exactly, the HF or the KS way, the temperature-dependence of their exchange energy has to be considered as correct. GGA functionals other than PBE can be assumed to exhibit a similarly wrong temperature dependence. Therefore the above discussed findings strongly indicate that ground-state GGA functionals, in general, treat the temperature-dependence of the free energy of electronic systems qualitatively wrongly, due to their lack of explicit temperature-dependence of the exchange energy. This would mean that ground-state GGA functionals are unsuitable for the treatment of the electronic structure of matter at high temperatures. It should, however, kept in mind, that in the EXX and HF calculations correlation was neglected and that the temperature-dependence of the correlation energy, at least to some extent, could counteract the temperature dependence of the exchange energy.

4.2 *Singularity Correction*

In this subsection we compare the two approaches considered in Sect. 3 to treat the integrable singularity that occurs in the integration in $\mathbf{k}$ space required in the evaluation of the exchange energy, the construction of the local multiplicative KS exchange potential, and the handling of the nonlocal HF exchange potential. The approach of Ref. [78] was generalized to the non-zero temperature case in Sect. 3 whereas the approach of Ref. [79] does not require a specific generalization

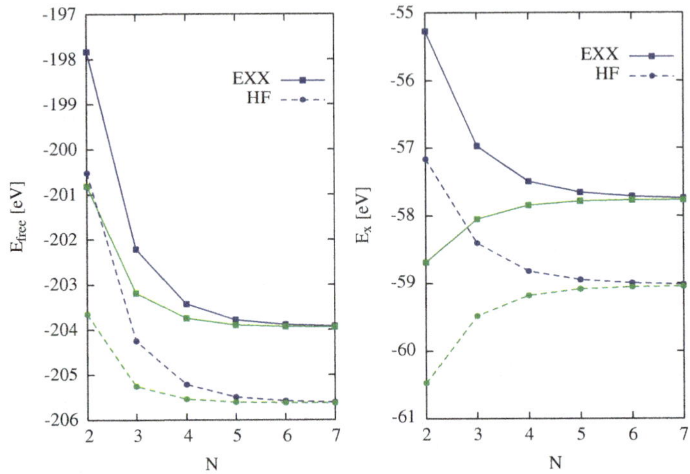

Fig. 8 Convergence of the free energy (*left panel*) and the exchange energy (panel on the *right*) of silicon at 10,000 K with **k** points for a treatment of the singularity of the exchange energy and the EXX KS and the HF exchange potentials according to this work generalizing Ref. [78], (*blue line*), and according to Ref. [79], (*green line*). The *solid lines* refer to EXX, the *dashed lines* to HF data. Regular **k** point meshes with $n \times n \times n$ points were used

to be applicable to non-zero temperatures. Free energies and band structures are considered. All data shown in this subsection are from calculations of silicon at a temperature of 10,000 K.

Figure 8 shows the convergence of the free energy and the exchange energy with the number of **k** points. Both treatments of the singularity converge rapidly to the same result for an increasing number of **k**-points. HF and EXX free energies and exchange energies exhibit a similar behavior because both methods share the same energy expression. Only the orbitals and occupation numbers entering that energy expression are different. For silicon, the treatment of the singularity according to Ref. [79] converges slightly faster than the method of this work generalizing Ref. [78]. This is not surprising because the scheme used in Ref. [79] is especially well-suited for highly symmetric systems like silicon with its face centered cubic lattice.

Next we consider the convergence with **k** points of the EXX band structure of silicon at 10,000 K with special emphasis on the region around the band gap, see Fig. 9. In the actual calculation of the EXX band structure, no singularities have to be dealt with because only the local multiplicative KS potential determined in the self-consistent EXX calculation preceding the band structure calculations is required but no integrations in **k** space have to be carried out. The KS potential determined in the self-consistent EXX calculation, however, depends on the treatment of the singularities in the **k** space integrations occurring in the course of the construction

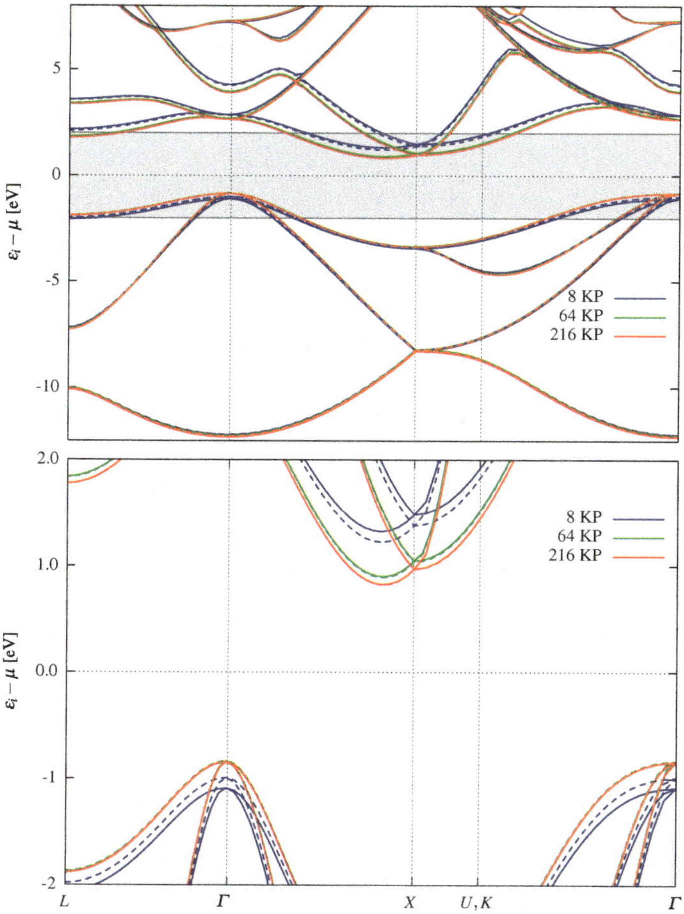

Fig. 9 Convergence of the exact-exchange band structure of silicon at 10,000 K with **k** points (KP). Results for regular **k** point grids with $2 \times 2 \times 2 = 8$, $4 \times 4 \times 4 = 64$, and $6 \times 6 \times 6 = 216$ **k** points are displayed. *Solid* and *dashed lines* refer to data obtained by treating the singularity in **k** space in the construction of the exact KS exchange potential according to this work generalizing Ref. [78] and according to Ref. [79], respectively

of the KS exchange potential. Figure 9 shows that both approaches to treat the singularities convergence very fast towards each other, with the bands below the Fermi level being less sensitive to smaller numbers of **k**-points than bands above the Fermi level.

Figure 10 shows the HF band structure of silicon at 10,000 K obtained with a treatment of the singularity in the nonlocal HF exchange potential according to Eq. (41) which follows from a generalization of the approach of Ref. [78].

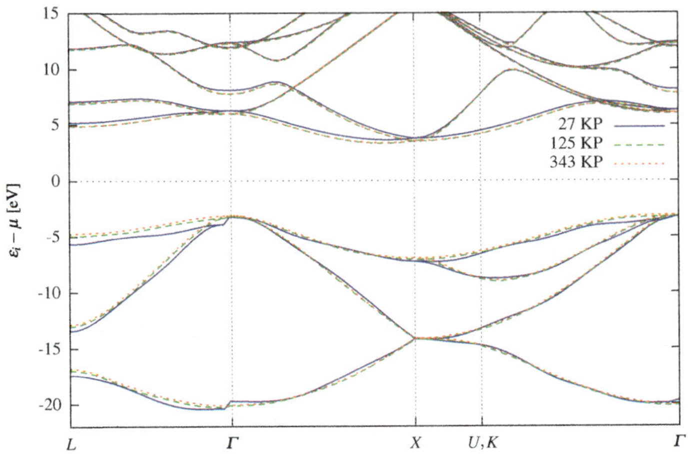

Fig. 10 Convergence of the HF band structure of silicon at 10,000 K with **k** points (KP) using a treatment of the singularity in the nonlocal HF exchange potential described in this work generalizing Ref. [78]. Results for regular **k** point grids with $3 \times 3 \times 3 = 27$, $5 \times 5 \times 5 = 125$, and $7 \times 7 \times 7 = 343$ **k** points are displayed

5 Concluding Remarks

Exact-exchange-only KS band structures were shown to exhibit a remarkably small dependence on temperature. The shape of the band structures of silicon and aluminum, chosen as representative examples for metallic and semi-conducting materials, even at temperatures of 50,000 K is virtually unaltered compared to the shape at 300 K. The only effect of increasing temperature on exact-exchange-only KS band structures is an almost uniform shift towards higher energies with respect to the Fermi level. The same behavior with temperature was found for PBE band structures, considered as representative examples for GGA bands structures. HF band structures, on the other hand, were found to exhibit a more complex dependence on temperature. Besides a shift towards higher energies we also observed changes in the shape of bands in the vicinity of the Fermi energy.

The free energy, considered for the example of aluminum, showed a clear dependence on temperature. In the exact-exchange-only KS case the free energy increased by roughly 4 eV per unit cell if going from 300 to 50,000 K for a lattice constant fixed at the experimental value at ambient conditions. The corresponding HF free energy exhibited a weaker increase of only 1 eV. The PBE free energy of aluminum, on the other hand, showed a decrease with temperature of about 1 eV.

Considering the contributions to the free energy we found that those parts that directly depend of occupation numbers, i.e., the kinetic energy, the exchange energy, and the product $-TS$ of temperature and entropy, exhibit a strong dependence with temperature. The behavior with temperature of kinetic energy and the product $-TS$ is opposite to that of the exchange energy, the former decrease with temperature the latter increases. On the other hand, those contributions to the free energy that are evaluated via the electron density and only indirectly depend on temperature through the electron density, i.e., the Coulomb energy and the interaction energy with the potential of the nuclei, exhibit only a very small temperature dependence. The PBE functional correctly yields the temperature dependence of those contributions calculated exactly in a ground-state GGA approach, i.e, kinetic energy, $-TS$, Coulomb energy, and interaction energy with the nuclei but completely fails to capture the temperature dependence of the exchange energy because the direct dependence of the exchange energy on the occupation numbers is not taken into account. This indicates that *ground-state GGA exchange functionals are unsuitable for higher temperatures*. Even if the correlation energy is likely to counteract the temperature dependence of the exchange energy to some extent, this means that the use of ground-state GGA functionals for the treatment of electronic structures of matter at high temperatures seems to be highly questionable. This is particularly true if matter at lower densities than the ambient density is considered because in this case the errors in the exchange energy by GGA functionals become more pronounced while it remains to be seen whether there is a significant cancelling effect of correlation.

We have shown how the integrable singularities occurring in the **k** space integrations for the exchange energy can be handled in the case of non-zero temperatures. Compared to the zero temperature case the regularization of the singular integrand is a bit more involved because it depends on occupation numbers in the non-zero temperature case. The computational effort of the regularization, however, remains negligible and thus poses no problem.

Acknowledgements We thank Sam Trickey for helpful comments and a careful reading of the manuscript. Financial support by the Deutsche Forschungsgemeinschaft through the Cluster of Excellence "Engineering of Advanced Materials" is gratefully acknowledged.

References

1. R.G. Parr, W. Yang, *Density-Functional Theory of Atoms and Molecules* (Oxford University Press, Oxford, 1989)
2. R.M. Dreizler, E.K.U. Gross, *Density Functional Theory* (Springer, Heidelberg, 1990)
3. W. Koch, M.C. Holthausen, *A Chemist's Guide to Density Functional Theory* (Wiley-VCH, New York, 2000)
4. N.D. Mermin, Phys. Rev. **137**, A1441 (1965)
5. M.V. Stoitsov, I.Z. Petkov, Ann. Phys. **121**, 121 (1988)
6. V.V. Karasiev, T. Sjostrom, S.B. Trickey, Phys. Rev. E **86**, 056704 (2012)
7. A.D. Becke, J. Chem. Phys. **98**(2), 1372 (1993)
8. A.D. Becke, J. Chem. Phys. **98**(7), 5648 (1993)
9. A. Seidl, A. Görling, P. Vogl, J.A. Majewski, M. Levy, Phys. Rev. B **53**, 3764 (1996)
10. A. Görling, M. Levy, J. Chem. Phys. **106**, 2675 (1997)
11. R.T. Sharp, G.K. Horton, Phys. Rev. **90**, 317 (1953)
12. J.D. Talman, W.F. Shadwick, Phys. Rev. A **14**, 36 (1976)

13. M. Städele, J.A. Majewski, P. Vogl, A. Görling, Phys. Rev. Lett. **79**, 2089 (1997)
14. M. Städele, M. Moukara, J.A. Majewski, P. Vogl, A. Görling, Phys. Rev. B **59**, 10031 (1999)
15. A. Görling, Phys. Rev. Lett. **83**, 5459 (1999)
16. S. Ivanov, S. Hirata, R.J. Bartlett, Phys. Rev. Lett. **83**, 5455 (1999)
17. A. Görling, J. Chem. Phys. **123**, 062203 (2005)
18. S. Hirata, S. Ivanov, I. Grabowski, R. Bartlett, K. Burke, J.D. Talman, J. Chem. Phys. **115**, 1635 (2001)
19. S. Hamel, M.E. Casida, D.R. Salahub, J. Chem. Phys. **116**, 8276 (2002)
20. D.R. Rohr, O.V. Gritsenko, E.J. Baerends, J. Mol. Struct. (THEOCHEM) **762**, 193 (2006)
21. V.K. Staroverov, G.E. Scuseria, E.R. Davidson, J. Chem. Phys. **124**, 141103 (2006)
22. A. Görling, A. Heßelmann, M. Jones, M. Levy, J. Chem. Phys. **128**, 104104 (2008)
23. A. Heßelmann, A. Görling, Chem. Phys. Lett. **455**, 110 (2008)
24. T. Kotani, J. Phys. Condens. Matter **10**, 9241 (1998)
25. F. Furche, Phys. Rev. A **64**, 195120 (2001)
26. T. Miyake, F. Aryastiawan, T. Kotani, M. van Shilfgaarde, M. Usuda, K. Terakura, Phys. Rev. B **66**, 245103 (2002)
27. M. Fuchs, X. Gonze, Phys. Rev. B **65**, 235109 (2002)
28. Y.M. Niquet, M. Fuchs, X. Gonze, Int. J. Quantum Chem. **101**, 635 (2004)
29. M. Fuchs, Y.M. Niquet, X. Gonze, K. Burke, J. Chem. Phys. **122**, 094116 (2005)
30. H. Jiang, E. Engel, J. Chem. Phys. **127**, 184108 (2007)
31. F. Furche, J. Chem. Phys. **129**, 114105 (2008)
32. H.V. Nguyen, S. de Gironcoli, Phys. Rev. B **79**, 205114 (2009)
33. A. Grüneis, M. Marsman, J. Harl, L. Schimka, G. Kresse, J. Chem. Phys. **131**, 154115 (2009)
34. J. Toulouse, I.C. Gerber, G. Jansen, A. Savin, J.G. Angyan, Phys. Rev. Lett. **102**, 096404 (2009)
35. B.G. Janesko, T.M. Henderson, G.E. Scuseria, J. Phys. Chem. **130**, 081105 (2009)
36. M. Hellgreen, U. von Barth, J. Chem. Phys. **131**, 044110 (2009)
37. H. Eshuis, J. Yarkony, F. Furche, J. Chem. Phys. **132**, 234114 (2010)
38. J. Harl, G. Kresse, Phys. Rev. Lett. **103**, 056401 (2009)
39. J. Harl, L. Schimka, G. Kresse, Phys. Rev. B **81**, 115126 (2010)
40. H.V. Nguyen, G. Galli, J. Chem. Phys. **132**, 044109 (2010)
41. D. Lu, H.V. Nguyen, G. Galli, J. Chem. Phys. **133**, 154110 (2010)
42. W. Zhu, J. Toulouse, A. Savin, J.G. Angyan, J. Chem. Phys. **132**, 244108 (2010)
43. J. Toulouse, W. Zhu, J.G. Angyan, A. Savin, Phys. Rev. A **82**, 032502 (2010)
44. J. Paier, B.G. Janesko, T.M. Henderson, G.E. Scuseria, A. Grüneis, G. Kresse, J. Chem. Phys. **132**, 094103 (2010). Erratum: J. Chem. Phys. 133 (2010) 179902
45. M. Hellgren, U. von Barth, J. Chem. Phys. **132**, 044101 (2010)
46. A. Heßelmann, A. Görling, Mol. Phys. **108**, 359 (2010)
47. A. Heßelmann, A. Görling, Phys. Rev. Lett. **106**, 093001 (2011)
48. A. Heßelmann, A. Görling, Mol. Phys. **109**(21), 2473 (2011)
49. X. Ren, A. Tkatchenko, P. Rinke, M. Scheffler, Phys. Rev. Lett. **106**, 153003 (2011)
50. W. Klopper, A.M. Teale, S. Coriani, T.B. Pedersen, T. Helgaker, Chem. Phys. Lett. **510**, 147 (2011)
51. H. Eshuis, F. Furche, J. Phys. Chem. Lett. **2**, 983 (2011)
52. H. Eshuis, J.E. Bates, F. Furche, Theor. Chem. Acc. **131**, 1 (2012)
53. X. Ren, P. Rinke, C. Joas, M. Scheffler, J. Mater. Sci. **47**, 7447 (2012)
54. P. Verma, R.J. Bartlett, J. Chem. Phys. **136**, 044105 (2012)
55. P. Bleiziffer, A. Heßelmann, A. Görling, J. Chem. Phys. **136**(13), 134102 (2012)
56. D.C. Langreth, J.P. Perdew, Solid State Commun. **17**, 1425 (1975)
57. D.C. Langreth, J.P. Perdew, Phys. Rev. B **15**, 2884 (1977)
58. M. Betzinger, C. Friedrich, S. Blügel, A. Görling, Phys. Rev. B **83**, 045105 (2011)
59. M. Betzinger, C. Friedrich, A. Görling, S. Blügel, Phys. Rev. B **85**, 245124 (2012)
60. T. Grabo, T. Kreibich, S. Kurth, E.K.U. Gross, *Strong Coulomb Correlations in Electronic Structure: Beyond the Local Density Approximatio* (Gordon & Breach, Tokyo, 1998)
61. Q. Wu, W. Yang, J. Theor. Comput. Chem. **2**, 627 (2003)

62. A. Heßelmann, A.W. Götz, F. Della Sala, A. Görling, J. Chem. Phys. **127**, 054102 (2007)
63. R.A. Lippert, N.A. Modine, A.F. Wright, J. Phys. Condens. Matter **18**, 4295 (2006)
64. M. Greiner, P. Carrier, A. Görling, Phys. Rev. B **81**, 155119 (2010)
65. M. Levy, J.P. Perdew, V. Sahni, Phys. Rev. A **30**, 2745 (1984)
66. J.P. Perdew, R.G. Parr, M. Levy, J.L. Balduz, Phys. Rev. Lett. **49**, 1691 (1982)
67. J.P. Perdew, M. Levy, Phys. Rev. Lett. **51**, 1884 (1983)
68. L.J. Sham, M. Schlüter, Phys. Rev. Lett. **51**, 1888 (1983)
69. R.W. Godby, M. Schlüter, L.J. Sham, Phys. Rev. Lett. **56**, 2415 (1986)
70. K. Schönhammer, O. Gunnarson, J. Phys. C **20**, 3675 (1987)
71. W. Knorr, R.W. Godby, Phys. Rev. Lett. **68**, 639 (1992)
72. M. Grüning, A. Marini, A. Rubio, Phys. Rev. B **74**, 161103 (2006)
73. M. Grüning, A. Marini, A. Rubio, J. Chem. Phys. **124**, 154108 (2006)
74. F. Della Sala, A. Görling, J. Chem. Phys. **115**, 5718 (2001)
75. T. Sjostrom, F.E. Harris, S.B. Trickey, Phys. Rev. B **85**, 045125 (2012)
76. F. Gygi, A. Baldereschi, Phys. Rev. B **34**, 4405 (1986)
77. B. Wenzien, G. Cappellini, F. Bechstedt, Phys. Rev. B **51**, 14701 (1995)
78. P. Carrier, S. Rohra, A. Görling, Phys. Rev. B **75**, 205126 (2007)
79. J. Spencer, A. Alavi, Phys. Rev. B **77**, 193110 (2008)
80. N.D. Mermin, Ann. Phys. **21**, 99 (1963)
81. J. Sokoloff, Ann. Phys. **45**, 186 (1967)
82. J.P. Perdew, K. Burke, M. Ernzerhof, Phys. Rev. Lett. **77**, 3865 (1996)
83. M. Moukara, M. Städele, J.A. Makewski, P. Vogl, A. Görling, J. Phys. C **12**, 6783 (2000)
84. E. Engel, A. Höck, R.N. Schmid, R.M. Dreizler, N. Chetty, Phys. Rev. B **64**, 125111 (2001)

Quantum Monte Carlo Techniques and Applications for Warm Dense Matter

Ethan Brown, Miguel A. Morales, Carlo Pierleoni, and David Ceperley

Abstract The Quantum Monte Carlo (QMC) method is used to study physical problems which are analytically intractable due to many-body interactions and strong coupling strengths. This makes QMC a natural choice in the warm dense matter (WDM) regime where both the *Coulomb coupling parameter* $\Gamma \equiv e^2/(r_s k_B T)$ and the *electron degeneracy parameter* $\Theta \equiv T/T_F$ are close to unity. As a truly first-principles simulation method, it affords superior accuracy while still maintaining reasonable scaling, emphasizing its role as a benchmark tool.

Here we give an overview of QMC methods including diffusion MC, path integral MC, and coupled electron-ion MC. We then provide several examples of their use in the WDM regime, reviewing applications to the electron gas, hydrogen plasma, and first row elements. We conclude with a comparison of QMC to other existing methods, touching specifically on QMC's range of applicability.

E. Brown
Lawrence Livermore National Laboratory, 7000 East Ave., Livermore, CA 94550, USA

Department of Physics, University of Illinois Urbana-Champaign, 1110 W. Green St. Urbana, IL 61801-3080, USA
e-mail: brown122@illinois.edu

M.A. Morales
Lawrence Livermore National Laboratory, 7000 East Ave., Livermore, CA 94550, USA
e-mail: moralessilva2@llnl.gov

C. Pierleoni
Dipartimento di Scienze Fisiche e Chimiche, Universita dell'Aquila and CNISM, UdR dell'Aquila, a V. Vetoio 10, Loc. Coppito, I-67100 L'Aquila, Italy
e-mail: carlo.pierleoni@aquila.infn.it

D. Ceperley (✉)
Department of Physics, University of Illinois Urbana-Champaign, 1110 W. Green St., Urbana, IL 61801-3080, USA
e-mail: ceperley@illinois.edu; ceperley@ncsa.uiuc.edu

F. Graziani et al. (eds.), *Frontiers and Challenges in Warm Dense Matter*, Lecture Notes in Computational Science and Engineering 96,
DOI 10.1007/978-3-319-04912-0_5,

1 Introduction

With the development of high performance computing over the last few decades, simulation has become a ubiquitous many-body physics tool, see, for example, some of the other contributions in this volume. Most simulations are based on a classical description of atoms and molecules and use either a molecular dynamics (MD) or a Monte Carlo (MC) algorithm.

However, the microscopic description of warm dense matter (WDM) poses a particular difficulty, since classical mechanics is far from adequate. It is not convenient to describe the system as a perturbation from the ground state, that is as a sum over electronic excitations, since there are so many states. At a sufficiently high temperature, one needs a method that treats electrons and other quantum particles as particles and not as delocalized wavefunctions. Imaginary time path integrals provide a particular fortuitous formalism, since they supply a direct mapping of the quantum system into a classical system and reduce to the classical limit at high temperature. A big conceptual difficulty preventing the straightforward use of the path integral method is the issue of how to map fermion statistics into a probability. As we discuss below, this has an "in principle" solution: the restricted path integral method which, however, requires an ansatz to treat general fermion systems.

In this short review, we briefly describe three quantum Monte Carlo (QMC) methods, including diffusion MC, path integral MC, and coupled electron-ion MC, and their application to several WDM systems. We then close with a summary of strengths and weakness of the QMC methods.

1.1 Diffusion Monte Carlo

Diffusion Monte Carlo (DMC), along with the simpler Variational Monte Carlo (VMC), are two of the most popular stochastic methods employed in ground-state calculations of the Schrödinger equation for continuous systems. Even though they are zero temperature methods, they can be used in conjunction with other methods to treat regimes not dominated by electronic excitations [1, 2].

1.1.1 Formalism

As the name suggests, Variational Monte Carlo (VMC) is based on the variational principle of quantum mechanics. It states that the expectation value of the Hamiltonian, with respect to any trial wave function Ψ_T, will be a minimum for the exact ground state wave function:

$$E[\Psi_T] = \frac{\int \Psi_T^* \hat{H} \Psi_T d\mathbf{r}}{\int |\Psi_T|^2 d\mathbf{r}} = \int \pi(\mathbf{r}) E_L(\mathbf{r}) d\mathbf{r},$$

$$\pi(\mathbf{r}) = \frac{|\Psi_T(\mathbf{r})|^2}{\int |\Psi_T|^2 d\mathbf{r}}, \qquad E_L(\mathbf{r}) = \frac{\hat{H}\Psi_T(\mathbf{r})}{\Psi_T(\mathbf{r})}, \tag{1}$$

where $\mathbf{r}$ is a $3N_e$-dimensional vector. The functional, $E[\Psi_T]$, provides an upper bound to the ground state energy of the system. The basic problem is finding flexible trial wave functions that are good approximations to the ground state. This typically requires some intuition and knowledge of the system.[1]

The Diffusion Monte Carlo (DMC) method [3], on the other hand, is based on the connection between quantum mechanics and classical diffusion. The time-dependent Schrödinger equation in imaginary time, β, becomes:

$$\frac{\partial \Psi(\mathbf{r},\beta)}{\partial \beta} = \lambda \nabla^2 \Psi(\mathbf{r},\beta) - V(\mathbf{r})\Psi(\mathbf{r},\beta). \tag{2}$$

Expanding $\Psi(\mathbf{r},\beta)$ in the eigenstates of the Hamiltonian it is easy to see that:

$$\begin{aligned}\Psi(\mathbf{r},\beta) &= \sum_i a_i \phi_i(\mathbf{r}) e^{-\beta\epsilon_i}, \\ &\propto a_0\phi_0(\mathbf{r}) + \sum_i a_i \phi_i(\mathbf{r}) e^{-\beta(\epsilon_i - \epsilon_0)}.\end{aligned} \tag{3}$$

As β gets large, all exited states are exponentially suppressed. Imaginary time propagation projects the trial wave function to the ground state of the system.

The standard DMC algorithm is obtained by considering the mixed distribution $f(\mathbf{r},\beta) = \Psi(\mathbf{r},\beta)\Psi_T(\mathbf{r})$, where Ψ_T is any given trial wave function. If Eq. (2) is multiplied by Ψ_T and a few simple manipulations are performed, the partial differential equation for $f(\mathbf{r},\beta)$ is given by:

$$\begin{aligned}\frac{\partial f(\mathbf{r},\beta)}{\partial \beta} &= \hat{L} f(\mathbf{r},\beta), \\ &= \lambda\nabla^2 f(\mathbf{r},\beta) - \lambda\nabla\cdot[f(\mathbf{r},\beta)\mathbf{F}(\mathbf{r})] + [E_T - E_L(\mathbf{r})]\, f(\mathbf{r},\beta),\end{aligned} \tag{4}$$

where we introduced an overall shift in the energy E_T. $\mathbf{F}$ is the quantum force defined by $\mathbf{F} = \nabla \ln|\Psi|^2$ and $E_L(\mathbf{r})$ is the local energy. Equation (4) can be mapped to a drift-diffusion random walk with a branching (death/birth) process. The standard DMC algorithm combines a propagation of a population of 'walkers' using a Langevin equation with a branching step. After each Langevin step, each walker gains a weight in the population given by:

$$\omega_i(t) = e^{-\frac{\tau}{2}(E_L(\mathbf{r}_i(t)) + E_L(\mathbf{r}_i(t-\tau)) - 2E_T)}, \tag{5}$$

[1] See Sect. 1.4 for more details on wave functions and optimization methods.

so that at time step k, the total weight of walker i is given by:

$$\omega_i(t_k) = \prod_{j=1}^{k} \omega_i(t_j), \tag{6}$$

and averages are obtained from weighted sums:

$$< O >= \frac{\sum_{i=1}^{N} \sum_{p=1}^{N_{pop}} \omega_p(t_i) O(\mathbf{r}_p(t_i))}{\sum_{i=1}^{N} \sum_{p=1}^{N_{pop}} \omega_p(t_i)}. \tag{7}$$

This algorithm is typically called *pure* DMC.

Since the multiplicative nature of the weights leads to very large fluctuations and potential instabilities in the time propagation, a branching scheme is introduced where instead of storing weights during the simulation, walkers are stochastically replicated (or killed) using

$$N_{add}(t) = INT\left[e^{-\frac{\tau}{2}(E_L(\mathbf{r}_i(t))+E_L(\mathbf{r}_i(t-\tau))-2E_T)} + \gamma\right] \tag{8}$$

where γ is a uniformly distributed random number in [0, 1] and *INT* is the integer part of its argument. The role of E_T is now apparent. By appropriately choosing E_T, the population of walkers can be stabilized. In modern implementations, a combination of both weights and branching is used to improve efficiency [4]. Weights are maintained for a few steps, but their fluctuations are controlled by branching when they get outside of specified bounds.

1.1.2 Fermions

As mentioned previously, we must take special care to treat the antisymmetric nature of wave functions for fermions. Since the electronic Hamiltonian is invariant to particle exchange, the methods described above can be applied to fermions as long as we restrict them to the antisymmetric sector of the Hilbert space. The ground state wave function corresponds to the lowest antisymmetric eigenstate of the electronic Hamiltonian, which is not the global minimum of the entire Hilbert space. This means that any method that does not explicitly limit solutions to antisymmetric functions will converge to the incorrect result (e.g. to a bosonic state).

An antisymmetric wave function will contain both positive and negative regions in configuration space (assuming a real wavefunction). Since, a probability distribution has to be non-negative, we must include the signs in the estimators. This causes a severe problem because the normalization of the resulting average will be proportional to the difference between the positive and negative volumes in configuration space. Particle symmetry implies that those volumes must be equal and, since the sampling is bosonic, the normalization will go to zero as the simulation time increases.

This can be seen if we assume that we sample the states according to the absolute value of the original distribution. In this case, the estimator for the energy, for example, becomes:

$$< E >= \frac{\sum_i \omega(\mathbf{r}_i) E_L(\mathbf{r}_i)}{\sum_i \omega(\mathbf{r}_i)}, \tag{9}$$

where $\omega(\mathbf{r}_i)$ is $\{+1, -1\}$ depending on the sign of the wave function. As the number of samples increases, both numerator and denominator approach zero, and the signal/noise ratio decreases. It can be shown that the efficiency of this direct approach decreases exponentially with the number of particles [5]. This exponential scaling with the number of particles is known as the *Fermion Sign Problem.*

There is currently no exact solution to the sign problem for general systems. The standard approach to perform stable simulations is known as the fixed-node approximation [6].[2] In this approximation, the simulation is restricted to a nodal pocket of the trial wave function, and walkers are not allowed to cross nodes. The resulting energy represents an upper bound to the exact ground state energy, becoming exact when the nodes of the trial wave function coincide with those of the exact solution. The fixed node method combined with a projection technique, such as DMC, produces the best energy consistent with the chosen nodal structure. This makes results, in principle, sensitive to the nodal structure. In practice, results are found to be very accurate.

1.2 *Path Integral Monte Carlo*

Next we briefly describe path integral Monte Carlo (PIMC), which is similar to DMC but can treat systems at non-zero temperatures; a many-body density matrix replaces the trial wave function. For a more complete overview of the method and its application to fermion systems, see Refs. [5] and [7] respectively.

1.2.1 Path Integrals

To begin, we define the solution to the Bloch equation $d\rho/d\beta = -\hat{H}\rho$, the many particle density matrix, to be

$$\rho(R, R'; \beta) = \langle R \mid e^{-\beta \hat{H}} \mid R' \rangle \tag{10}$$

[2] Note that this assumes real wave functions. If otherwise, the fixed-phase approximation may be used [4].

where $R \equiv (r^{(1)}, \ldots, r^{(N)})$ with $r^{(i)}$ specifying the spacial coordinates of the ith of N particles, and $\beta \equiv 1/k_B T$, the inverse temperature. The quantum statistical partition function is defined as the trace of the density matrix,

$$Z(\beta) = Tr(\rho) = \int dR \langle R \mid e^{-\beta \hat{H}} \mid R \rangle = \int dR \rho(R, R; \beta). \tag{11}$$

The expectation value of any observable may be computed from this definition as

$$\langle \hat{\mathscr{O}} \rangle = Tr(\hat{\mathscr{O}} \rho)/Z = Tr(\hat{\mathscr{O}} \rho)/Tr(\rho). \tag{12}$$

Using the product property of the density matrix M times, such that $\beta = M\tau$, we discretize the partition function integral. The partition function becomes

$$Z(\beta) = \int \prod_{i=0}^{M-1} d^d R_i \rho(R_0, R_1; \tau) \rho(R_1, R_2; \tau) \ldots \rho(R_{M-1}, R_0; \tau). \tag{13}$$

Thus, we have effectively reduced the problem of finding a low temperature density matrix to one of finding the product of many high temperature density matrices. For each, we define an action

$$S(R_i, R_j; \tau) \equiv -ln[\rho(R_i, R_j; \tau)] \tag{14}$$

which may be broken into kinetic and potential parts. For systems with a long-range interaction, a variant of the Ewald summation technique is used [8]. Finally, in order to properly account for the particle statistics of the simulated system, we must sum over permutations $\mathscr{P}$, giving

$$Z(\beta) = \frac{1}{N!} \sum_{\mathscr{P}} (\pm 1)^{\mathscr{P}} \int_{R \to \mathscr{P} R} dR_t e^{-S[R_t]} \tag{15}$$

where the action S represents the path starting at R and ending at $\mathscr{P} R$.

1.2.2 Restricted Paths

Once again, with fermions negative terms enter in this sum, leading to a sign problem. As was done in the previous discussion of DMC, one way to circumvent this issue is to impose a nodal constraint [6]. We define the *nodal surface* $\Upsilon_{R_\star \beta}$ for a given point $R_\star$ and inverse temperature β to be

$$\Upsilon_{R_\star \beta} = \{R \mid \rho(R, R_\star; \beta) = 0\} \tag{16}$$

which is a $(dN - 1)$-dimensional manifold in dN-dimensional configuration space. Here, $R_\star$ is dubbed the *reference point*, as it is needed to define the nodal surfaces. Inside a nodal cell, by definition the sign of the density matrix is uniform. Using Dirichlet boundary conditions, we may solve the Bloch equation within each nodal cell. We define the *reach* $\Gamma_\beta(R)$ as the set of all continuous paths R_τ, for which $\rho(R_\tau, R_\star, \beta) \neq 0$ for all intermediate τ $(0 < \tau \leq \beta)$, i.e. a node-avoiding path.

$$\Gamma_\beta(R) = \{\gamma : R \to R' \mid \rho(R, R_\tau; \beta) \neq 0\}. \tag{17}$$

It is clear then that all paths contributing to the Bloch equation solution must belong to this reach. For all diagonal contributions, odd permutations must cross a node an odd number of times and thus are not allowed by this constraint.[3] In fact, they are exactly cancelled by all paths of node-crossing even permutations. This leaves us with the following expression for the density matrix,

$$\rho(R, R; \beta) = \frac{1}{N!} \sum_{\mathscr{P}, even} \int_{\gamma: R \to \mathscr{P}R}^{\gamma \in \Gamma_\beta(R)} dR_\tau e^{-S[R_\tau]}. \tag{18}$$

We have thus turned the sign-full expression for the density matrix into one which includes only terms of a single sign, allowing efficient computation. However, because ρ appears on both sides of Eq. (18), this requires a priori knowledge of the density matrix nodal structure, which is generally unknown. To escape this self-consistency issue, an ansatz density matrix that approximates the actual nodal structure, is introduced. This ansatz is a solution to the Bloch equation inside the trial nodal cells and obeys the correct initial conditions, providing an exact solution only when its nodes are the true nodes of the density matrix. This method is often called *restricted* PIMC (RPIMC).

For spinless fermions, it is often simplest to write down the anti-symmetrized density matrix as the Slater determinant of single-particle distinguishable density matrices, $\rho(R, R_\star; \beta) = \frac{1}{N!} \det \rho_{ij_\star}$ where

$$\rho_{ij_\star} = (4\pi\lambda\beta)^{-d/2} e^{-\frac{(r_i - r_{j\star})^2}{4\lambda\beta}}. \tag{19}$$

Generally, we expect this approximation to be best at high temperature and when correlation effects are weak. Furthermore, due to the constraint of translational invariance, free particle nodes are quite reasonable for homogeneous systems, specifically in the WDM regime. As a simple illustration, Fig. 1 shows the exact, the ground state, and the free particle nodal structures for a system of three particles in a two dimensional harmonic trap. One can see that in the classical limit, the exact nodes match those of free particles, while in the $T \to 0$ limit, the nodes are that of the ground state.

[3] Note, for non-diagonal elements the sum over odd permutations must be retained.

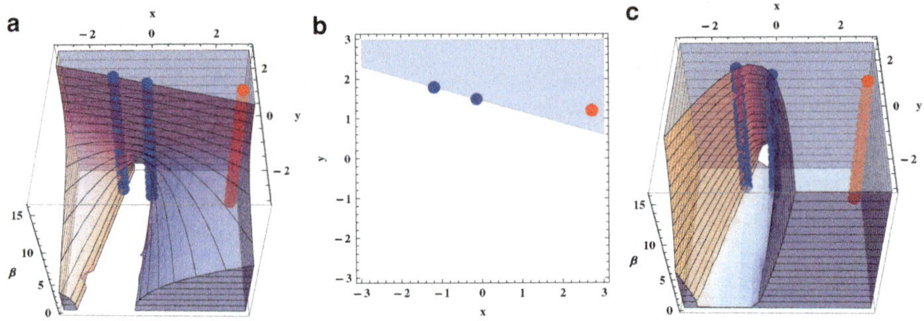

Fig. 1 An example nodal surface for three particles in a two dimensional harmonic trap. The nodal line cuts through two of the particles (*blue*), defining the reach of the third (*red*). Here all time slices occupy the same point in space, showing how the exact nodal structure (**a**) is the precise extrapolation from the ground state nodal structure (**b**) to those of free particles (**c**). The third axis in (**a**) and (**c**) represents β, the inverse temperature

1.2.3 Possible Sources of Error

We now discuss some of the sources of error. The largest is caused by the finite-size of the simulation cell. This error may be controlled either by performing multiple-sized simulations and extrapolating to the thermodynamic limit or by using an analytic correction. For homogeneous systems with long-range interactions, we can use the finite temperature extension of the analytic correction given in Ref. [9] as we discuss in Sect. 1.5.

The second largest error comes from the discretization of imaginary time by τ. While the naive Trotter breakup of $\exp[-\hat{S}] \approx \exp[-\hat{T}]\exp[-\hat{V}]$ has error $\sim\tau^2$, one can often do better. Specifically, we employ the *pair product approximation* for the action which has error $\sim\lambda\tau^3$ [5]. For fermions a time-step error arises from the nodal constraint when paths cross and recross a node within τ. Though these contributions may be difficult to separate, one can always extrapolate to the $\tau \to 0$ limit by adding more time slices. The third controllable error is due to the statistical fluctuations of the Monte Carlo algorithm itself. However, as in DMC, one may run the simulation longer to provide smaller errors on measured quantities.

The final source of error, the nodal error, is the most problematic since it is uncontrollable. As mentioned, there is no a priori way to know the exact nodal structure. PIMC's variational principle is through the free energy, as opposed to the internal energy in DMC. Thus one possible solution is parameterize the nodal ansatz, and then minimize the free energy by varying the parameters. However, the free energy is not usually easily obtainable; it may require a thermodynamic integration. Nevertheless, in the systems mentioned below, estimates suggest that the nodal error arising from the free-particle ansatz to be smaller than the statistical error.

1.3 Coupled Electron Ion Monte Carlo

In applying QMC methods to WDM, systems of nuclei and electrons must be considered. The large nucleon-electron mass ratio implies a wide separation of time scales. Because of this, first principles methods based on Density Functional Theory (DFT), find the adiabatic, or Born-Oppenheimer (BO), approximation to be very useful even for the lightest element, hydrogen. Conversely, the QMC methods described so far ignore this separation and treat nuclei and electrons on equal footing, causing difficulties. The imaginary time step of the path integral representation (both in DMC and PIMC) is imposed by the light electron mass. In DMC this means that nuclear "dynamics" (the speed of sampling configuration space) is much slower than electron "dynamics" requiring very long (and time consuming) trajectories. In PIMC the separation of time scales presents itself as a separation in the regions where thermal effects are relevant: in high pressure hydrogen for instance nuclear quantum effects becomes relevant below $\sim$1,000–5,000 K where electrons are, to a very good approximation, in their ground state. Performing PIMC in this region of temperatures requires very long electronic paths causing a slowing down of the exploration of configuration space and effectively limiting the ability of PIMC to perform accurate calculations at low temperatures.

To study this interesting temperature region, a QMC method based on the BO approximation, the Coupled Electron-Ion Monte Carlo (CEIMC) has been developed [1]. In CEIMC a Monte Carlo calculation for finite temperature nuclei (either classical or quantum represented by path integrals) is performed using the Metropolis method with the BO energy obtained by a separate QMC calculation for ground state electrons. CEIMC has been extensively reviewed in Refs. [1, 10]. Here, we only briefly report the main technical features of the method.

1.3.1 Penalty Method

In CEIMC the difference of BO energies of two nearby nuclear configurations in a MC attempted step, as obtained by an electronic QMC run, is affected by statistical noise which, if ignored, results in a biased nuclear sampling. To cope with this situation either the statistical noise needs to be reduced to a negligible value by long electronic calculations (very inefficient), or the Metropolis acceptance/rejection scheme needs modifications to cope with noisy energy differences. The latter strategy is implemented in the Penalty Method [11] which enforces detailed balance to hold on average over the noise distribution. The presence of statistical noise causes an extra rejection with respect to the noiseless situation. An extra "penalty" defined as the variance of the energy difference over the square of the physical temperature is added to energy differences. Therefore running at lower temperatures requires a reduced variance to keep an acceptable efficiency of the nuclear sampling. Small variances can be obtained if correlated sampling is used to compute the energy of the two competing nuclear configurations. In an attempted nuclear MC step, a single electronic run is performed with a trial wave function which is a linear combination of the wave functions of the two nuclear configurations considered. The BO energy of the two nuclear configurations is obtained by a reweighting procedure which provides energy differences with a much reduced variance with respect to performing two independent electronic runs if the "distance" between the two nuclear configurations is limited (i.e. the overlap between the trial wave functions of the two configurations is large) [12]. This strategy allows an efficient sampling of nuclear configuration space for high pressure hydrogen and helium down to temperature as low as $\sim$100–200 K.

1.3.2 PIMC

When nuclear quantum effects are included using a path integral representation (see Sect. 1.2), the relevant inverse temperature in the penalty method is the imaginary time discretization step τ, so that no loss of efficiency is experienced when lowering the temperature (i.e. taking longer paths). For quantum protons in high pressure hydrogen, CEIMC can be used to efficiently study systems at temperatures as low as $\sim$50 K. In the present implementation of nuclear quantum effects in CEIMC, we introduce an effective pair potential between nuclei and use the pair density matrix (see Sect. 1.2). The residual difference between the effective pair potential and the BO energy is considered at the primitive approximation level of the Trotter break-up of the proton propagator [1]. In high pressure hydrogen ($r_s = 1.40$) it is found that with this strategy, $\tau^{-1} \simeq 4{,}500$ K is enough to reach convergence of the thermodynamics properties, which allows to study systems at low temperature with a limited number of time slices ($\leq$100).

1.3.3 VMC vs RQMC

The main ingredient of CEIMC is the electronic QMC engine used to compute the BO energy. As mentioned a very important aspect for the efficiency of CEIMC is the noise level which is related to the variance of the local energy. In ground state QMC (see Sect. 1.1) the "zero variance principle" applies: if the trial wave function is an eigenfunction of the Hamiltonian, the local energy is no longer a function of the electronic coordinates and a single calculation provides the exact corresponding eigenvalues. Therefore by improving the trial wave function and approaching the exact ground state, the variance of the local energy decreases to zero. In connection with CEIMC, this is important not only for the accuracy of the BO energy but also for the efficiency of the nuclear sampling since the extra rejection due to the noise is reduced for a more accurate trial wave function. For high pressure hydrogen we have devised a quite accurate trial function of the Slater-Jastrow, single determinant, form. The Slater determinant is built with single electron orbitals, obtained by a self-consistent DFT solution, expressed in terms of quasi-particle coordinates with a backflow transformation. The Jastrow part has an electron-proton and electron-electron RPA Jastrow plus two-body and three-body numerical terms [13, 14].

To go beyond VMC accuracy in CEIMC we have implemented the so called Reptation QMC method (RQMC) [1, 15]. The theoretical basis of RQMC and DMC are the same, however their implementations are quite different. In DMC a generalized drift-diffusion-source equation is solved which represents the Schrödinger equation in imaginary time. An initial population of replicas of the system is propagated in imaginary time, and the projected wave function at a given time is represented by the population of replicas at that time. In RQMC a single replica is considered, and the entire evolution of the system is represented by an imaginary time path which, as in PIMC, is kept in the memory of the computer. The problem is therefore transformed into sampling the configuration space of such paths which can be accomplished using Metropolis Monte Carlo. Expectations of the ground state wave function are calculated in the middle of the path [1]. Since RQMC uses an explicit representation of the statistical weight of each path, the reweighting procedure needed for estimating energy differences in CEIMC can be more easily applied in RQMC than in DMC; this is the reason for our choice of RQMC in CEIMC.

1.4 Improved Wavefunctions and Density Matrices

The simplest and most common trial wave function used in QMC calculations is the Slater-Jastrow form, $\Psi_T(\mathbf{r}) = D^{\uparrow} D^{\downarrow} e^{-J}$, where $D^{\uparrow}$ is a Slater determinant of single particle states and J is a Jastrow factor. The orbitals in the determinant are typically obtained from a mean-field method, while the Jastrow term incorporates correlation directly and typically includes terms involving electron-ion, electron-electron, and electron-electron-ion terms. Over the last decade, many wave function forms have been explored. Some of these include: Pfaffians [16], generalized valence bond (GVB) [17], antisymmetrized geminal power (AGP) [18], backflow transformation [19], and multi-determinant expansions [20–23]. In particular, the multideterminant form has shown great promise in molecular calculations, leading to wave functions with controlled accuracy capable of reaching chemically accurate results, and competing with the best methods in quantum chemistry. This wave function has the form: $\Psi_T(\mathbf{r}) = e^{-J} \sum_k c_k D_k^{\uparrow} D_k^{\downarrow}$, the linear parameters, c_k, as well as all the variational parameters in the Jastrow are optimized using VMC.

Over the last decade, robust and efficient wave function optimization methods have been developed [24, 25]. Among the most notable is the linear optimization method of Umrigar and co-workers [26]. In this case, the Hamiltonian is diagonalized in the sub-space obtained by linearizing the wave function derivatives with respect to its variational parameters. For linear parameters, as in the multi-determinant expansion, the resulting sub-space is complete, and the solution to the eigenvalue equations is the lowest energy solution. Rescaling of the lowest energy eigenvector is allowed when non-linear parameters are included, either by a line minimization, controlling the normalization, or any other criterion, to speed the convergence to the minimum. Typically, fewer than ten iterations are required, each using an increasing number of Monte Carlo steps. The continuous development of robust wave function optimization methods with the implementation and use of flexible trial wave function forms has lead to a large improvement in the accuracy of QMC methods over the last several years.

Density matrices may be improved in the same fashion, though all proper breakups of the action will converge to the exact density matrix given small enough time step τ. Several approximations which include higher-order electronic correlations have been proposed, though generally the pair product approximation is chosen as the best compromise between accuracy and efficiency [5]. For fermions, however, there is still a need to devise better nodal density matrices. At temperatures above zero, QMC methods obey a variational principle in the free energy (as opposed to the internal energy at zero-temperature). To this end, there have been some recent efforts to improve the nodal ansatz used in PIMC by minimizing the free energy [27]. The inclusion of backflow effects has also been proposed, though not tested, as a possible way forward. The literature discusses several recent attempts to improve nodal surfaces [27, 28].

1.5 Finite Size Effects and Extrapolations

One of the main approximations in simulations is to reach the thermodynamic limit by doing simulations with a microscopic system. Simulations are limited today to using fewer than several thousand electrons; typically this is used to model a physical system which contains enormously more electrons. There has been significant progress on this during the past decade. For WDM these methods are particularly appropriate since particle correlations are, for a large part, local, and if the correct boundary conditions are applied, a supercell can mimic an infinite system. To minimize the finite size errors, periodic boundary conditions are used. The dependence of the energy on the size of the simulation cell is often the largest systematic error in QMC calculations. Finite size effects are larger in QMC than in mean field methods because electrons are explicitly represented.[4] Fortunately, they can be minimized by several methods. We will discuss a recent analysis of finite system effects, first of the kinetic energy, and second of the potential energy.

If calculations are done with periodic boundary conditions, the largest effect for systems with a Fermi surface at low temperture comes from the kinetic energy, arising from the discontinuous filling of the shells in k-space. In the original calculations of the HEG, calculations were done with a sequence of closed shells, and Fermi liquid theory was used to extrapolate to the thermodynamic limit [29]. Although this works well for homogeneous systems with small unit cells, it does not

[4] All exact methods that work with finite systems share this difficulty.

work as well for more complex systems, having larger cells, or in a liquid. General boundary conditions on the phase are:

$$\Psi(\mathbf{r}_1 + \mathbf{L}, \mathbf{r}_2, \ldots) = e^{i\theta}\Psi(\mathbf{r}_1, \mathbf{r}_2, \ldots) \tag{20}$$

where $\mathbf{L}$ is a lattice vector of the supercell. If the twist angle θ is averaged over, (twist averaged boundary conditions or TABC), most single-particle finite-size effects, arising from shell effects in filling the plane wave orbitals, are eliminated [30]. In fact, within the grand canonical ensemble, there are no single-particle kinetic finite size effects. One can understand this in terms of the momentum distribution since the kinetic energy per electron may be written as

$$T_N/N = \frac{(2\pi)^3}{2\Omega}\sum_{\mathbf{k}} k^2 \rho_N(k), \qquad T_\infty = \frac{1}{2}\int d\mathbf{k} k^2 \rho_\infty(k) \tag{21}$$

where ρ_N and ρ_∞ are the momentum distribution for an N electron and infinite systems respectively. Twist averaging gets rid of the largest approximation, namely the "sampling" of $\mathbf{k}$-space, all the while still using $\rho_N(\mathbf{k})$.

The potential energy per electron is the integral over the two particle correlation function, conveniently expressed in k-space in terms of the structure factor (S_k):

$$V_N = \sum_{\mathbf{k}} V_N(\mathbf{k}) S_N(k), \qquad V_\infty = \int d\mathbf{k} V_\infty(\mathbf{k}) S_\infty(k) \tag{22}$$

Here $V_N(k) = V_\infty(k) = 4\pi k^{-2}$ is the Fourier transform of the Coulomb interaction, and we have assumed that the Ewald image procedure is used to treat the long-range interaction in a periodic system. As with the expression for the kinetic energy, we assume that $S_N(\mathbf{k}) = S_\infty(\mathbf{k})$. However, twist averaging does not convert the sum in Eq. (22) to an integral since $S(k)$ is a two-particle quantity; the values of $\mathbf{k}$ that can be computed are all in the reciprocal lattice of the supercell no matter what the twist angle. The dominant finite size error comes from the contribution near $\mathbf{k} = 0$. For small values of $\mathbf{k}$ we have that $S(\mathbf{k}) = ck^2$. Integrating over the sphere $k < \pi/L$ (assuming a cubic supercell) one gets a correction $4\pi k_0^3 c$. This procedure was introduced in Ref. [31], and thorough analysis and comparison with other methods was performed in Ref. [32].

2 Applications

With the above methods, the QMC community has been able to study several systems in WDM. Here we provide recent calculations of the electron gas, hydrogen, helium, carbon, and water, and address their importance to WDM research.

2.1 Electron Gas

The one-component plasma (OCP), a fundamental many-body model, consists of a single species of charged particles immersed in a rigid neutralizing background. For electrons, the OCP is a model of simple metals and is often called the homogeneous electron gas (HEG) or *jellium*. It is customary to define the natural length scale $r_s a_B \equiv (3/4\pi n)^{1/3}$ and energy scale $Ry = e^2/2a_B$, where n is the system density. When r_s, the *Wigner-Seitz radius*, is small (high density), the kinetic energy term dominates and the system becomes qualitatively similar to a non-interacting gas. At low density ($r_s \to \infty$), the potential energy dominates and the system is predicted to form a Wigner crystal [33]. In 3D at intermediate densities, a partially polarized state is predicted to emerge [34, 35].

Over the past few decades QMC calculations of the ground state HEG examined each of these phases [3, 36]. In addition to determining phase boundaries, the results of these studies have proven invaluable in the rigorous parameterization of functionals in ground state density functional theory (DFT) [37]. In the opposite limit, the classical OCP, whose equation of state depends only on the *Coulomb coupling parameter* $\Gamma \equiv e^2/(r_s k_B T)$, has been studied extensively [38–40]. However, the accuracy of these results quickly deteriorates as the temperature is lowered and quantum correlations play a greater role [41]. This breakdown is most apparent in the WDM regime where both Γ and the *electron degeneracy parameter* $\Theta \equiv T/T_F$ are close to unity.

There have been several attempts to tackle this strongly interacting regime through both analytics (diagrammatic techniques, perturbative expansions, RPA, etc.) and numerics (DFT, hypernetted-chain calculations, etc.). However, without a reliable first principles benchmark, these efforts have gone unverified, except in the weakly-interacting limits. Reference [42], using the RPIMC method (see Sect. 1.2), provides such accurate, first-principles thermodynamic data for the $3D$ HEG throughout the WDM regime for both the fully spin-polarized $\xi = 1$ and unpolarized $\xi = 0$ systems. This reference makes connections to both previous semi-classical and ground-state studies, as well as matches on to *exact*, signful calulations where possible. In the proceeding, we briefly touch on those results, as well as provide a new comparison to existing finite temperature energy parameterizations.

For densities ranging from $r_s = 1.0$ to 40.0 and temperatures ranging from $\Theta \equiv T/T_F = 0.0625$ to 8.0, energies, pair correlation functions, and structure factors were computed.[5] At each density, we observe a smooth convergence to previous semi-classical studies [38] at high temperature. In Fig. 2 we plot the total excess energy $(E_{tot} - E_0/E_0)$ where E_0 is the kinetic energy of the free Fermi gas, for the polarized system at all temperatures with $r_s = 4.0$ and 40.0. At the highest temperatures, the results match well with the purely classical Monte

[5]All data can be found in a repository hosted at http://github.com/3dheg/3DHEG

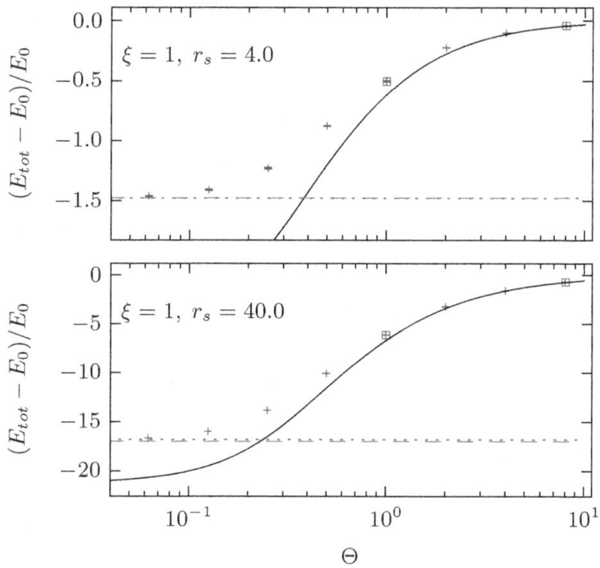

Fig. 2 (color online) Excess energies for $r_s = 4.0$ (*top*) and $r_s = 40.0$ (*bottom*) for the polarized state. For both densities, the high temperature results fall smoothly on top of previous Monte Carlo energies for the classical electron gas [38] (*solid line*). Differences from the classical Coulomb gas occur for $\Theta < 2.0$ for $r_s = 4.0$ and $\Theta < 4.0$ for $r_s = 40.0$. Simulations with the Fermion sign (*squares*) confirm the fixed-node results at $\Theta = 1.0$ and 8.0. The zero temperature limit (*dotted line*) smoothly extrapolates to the ground-state QMC results of Ceperley-Alder [3] (*dashed line*)

Carlo results of Ref. [38] (solid line). For a few select points, the much more time-consuming but more accurate, signful PIMC simulation (squares) was done. These points which are essentially exact, i.e. without possible nodal error, match well with fixed-node results. Finally, we know from Fermi liquid theory the low-temperature gas should have a linear form for the heat capacity, and therefore a quadratic form for the internal energy. Thus for each density we fit the low-temperature points to a quadratic function and extrapolate to $T = 0\,\mathrm{K}$. Figure 2 shows the extrapolated results (dotted line) match well with the zero temperature QMC results of Ceperley-Alder [3] (dashed line).

In Ref. [42], was found similar limiting agreement in both the calculated structure factors and pair correlation functions. Through this comparison against existing numerical and analytical data, we conclude the free-particle nodal approximation performs well for the densities studied. Further investigation is needed at even smaller values of r_s and lower temperatures in order to determine precisely where this approximation begins to fail. Such studies will necessarily require algorithmic improvements, however, because of difficulty in sampling paths at high density and low temperature [43].

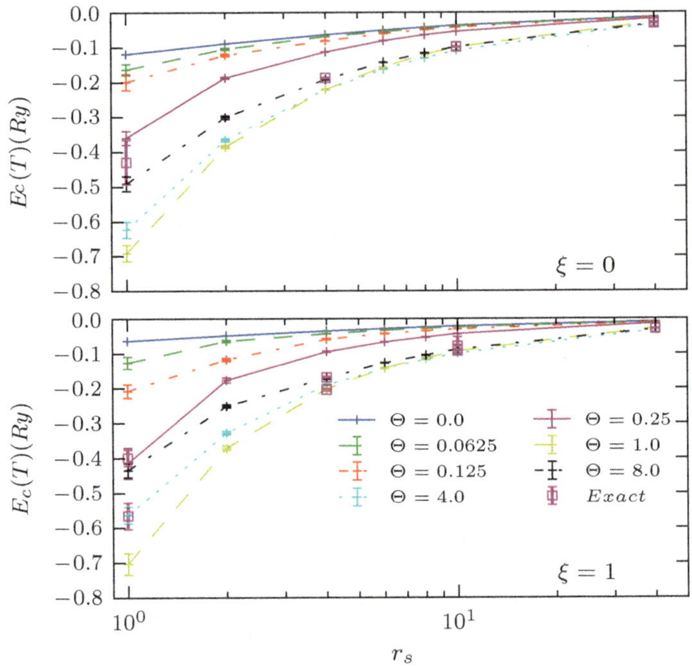

Fig. 3 (color online) Correlation energy $E_c(T)$ of the $3D$ HEG at several temperatures and densities for the unpolarized (*top*) and fully spin-polarized (*bottom*) states. *Exact* (signful) calculations (*squares*) confirm the fixed-node results where possible. For comparison, we plot the $\Theta = 0.0$ correlation energy used in LDA DFT calculations

Finally, we have evaluated the exchange-correlation energy E_{xc}, an essential quantity in any DFT formulation, defined as

$$E_{xc}(T) \equiv E_{tot}(T) - E_0(T) \tag{23}$$

As is customary, we further break up E_{xc} into exchange and correlation parts,

$$E_{xc}(T) = E_x(T) + E_c(T) \tag{24}$$

where $E_x(T)$ is the Hartree-Fock exchange energy for a free Fermi gas at temperature T.

By calculating $E_{tot}(T)$ through RPIMC simulations $E_c(T)$ was determined at all studied densities and spin-polarizations. As one can see in Fig. 3, correlation effects increase both with density (smaller r_s) and temperature up to the Fermi temperature T_F. Above this temperature, the electron gas begins to be less correlated. This represents the point at which electron screening is a dominant effect, the interaction becomes effectively short-ranged, and the Debye-approximation becomes relatively

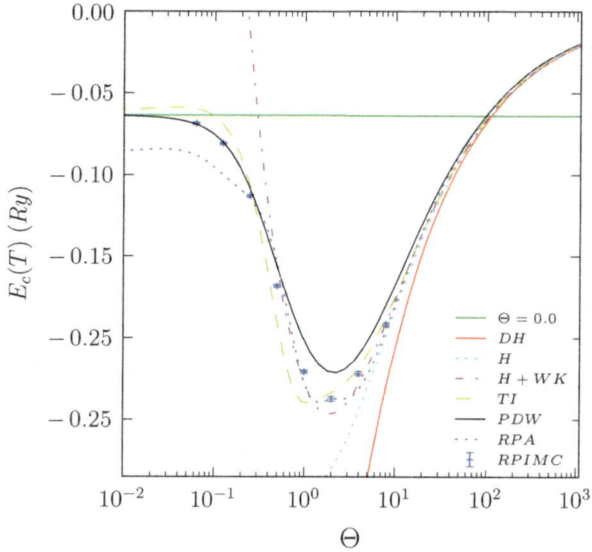

Fig. 4 (color online) Correlation energy $E_c(T)$ of the $3D$ HEG at $r_s = 4.0$ for the unpolarized $\xi = 0$ state from the above calculations (RPIMC) and several previous parameterizations. The latter include Debye-Huckel (DH), Hansen (H), Hansen+Wigner-Kirkwood (H+WK), random phase approximation (RPA), Tanaka and Ichimaru (TI), and Perrot and Dharma-wardana (PDW). We also include the ground state $\Theta = 0.0$ result for comparison

accurate [44]. As the density increases, the relative temperature at which this occurs decreases. At $r_s = 1.0$ the maximal effect of interactions occurs very near T_F, $\Theta = 1$. Most notably, we see a departure from the $T = 0$ K correlation energy used ubiquitously in both ground state and finite temperature LDA DFT calculations. This discrepancy is significant throughout the WDM regime, calling into question the use of ground state correlation functionals at such temperatures and densities.

We conclude with a brief comparison of these calculations to previous estimations of the $3D$ HEG correlation energy. Such parameterizations generally fall into two categories: those which extend down from the classical regime and those which assume some interpolation between the $T = 0$ and high-T regimes. From the former group, in Fig. 4, we plot E_c coming from Debye-Huckel (DH) theory which assumes short-ranged interactions and the classical Monte Carlo simulations of Hansen et al. of the full Coulomb system both with Wigner-Kirkwood corrections (H+WK) and without (H) [40, 44]. Clearly these methods do not perform well in the quantum regime below the Fermi temperature since they lack quantum exchange.

The random phase approximation (RPA), from the latter group, is a reasonable approximation in the low-density, high-temperature limit (where it reduces to DH) and the low-temperature, high-density limit, since these are both weakly interacting regimes. Its failure, however, is most apparent in its estimation of the equilibrium, radial distribution function $g(r)$ which becomes negative for stronger coupling. Extensions of the RPA into intermediate densities and temperatures have largely focused on constructing local-field corrections (LFC) through interpolation since diagrammatic resummation techniques often become intractable in strongly-coupled regimes. Singwi et al. [45] introduced one such strategy relying on two assumptions. First, they use the static polarization-potential approximation allowing one to write the LFC $G(k, z) \equiv G(k, z = 0) \equiv G(k)$. Next they assume the two-particle distribution function is a function of the $n(r)$ and $g(r)$ which allows a self-consistent solution for $G(k)$. Tanaka and Ichimaru [46] (TI) extended this method to finite temperatures and provided the shown parameterization of the $3D$ HEG correlation energy. A similar method (not shown) by Dandrea et al. uses the Vashista-Singwi LFC [47] to interpolate between the high- and low-temperature limits. Both methods appear to perform marginally better than the RPA at all temperatures, though both still fail to produce a positive-definite $g(r)$ at values of $r_s > 2$.

A third, more recent approach introduced by Perrot and Dharma-wardana (PDW) [48] relies on a classical mapping where the distribution functions of a classical system at temperature T_{cf}, solved for through the hypernetted-chain equation, reproduce those for the quantum system at temperature T. In a previous work, PDW showed such a temperature T_q existed for the classical system to reproduce the correlation energy of the quantum sytem at $T = 0$ [49]. To extend this work to finite temperature quantum systems, they use the simple interpolation formula $T_{cf} = \sqrt{T^2 + T_q^2}$. This interpolation is clearly valid in the low-T limit where Fermi liquid theory gives the quadratic dependence of the energy on T. Further in the high-T regime, T dominates over T_q as the system becomes increasingly classical. The PDW line in Fig. 4 clearly matches well with the RPIMC results in these two limits. It is not surprising, however, that in the intermediate temperature regime, where correlation effects are greatest, the quadratic interpolation fails. A contemporary, but similar approach by Dutta and Dufty [50] uses the same classical mapping as PDW which relies on matching the $T = 0$ pair correlation function instead of the correlation energy. While we expect this to give more accurate results near $T = 0$, we would still expect a breakdown of the assumed Fermi liquid behavior near the Fermi temperature.

In a recent publication, we have fitted an analytic form to the exchange-correlation energy data [51]. In doing so, we constructed a Padé approximant which collapses to Debye-Hückel theory in the high-temperature, low-density limit. Likewise, the zero-temperature limit matches the numerical results of ground-state quantum Monte Carlo, as well as RPA results in the high-density limit.

2.2 Hydrogen

Hydrogen is the simplest element of the periodic table and also the most abundant element in the Universe. Together with its isotopes, deuterium and tritium, it occupies a special place in WDM since it is particularly relevant in astrophysics, planetary physics and nuclear energy applications.

Because of its simple electronic structure, it has been instrumental in the development of quantum mechanics and remains important for developing ideas and theoretical methods. QMC is not an exception, and a DMC investigation of the ground state of hydrogen across the pressure-induced molecular dissociation regime appeared as early as 1987 [29], followed by more recent and accurate studies [52, 53]. A few years later RPIMC was developed to investigate the WDM regime of hydrogen [54–56] and to study the primary and secondary Hugoniots [57, 58]. RPIMC predictions for the principal Hugoniot of deuterium were first in disagreement with pulsed laser-produced shock compression experiments [59–61], but were later confirmed by magnetically generated shock compression experiments at the Z-pinch machine [62–67] and by converging explosive-driven shock waves techniques [68, 69]. A general agreement between RPIMC and First Principle Molecular Dynamics (FPMD) predictions is also observed [10] except at the lowest temperatures that could be reached by RPIMC (~10,000 K). There the Hugoniot from FPMD is slightly softer than from RPIMC (see Fig. 13 of Ref. [10]).[6] Lower temperatures cannot be easily reached by RPIMC without reducing the level of accuracy. However, most of the interesting phenomena like molecular dissociation under pressure, metallization, solid-fluid transition, a possible liquid-liquid phase transition and its interplay with melting, the various crystalline phases and the transition to the atomic phases [10], occur at lower temperature out of the reach of RPIMC. Investigating this regime by QMC methods has been the main motivation in developing CEIMC. The other motivation is to benchmark the much more developed (and less demanding) alternative theoretical method, namely FPMD based on DFT. Indeed the numerical implementation of DFT is based on approximations (the exchange-correlation functional) the accuracy of which can only be established against experiments or, better, against more accurate theories. As mentioned earlier, QMC is based on the variational principle and therefore has an internal measure of accuracy.

CEIMC has been applied to investigate the WDM regime and benchmark FPMD [14, 70, 71] and to investigate the Liquid-Liquid phase transition (LLPT) region in hydrogen [72, 73]. The emerging picture is that a weak first-order phase transition occurs in hydrogen between a molecular-insulating fluid and a metallic-mostly monoatomic fluid. At higher temperature, molecular dissociation and metallization occur continuously. However the precise location of the transition line and the critical point are still matter of debate since several levels of the theory provide different locations. Within FPMD-DFT the location of the transition line depends strongly on the exchange-correlation functional employed and on whether classical or quantum protons are considered [74, 75]. Transition lines from the PBE and vdW-DF2 approximations differ by roughly 200–250 GPa, the PBE one being located at lower pressure. The PBE melting line with quantum protons is not in agreement with experiments, which highlights the failure of the PBE approximation when employed together with the quantum description of the nuclei. On the other hand, optical properties for the vdW-DF2 approximation are in agreement with experiments supporting the use of this functional for hydrogen in the WDM regime. The LLPT line from CEIMC lies in between the lines from PBE and vdW-DF2 functionals

[6] Note that both theories are compatible with experiments because of the large uncertainty of the latter.

[72, 73]. However, those calculations were limited to classical protons and nuclear quantum effects are still to be considered.

Note that at higher temperatures and pressures, when molecules are entirely dissociated, the system is in the metallic state and protons can be considered classical particles, PBE and CEIMC have been found to be in very good agreement [71].

Our current understanding of the metallic state and metallization phenomena in high pressure hydrogen is mostly based on single electron theory [76]. There have been some effort to compute the DC conductivity of hydrogen in the WDM regime using DFT [77, 78], however to the best of our knowledge there has only been a single attempt by a QMC based methods [79]. Protonic configurations at finite temperature are generated by CEIMC while the Correlation Function QMC [80, 81] is employed to compute the low energy many-body excitation spectrum of the liquid. Combining the excitation energies with the Green-Kubo formula, the electrical conductivity of hydrogen is obtained based entirely on QMC; these calculations do not suffer from self-interaction errors but suffer from other limitations, notably the numerical difficulty in obtaining accurate properties of excited states because of the QMC sign problem, and large finite cell size effects. However, good agreement with the limited data from shock experiments measurements was obtained [82].

Much more effort is needed in developing QMC method for WDM, in particular to explore the sensitivity of CEIMC to orbitals from DFT with different functionals and to further develop the method for transport and dynamical properties.

2.3 Helium, Carbon, and Water

While the application of quantum Monte Carlo methods in the WDM regime has been focused mainly on hydrogen, there has been recent interest in the extension of these methods to other materials in the periodic table. The first application of PIMC to elements beyond hydrogen involved the study of the phase diagram and equation of state of helium [83] and helium-hydrogen mixtures [84]. Similar to the case of pure hydrogen, the properties of hydrogen-helium mixtures are critical to the development of interior models for giant planets, prompting previous DFT efforts [85]. Combining DFT and PIMC calculations, Militzer [86] was able to generate an equation of state for helium purely based on first-principles calculations, covering many orders of magnitude in density and temperature. This work represents a big step forward in the understanding of this material at extreme conditions and is an important piece in the development of next-generation planetary models.

The PIMC method was recently extended to other first row elements in the periodic table, including carbon and oxygen [87]. These calculations employed all-electron simulations of carbon and water at extreme conditions and covered temperature ranges from $10^4 - 10^9$ K. These represent the first PIMC calculations including core electrons and realistic materials. While some precautions must be taken when including core electrons, the work of Driver et al. shows a marked improvement in the description of these materials at extreme temperatures and leads the way for the extension of the method to other interesting materials. At low temperatures, these calculations compare favorably to previous finite-temperature DFT work for carbon [88, 89] and water [90, 91]. This work also shows the need to generalize and improve the nodal constraint used in path integral simulations at low temperatures, particularly when bound states appear. With the development of optimization techniques for density matrices and the use of more complicated ansatz for the nodal constraint, the method can become the standard simulation technique at finite temperatures, applicable across the entire periodic table.

3 Discussion

We conclude this review with a quick comparison of QMC to other existing methods followed by a brief summary of some of the positive and negative aspects of QMC methods for the simulation of WDM.

3.1 Comparison of Methods

As discussed throughout this chapter, several simulation techniques exist that can be used in the study of materials in the WDM regime. Since most methods are based on very different approximations, which work better in different regimes, it is important to understand the regimes of applicability of methods as well as the expected accuracy as a function of both temperature and density. For example, methods based on density functional theory (DFT) are typically accurate at low temperatures, but are limited to temperatures below $10^4 - 10^5$ K due to computational difficulties above this regime. This difficulty may be overcome using an orbital-free DFT, however this is often done at the expense of accuracy [92, 93]. Path integral methods, on the other hand, are very accurate at high temperatures, but the accuracy is limited as the temperature is lowered. Coupled Electron-Ion Monte Carlo (CEIMC), while being accurate also at low temperatures, becomes inefficient in the limit of very low temperatures. A combination of all three methods can then be used to cover very wide ranges in density and temperature, and at the same time producing accurate properties in an efficient way. As a consequence, considerable work has been performed in order to compare the accuracy and limitations of these methods in the regimes where they overlap.

In the case of helium, a recent study compared the EOS between DFT-based BOMD and PIMC methods [86]. Excellent agreement between the two methods is obtained at temperatures above 80,000 K. This work is particularly interesting because it establishes the importance of the electronic temperature in DFT-MD calculations. In the case of hydrogen at Mbar pressures, several comparisons between DFT-MD, CEIMC, and PIMC have been made over the years [2, 70, 74]. In

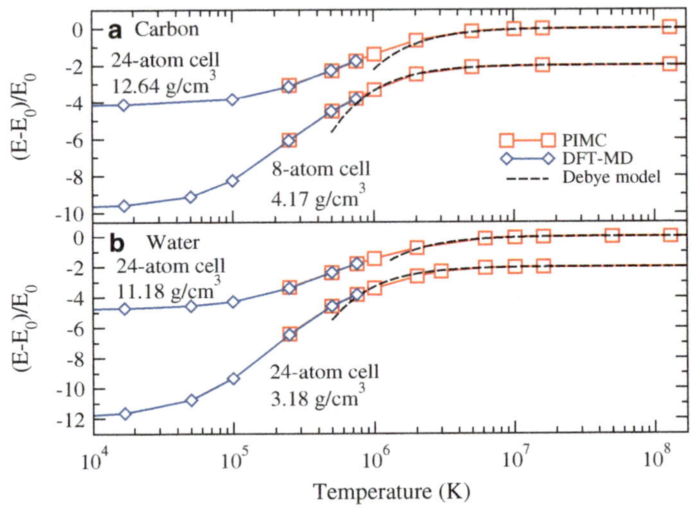

Fig. 5 A direct comparison of the excess energy in (**a**) carbon and (**b**) water, calculated through DFT-MD, PIMC, and Debye-Huckel theory. From this plot, one can conclude that PIMC may act as a bridge between the two quicker computational methods [87]

particular, CEIMC calculations have been used to benchmark DFT-MD simulations at high pressure in the development of equation of state models for hydrogen from first-principles [23, 74].

The recent work of Driver et al. [87] on all-electron simulations of water and carbon using path integral methods also offers a comparison to DFT-MD over an extended temperature range (see Fig. 5). Similar to the case of hydrogen and helium, a reasonable agreement between the two methods is observed at temperatures above 200,000 K. This result serves to emphasize PIMC's role as an intermediary between DFT and classical methods such as hyper-netted chain (HNC) or Debye-Huckel theory.

3.2 Positives

In the spirit of this chapter being used as reference, we conclude by listing the specific areas where QMC performs well and where QMC still needs improvement. We begin with the things QMC is known to do well:

- Strong correlation is not a problem for QMC, in fact strong correlation is when QMC works best, e.g. in a Wigner crystal. Also there is no problem with the self-interaction, e.g. the formation of hydrogen molecules is exactly captured. For the singlet state of an isolated molecule, there is also no fermion sign problem.

- Because of its high accuracy, QMC usually serves as a benchmark for other methods. As an example, QMC provides the exchange-correlation energy of the electron gas at zero temperature. And, as discussed in Sect. 2.1, QMC can now provide high temperature correlation energy for the electron gas. Other properties besides the energy are possible to compute, e.g. correlation functions and static response functions.
- QMC is the most general method of simulating quantum systems. For example, one can add nuclear zero point motion of the ions, or include other quantum particles such as positrons, study bose condensation, all within the same framework.
- Adding additional randomness typically does not cause a large increase in computer resources. Adding randomness simply increases the dimensionality of the integrals by a small amount, while the computational efficiency depends on the variance and not on the number of integration variables. For other methods, where the quantum mechanics is computed deterministically, this is not true. This means for problems where randomness is part of the problem, QMC methods are not as expensive. Some examples of this averaging are twist average boundary conditions to reduce finite size effects (see Sect. 1.5), a liquid as treated with CEIMC (see Sect. 2.2), and proton zero point motion (see Sect. 2.2).
- QMC has reasonable scaling versus number of particles. The expensive part of scaling is the fermion determinant that takes N^3 operations if the matrix is full. However at high temperature the matrix elements have a range of the thermal deBroglie wavelength. For example, at the Fermi temperature, each row and column have only on the order of 10 elements with a value greater than 0.01. In this case $O(N)$ or $O(N^2)$ scaling should be possible.
- Perhaps most importantly, imaginary time path integrals give new understanding of quantum statistical mechanics. A notable example is the winding number relation to superfluidity [5]. There is a lot of work that needs to be be done in the direction to find new relations between properties of restricted paths and quantum observables.

3.3 Negatives

Finally, we list the issues that are currently holding up the applications of QMC to problems in WDM:

- There are a number of convergence issues: the algorithms are mathematically well-posed, but become inefficient as the needed limit is taken. For example, there are finite size effects (taking the simulation box size to infinity) (see Sect. 1.5). We do not view this problem as critical for simple systems, since the correlation length in WDM is not that large, but it becomes more difficult for mixtures. Another example of a convergence issue is the *Trotter error* or time step error (see Sect. 1.2). This also is not a practical limitation, but more efficient actions would increase the overall efficiency. Most important are ergodic

problems of the Markov chain for restricted paths at low temperatures, aptly dubbed *reference point freezing* [57]. We do not know a solution to this problem, but also, we do not know if it is a fundamental problem.
- Imaginary time path intergrals can calculate imaginary time Green's functions. There are several problems in using this to find out dynamical response properties. First of all, this is equilibrium response only, true many-body dynamics is yet to be developed. Some linear response is possible, but it is not known precisely what are its limitations. Even within equilibrium linear response theory, there are two serious problems. First, the restricted path method for solving the sign problem changes the dynamics, even with the correct nodes. Second, there is the issue of imaginary to real time conversion, though for disordered systems in WDM, this should not be the biggest problem.
- For ground state problems, methods to remove core electrons using non-local pseudopotentials have been developed. However with path integrals, only all electron calculations have been done, see Sect. 2.3. Core electrons have problems for non-local pseudopotentials with restricted paths that have not been resolved. It is not known yet whether this is a serious problem or if pseudopotentials developed at $T = 0$ would work in the WDM regime.
- The final, but most fundamental problem is the *fermion sign problem*. This is a well-studied problem with as yet no known general solution (in fact it has been shown that there exist quantum glass states that are NP-complete [94]). Above the Fermi temperature the nodes should be dominated by the free particle nodes, and perturbation theory should take us to lower temperatures. Also, as mentioned, a variational principle should be able to guide us to lower temperatures as it does at zero temperature. Little work has been done along these lines in practice.

The problems listed above are opportunities to find solutions at the intersection of applied mathematics and many-body theory. None of these issues, beyond the sign problem in a specific spin system, have been shown to be fundamental and hence insurmountable. Because of this, we foresee QMC continuing to be a central tool as WDM research moves forward.

Acknowledgements DC is supported by DOE DE-NA0001789. EB is supported by DOE DE-AC52-07NA27344, LDRD 10-ERD-058 and the LLNL Lawrence Scholar program. CP is supported by the Italian Institute of Technology (IIT) under the SEED project grant number 259 SIMBEDD Advanced Computational Methods for Biophysics, Drug Design and Energy Research. This work was performed in part under the auspices of the US DOE by LLNL under Contract DE-AC52-07NA27344.

References

1. C. Pierleoni, D.M. Ceperley, Lect. Notes Phys. **714**, 641 (2006)
2. M.A. Morales, C. Pierleoni, D.M. Ceperley, Phys. Rev. E **81**, 021202 (2010)
3. D.M. Ceperley, B.J. Alder, Phys. Rev. Lett. **45**, 566 (1980)
4. J. Kolorenč, L. Mitas, Rep. Prog. Phys. **74**(2), 026502 (2011)

5. D.M. Ceperley, Rev. Mod. Phys. **67**, 279 (1995)
6. D.M. Ceperley, J. Stat. Phys. **63**, 1237 (1991). doi:10.1007/BF01030009
7. D.M. Ceperley, Path integral Monte Carlo methods for fermions, in *Monte Carlo and Molecular Dynamics of Condensed Matter Systems*, ed. by K. Binder, G. Ciccotti (Editrice Compositori, Bologna/Italy, 1996)
8. V. Natoli, D.M. Ceperley, J. Comput. Phys. **117**(1), 171 (1995)
9. S. Chiesa, D.M. Ceperley, R.M. Martin, M. Holzmann, Phys. Rev. Lett. **97**, 076404 (2006)
10. J. McMahon, M. Morales, C. Pierleoni, D.M. Ceperley, Rev. Mod. Phys. **84**, 1607 (2012)
11. D.M. Ceperley, M. Dewing, J. Chem. Phys. **110**, 9812 (1999)
12. D. Ceperley, M. Dewing, C. Pierleoni, Lect. Notes Phys. **605**, 473 (2002)
13. M. Holzmann, D.M. Ceperley, C. Pierleoni, K. Esler, Phys. Rev. E **68**(046707), 1 (2003)
14. C. Pierleoni, K. Delaney, M. Morales, D. Ceperley, M. Holzmann, Comput. Phys. Commun. **179**(1–3), 89 (2008)
15. S. Baroni, S. Moroni, Phys. Rev. Lett. **82**(24), 4745 (1999)
16. M. Bajdich, L. Mitas, G. Drobný, L.K. Wagner, K.E. Schmidt, Phys. Rev. Lett. **96**, 130201 (2006)
17. A.G. Anderson, W.A. Goddard, J. Chem. Phys. **132**(16), 164110 (2010)
18. M. Casula, C. Attaccalite, S. Sorella, J. Chem. Phys. **121**(15), 7110 (2004)
19. Y. Kwon, D.M. Ceperley, R.M. Martin, Phys. Rev. B **58**, 6800 (1998)
20. M. Bajdich, M.L. Tiago, R.Q. Hood, P.R.C. Kent, F.A. Reboredo, Phys. Rev. Lett. **104**, 193001 (2010)
21. P. Seth, P.L. Rios, R.J. Needs, J. Chem. Phys. **134**(8), 084105 (2011)
22. B.K. Clark, M.A. Morales, J. McMinis, J. Kim, G.E. Scuseria, J. Chem. Phys. **135**(24), 244105 (2011)
23. M.A. Morales, J. McMinis, B.K. Clark, J. Kim, G.E. Scuseria, J. Chem. Theory Comput. **8**(7), 2181 (2012)
24. C.J. Umrigar, J. Toulouse, C. Filippi, S. Sorella, R.G. Hennig, Phys. Rev. Lett. **98**, 110201 (2007)
25. J. Toulouse, C.J. Umrigar, J. Chem. Phys. **128**(17), 174101 (2008)
26. J. Toulouse, C.J. Umrigar, J. Chem. Phys. **126**(8), 084102 (2007)
27. S.A. Khairallah, J. Shumway, E.W. Draeger, ArXiv e-prints (2011)
28. B. Militzer, E.L. Pollock, Phys. Rev. E **61**, 3470 (2000)
29. D.M. Ceperley, B.J. Alder, Phys. Rev. B **36**, 2092 (1987)
30. C. Lin, F.H. Zong, D.M. Ceperley, Phys. Rev. E **64**, 016702 (2001)
31. S. Chiesa, D.M. Ceperley, R.M. Martin, M. Holzmann, Phys. Rev. Lett. **97**, 076404 (2006)
32. N.D. Drummond, R.J. Needs, A. Sorouri, W.M.C. Foulkes, Phys. Rev. B (Condens. Matter Mater. Phys.) **78**(12), 125106 (2008)
33. E. Wigner, Phys. Rev. **46**, 1002 (1934)
34. F. Bloch, Z. Phys. **57**, 545 (1929)
35. F.H. Zong, C. Lin, D.M. Ceperley, Phys. Rev. E **66**, 036703 (2002)
36. D. Ceperley, Phys. Rev. B **18**, 3126 (1978)
37. S. Kümmel, L. Kronik, Rev. Mod. Phys. **80**, 3 (2008)
38. E.L. Pollock, J.P. Hansen, Phys. Rev. A **8**, 3110 (1973)
39. B. Jancovici, Phys. A Stat. Mech. Appl. **91**, 152 (1978)
40. J.P. Hansen, P. Vieillefosse, Phys. Lett. A **53**, 187 (1975)
41. M.D. Jones, D.M. Ceperley, Phys. Rev. Lett. **76**, 4572 (1996)
42. E.W. Brown, B.K. Clark, J.L. DuBois, D.M. Ceperley, Phys. Rev. Lett. **110**, 146405 (2013)
43. D.M. Ceperley, Phys. Rev. Lett. **69**, 331 (1992)
44. J.P. Hansen, Phys. Rev. A **8**, 3096 (1973)
45. K.S. Singwi, M.P. Tosi, R.H. Land, A. Sjölander, Phys. Rev. **176**, 589 (1968)
46. S. Tanaka, S. Ichimaru, J. Phys. Soc. Jpn. **55**, 2278 (1986)
47. R.G. Dandrea, N.W. Ashcroft, A.E. Carlsson, Phys. Rev. B **34**, 2097 (1986)
48. F. Perrot, M.W.C. Dharma-wardana, Phys. Rev. B **62**, 16536 (2000)
49. M.W.C. Dharma-wardana, F. Perrot, Phys. Rev. Lett. **84**, 959 (2000)

50. S. Dutta, J. Dufty, Phys. Rev. E **87**, 032102 (2013)
51. E.W. Brown, J.L. DuBois, M. Holzmann, D.M. Ceperley, Phys. Rev. B **88**, 081102 (2013)
52. V. Natoli, R.M. Martin, D.M. Ceperley, Phys. Rev. Lett. **70**, 1952 (1993)
53. V. Natoli, R.M. Martin, D.M. Ceperley, Phys. Rev. Lett. **74**, 1601 (1995)
54. C. Pierleoni, D. Ceperley, B. Bernu, W. Magro, Phys. Rev. Lett. **73**(16), 2145 (1994)
55. W. Magro, D. Ceperley, C. Pierleoni, B. Bernu, Phys. Rev. Lett. **76**(8), 1240 (1996)
56. C. Pierleoni, W.R. Magro, D. Ceperley, B. Bernu, in *Physics of Strongly Coupled Plasmas*, Binz/Rügen, ed. by W.D. Kraeft, M. Schlanges (World Scientific, 1996), pp. 11–26
57. B. Militzer, D. Ceperley, Phys. Rev. Lett. **85**(9), 1890 (2000)
58. B. Militzer, D. Ceperley, J. Kress, J. Johnson, L. Collins, S. Mazevet, Phys. Rev. Lett. **87**(27), 275502 (2001)
59. L. DaSilva, P. Celliers, G. Collins, K. Budil, N. Holmes, T. Barbie Jr., B. Hammel, J. Kilkenny, R. Wallace, M. Ross, R. Cauble, Phys. Rev. Lett. **78**(3), 483 (1997)
60. G. Collins, L.D. Silva, P. Celliers, D. Gold, M. Foord, R. Wallace, A. Ng, S. Weber, K. Budil, R. Cauble, Science **281**(5380), 1178 (1998)
61. P. Celliers, G. Collins, L.D. Silva, D. Gold, R. Cauble, R. Wallace, M. Foord, B. Hammel, Phys. Rev. Lett. **84**(24), 5564 (2000)
62. M. Knudson, D. Hanson, J. Bailey, C. Hall, J. Asay, W. Anderson, Phys. Rev. Lett. **87**(22), 225501 (2001)
63. M. Knudson, D. Hanson, J. Bailey, C. Hall, J. Asay, Phys. Rev. Lett. **90**(3), 35505 (2003)
64. M. Knudson, D. Hanson, J. Bailey, C. Hall, J. Asay, C. Deeney, Phys. Rev. B **69**(14), 144209 (2004)
65. J. Bailey, M. Knudson, A. Carlson, G. Dunham, M. Desjarlais, D. Hanson, J. Asay, Phys. Rev. B **78**(14), 144107 (2008)
66. D. Hicks, T. Boegly, P. Celliers, J. Eggert, S. Moon, D. Meyerhofer, G. Collins, Phys. Rev. B **79**, 014112 (2009)
67. M. Knudson, M. Desjarlais, Phys. Rev. Lett. **103**(22), 225501 (2009)
68. G. Boriskov, A. Bykov, R. Il'Kaev, V. Selemir, G. Simakov, R. Trunin, V. Urlin, A. Shuikin, W. Nellis, Phys. Rev. B **71**(9), 092104 (2005)
69. S. Grishechkin, S. Gruzdev, V. Gryaznov, M. Zhernokletov, R. Il'Kaev, I. Iosilevskii, G. Kashintseva, S. Kirshanov, S. Manachkin, V. Mintsev, JETP Lett. **80**(6), 398 (2004)
70. C. Pierleoni, D.M. Ceperley, M. Holzmann, Phys. Rev. Lett. **93**, 146402 (2004)
71. M.A. Morales, C. Pierleoni, D.M. Ceperley, Phys. Rev. E **81**(2), 021202 (2010)
72. M.A. Morales, C. Pierleoni, E. Schwegler, D.M. Ceperley, Proc. Natl. Acad. Sci. **107**(29), 12799 (2010)
73. E. Liberatore, M.A. Morales, D.M. Ceperley, C. Pierleoni, Mol. Phys. **109**, 3029–3036 (2011)
74. L. Caillabet, S. Mazevet, P. Loubeyre, Phys. Rev. B **83**, 094101 (2011)
75. M.A. Morales, J.M. McMahon, C. Pierleoni, D.M. Ceperley, Phys. Rev. Lett. **110**, 065702 (2013)
76. R.M. Martin, *Electronic Structure: Basic Theory and Practical Methods* (Cambridge University Press, Cambridge/New York, 2004)
77. B. Holst, M. French, R. Redmer, Phys. Rev. B **83**, 235120 (2011)
78. M. French, A. Becker, W. Lorenzen, N. Nettelmann, M. Bethkenhagen, J. Wicht, R. Redmer, Astrophys. J. Suppl. Ser. **202**(1), 5 (2012)
79. F. Lin, M. Morales, K. Delaney, C. Pierleoni, R.M. Martin, D.M. Ceperley, Phys. Rev. Lett. **103**, 256401 (2009)
80. D.M. Ceperley, B. Bernu, J. Chem. Phys. **89**, 6316 (1988)
81. B. Bernu, D.M. Ceperley, W.A. Lester, J. Chem. Phys. **93**, 552 (1990)
82. S. Weir, A. Mitchell, W. Nellis, Phys. Rev. Lett. **76**(11), 1860 (1996)
83. B. Militzer, Phys. Rev. Lett. **97**, 175501 (2006)
84. B. Militzer, J. Low Temp. Phys. **139**, 739 (2005)
85. W. Lorenzen, B. Holst, R. Redmer, Phys. Rev. Lett. **102**, 115701 (2009)
86. B. Militzer, Phys. Rev. B **79**, 155105 (2009)
87. K.P. Driver, B. Militzer, Phys. Rev. Lett. **108**, 115502 (2012)

88. N.A. Romero, W.D. Mattson, Phys. Rev. B **76**, 214113 (2007)
89. M.D. Knudson, M.P. Desjarlais, D.H. Dolan, Science **322**(5909), 1822 (2008)
90. T.R. Mattsson, M.P. Desjarlais, Phys. Rev. Lett. **97**, 017801 (2006)
91. M. French, T.R. Mattsson, N. Nettelmann, R. Redmer, Phys. Rev. B **79**, 054107 (2009)
92. V. Karasiev, S. Trickey, Comput. Phys. Commun. **183**(12), 2519 (2012)
93. T. Sjostrom, F.E. Harris, S.B. Trickey, Phys. Rev. B **85**, 045125 (2012)
94. M. Troyer, U.J. Wiese, Phys. Rev. Lett. **94**(17), 170201 (2005)

The Structure of Warm Dense Matter Modeled with an Average Atom Model with Ion-Ion Correlations

D. Saumon, C.E. Starrett, J.A. Anta, W. Daughton, and G. Chabrier

Abstract We present a new model of warm dense matter that represents an intermediate approach between the relative simplicity of "one-ion" average atom models and the more realistic but computationally expensive ab initio simulation methods. Physical realism is achieved primarily by including the correlations in the plasma that surrounds a central ion. The plasma is described with the Ornstein-Zernike integral equations theory of fluids, which is coupled to an average atom model for the central ion. In this contribution we emphase the key elements and approximations and how they relate to and expand upon a typical average atom model. Besides being relatively inexpensive computationally, this approach offers several advantages over ab initio simulations but also has a number of limitations. The model is validated by comparisons with numerical solutions for the pair distribution function of the ions from ab initio simulations for several elements and a wide range of plasma conditions. Simulations results are reproduced remarkably well and simpler limiting theories are recovered as well. This model has many potential applications to the calculation of properties of warm dense matter such as the equation of state and conductivities for a wide range of temperatures and densities.

D. Saumon (✉) • C.E. Starrett • W. Daughton
Los Alamos National Laboratory, P.O. Box 1663, Los Alamos, NM 87545, USA
e-mail: dsaumon@lanl.gov; starrett@lanl.gov; daughton@lanl.gov

J.A. Anta
Departamento de Sistemas Fisicos, Químicos y Naturales, Universidad Pablo de Olavide, Ctra. Utrera, km 1, 41013 Sevilla, Spain
e-mail: anta@upo.es

G. Chabrier
CRAL (UMR CNRS 5574), École Normale Supérieure de Lyon, 69364 Lyon Cedex 07, France
e-mail: chabrier@ens-lyon.fr

F. Graziani et al. (eds.), *Frontiers and Challenges in Warm Dense Matter*, Lecture Notes in Computational Science and Engineering 96,
DOI 10.1007/978-3-319-04912-0_6,

1 Introduction

Warm dense matter (WDM) is an "inner" frontier of research in material properties in the sense that it is circumscribed by regimes that can be described by well-understood theoretical approaches and well-known limits. For instance, neither condensed matter theory, which is primarily based on treating electrons at near-zero temperature in periodic structures, nor plasma physics theory, based on weak Coulomb coupling between the constitutive particles, are applicable to modeling WDM. More specifically, WDM is a partially ionized, globally neutral plasma characterized by moderate to strong coupling between all charged particles (ion-ion, ion-electron, and electron-electron) and partial electron degeneracy, where weakly bound states can cross into the continuum (pressure ionization). Depending on the material, this corresponds to temperatures of $\sim$1–100 eV and densities of $\sim$0.1–50 times the normal solid density. A general but admittedly vague definition of WDM, is that it occurs in the physical regime where "all the physics" is important. By definition, it is thus quite challenging to model.

Experimentally, WDM states are not particularly difficult to achieve nowadays as many experiments cross the WDM regime on their way to denser and, more often, hotter final states. The difficulty is in making a sample of WDM that is large enough, that is relatively long-lived and with a spatial homogeneity that allow meaningful probing of its properties. The nature of WDM also limits the effectiveness of well-established diagnostics of dense matter and plasmas and new methods need to be developed [1]. There is presently only a limited but growing amount of data on WDM states [2–10].

Despite the experimental and theoretical challenges it presents, WDM has long been of interest as it occurs naturally in the giant planets of the solar system (Jupiter, Saturn, Uranus and Neptune) [11, 12], in dense stars such as in the envelopes of white dwarf stars (the final stage of a star's life) [13] and in the many giant and exotic planets that have been discovered orbiting other stars in the last decade [14]. In the laboratory, WDM is a transient state of imploding capsules in inertial confinement fusion experiments and can be produced with pulsed-power platforms and high-power laser facilities. Modeling these physical systems typically requires a knowledge of the equation of state, opacity, conductivity, and diffusion coefficients of WDM.

The intrinsic interest of WDM as a hard physics problem, as well as its importance to other fields of research has led to many approximate models that have steadily increased in sophistication. One large and important class of models, known as average atoms (AA) models, has a long history and have proved to be very useful. An AA model is essentially a one-ion model that solves for the electronic structure (bound and free states) around a central nucleus embedded in a spherically averaged, homogeneous plasma outside the ion sphere. There are many variants of such models but they all assume spherical symmetry and that the resulting modeled ion, which typically has a fractional net charge, represents in some sense an average of the multiple ion configurations (ionic states) present in the plasma. The assumptions about the surrounding plasma effectively replace the environment of the central ion by a boundary condition at the radius of the ion sphere R defined by

$$\frac{4\pi R^3}{3} = \frac{1}{n_I^0} \tag{1}$$

where n_I^0 is the number density of ions. The system of equations defining an AA model (see Sect. 2 below) amounts to a self-consistent field problem in spherical symmetry and finite temperature. This is a conceptually straightforward problem which explains their early and long-lasting popularity to model WDM.[1] Through the boundary condition applied at the ion sphere radius, AA models can qualitatively, and to a fair extent, quantitatively, predict the electronic structure and charge of ions as a function of density and temperature [17, 18]. Despite their popularity and usefulness, average atom models account for the surrounding plasma only in the simplest possible way and ignore the correlations in the plasma, an important characteristic of WDM.

At the other end of the spectrum of complexity are computer simulations of WDM, based on ab initio methods such as Path Integral Monte Carlo (PIMC) and Quantum Molecular Dynamics (QMD) [19–25]. These methods take a more direct approach to solving the quantum many-body problem of WDM and use few approximations in simulating the properties of a mixture of nuclei and electrons in a simulation box. Bound and free states, radial and angular correlations, the formation of clusters, and multiple ion configurations naturally occur in such simulations. The theoretical appeal of these methods is tempered by their considerable computational cost even in view of the computing power available today. Simulations of higher-Z elements remain very onerous.

We present a new model of WDM that takes an intermediate path between simple average atom models and expensive simulations, with the goal of producing realistic material properties at a reasonable computational cost. The concept of the average atom is extended by including correlations between charged particles in the surrounding plasma. The structure of the central ion is thus solved in the field of the central nucleus and of the surrounding, correlated ions and electrons. This re-introduces the surrounding plasma correlations in a much more realistic fashion. By maintaining spherical symmetry in the formulation of the model, the computational cost remains modest. This opens the possibility of generating large tables of properties, particularly of the equation of state, that can be used in various applications. Similar models have been published in the past [26–29] but this new model is based on a more formal derivation and a higher level of internal consistency. Limitations of earlier models have been overcome, resulting in a model that is applicable to a broad range of physical regimes, from liquid metals to WDM and high-temperature plasmas.

---

[1]Achieving accurate numerical solutions is more challenging [15] but has progressed considerably [16].

We first briefly review a typical average atom model in Sect. 2 to set the context for its extension to include plasma correlations. The full model is described in Sect. 3, with illustrations of the key models quantities. The models has been applied to a wide range of elements, temperature and densities, and it has been validated by comparisons with ab initio simulations, primarily in terms of pair distribution functions (Sect. 4). The last section (Sect. 5) offers a summary and outlook. This contribution emphasizes the concepts that underlie the model and how it differs from previous similar efforts. A detailed presentation along with derivations are given in [30, 31], where additional numerical results can also be found.

2 Average Atom Models

An average atom model describes a single central nucleus of charge Z with Z electrons embedded in a radially and spherically smoothed plasma residing outside the ion sphere surrounding the nucleus. The surrounding plasma has number density n_I^0 and, like the central ion, is at a temperature $T = 1/\beta$. Average atom models invariably describe the electron cloud surrounding the nucleus with the finite temperature formulation of the Kohn-Sham density functional theory (DFT) [32–34]. The latter can be cast in the semi-classical Thomas-Fermi model of the electrons [35–37], a full quantum mechanical description with the Schrödinger equation [38], or the relativistic quantum mechanical Dirac equation [16, 18, 39]. For the clarity of the discussion and without loss of generality, hereafter we will consider only the quantum mechanical case (Schrödinger equation). The structure of a typical AA model and its key equations are shown in Fig. 1, which is a guide to the following discussion. The effective one-electron DFT Schrödinger equation is

$$[\hat{T} + V_{Ne}^{\text{eff}}(r)]\psi_i(r) = \epsilon_i \psi_i(r) \tag{2}$$

where $\hat{T}$ is the kinetic energy operator, $V_{Ne}^{\text{eff}}(r)$ is the effective one-electron potential energy, $\psi_i(r)$ is the one-electron radial wave function of state i with energy ϵ_i. The outer boundary condition is applied at $r \to \infty$ where $V_{Ne}^{\text{eff}} \to 0$ and a free particle solution is obtained. $\psi_i(r) \to 0$ at infinity for bound states. For continuum states, $\psi_i(r \to \infty)$ is required to match the free solution (spherical Bessel function). In practice, the outer boundary condition is applied at some large radius R_{max} that defines the edge of the computational boundary, where $\psi_i(R_{\text{max}}) = 0$ for bound states, and $\psi_i(R_{\text{max}})$ is matched to spherical Bessel functions for continuum states [40]. The potential is

$$V_{Ne}^{\text{eff}}(r) = -\frac{Z}{r} + \int \frac{n_e(r') - Z^* n_I(r')}{|\mathbf{r} - \mathbf{r}'|} d^3r' + V_{\text{xc}}^{ee}[n_e(r)] - V_{\text{xc}}^{ee}[n_e^0] \tag{3}$$

where $n_e(r)$ and $n_I(r)$ are the radial density profiles of electrons and ions around the central nucleus, respectively, Z^* is the ion charge in the average atom model,

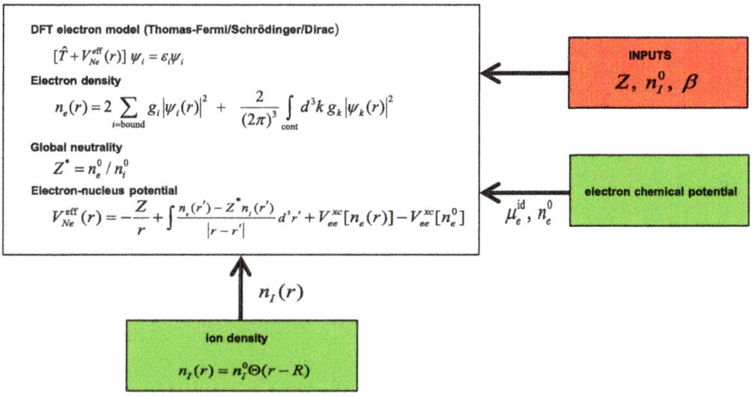

Fig. 1 Structure and key equations of a typical average atom model. The *red box* gives the input parameters to the model. *Green boxes* show model components where several choices are possible. *Arrows* show the flow of information between the key elements of the model. See text for details

and $V_{\rm xc}^{ee}$ is the electron exchange and correlation potential. The electron density is obtained by summing the density associated with the eigenfunctions of both bound and continuum states

$$n_e(r) = 2 \sum_{i=\text{bound}} g_i |\psi_i(r)|^2 + \frac{2}{(2\pi)^3} \int_{\text{cont}} d^3k \, g_k |\psi_k(r)|^2 \tag{4}$$

and the free electron density is

$$n_e^0 = \lim_{r\to\infty} n_e(r). \tag{5}$$

The g_i and g_k are Fermi occupation factors

$$g_i = \frac{1}{e^{\beta(\epsilon_i - \mu_e^{\rm id})} + 1}, \tag{6}$$

where $\mu_e^{\rm id}$ is the ideal chemical potential of the electrons. Global neutrality of the plasma requires that the ion charge be

$$Z^* = n_e^0 / n_I^0. \tag{7}$$

This essentially defines the self-consistent field problem of the AA model.

Two additional elements are needed to close this system of equations, neither of which is unique (green boxes in Fig. 1). One is the density profile of the ions surrounding the central nucleus, $n_I(r)$, which describes the surrounding plasma. In AA models, this is taken as a simple step function with a constant ion density outside the ion sphere:

$$n_I(r) = n_I^0 \Theta(r - R), \tag{8}$$

which describes a cavity (the ion sphere) in the plasma centered at the origin. This form was originally devised to approximate the periodic nature of solids in AA models [26, 41], by confining each average atom within the ion sphere[2] but it can also be interpreted as the ion pair distribution function in a dense plasma. With this choice for $n_I(r)$, the average atom is coupled to the external plasma only through the ion sphere radius R, which depends only on the density n_I^0. This choice of $n_I(r)$ implies that the external ions are not correlated with each other or with the central nucleus. It has the virtue of extreme simplicity but is clearly quite approximate and leaves out important physics.

The second element is the chemical potential μ_e^{id} of non-interacting electrons which determines the Fermi occupation factors g_i and g_k in the electron density (Eq. 6). This can be determined in several different ways. A very common practice [16, 18, 26, 39] is to use a local charge neutrality condition. For example

$$Z = \int_0^R n_e(\mu_e^{\text{id}}, r)\, d^3 r, \tag{9}$$

where the μ_e^{id} dependence of $n_e(r)$ is shown explicitly, ensures neutrality within the volume of the ion sphere. This relation uniquely determines μ_e^{id}. The free electron density that corresponds to $V_{Ne}^{\text{eff}}(r \to \infty) = 0$ then follows from

$$n_e^0 = \frac{\sqrt{2}}{\pi^2 \beta^{3/2}} \int \frac{x^{1/2}\, dx}{e^{x - \beta\mu_e^{\text{id}}} + 1}, \tag{10}$$

which also determines the ion charge Z^* (Eq. 7). It is well known that there is no rigorous definition of the ion charge in AA models [17, 18]. While Eq. (9) is intuitively sensible, there are other reasonable choices [17, 27, 28, 42, 43], such as requiring that $Z = N_b + n_e^0/n_I^0$, where N_b is the number of electrons bound to the central nucleus

$$N_b = 2 \int \sum_{i=\text{bound}} g_i |\psi_i(r)|^2\, d^3 r \tag{11}$$

(see Eq. 4). This is equivalent to requiring that the ion charge is that of the nucleus minus the number of bound electrons: $Z^* = Z - N_b$. It has been shown [18] that requiring that the AA model gives the same pressure whether obtained from the virial or thermodynamic routes gives a condition for μ_e^{id} that involves the electrostatic potential rather than the electron density:

---

[2] In this case the boundary condition is applied at $r = R$.

$$\int_0^\infty d^3r\, \Theta(r-R) V_{Ne}^{\mathrm{el}}(r) = 0, \tag{12}$$

where

$$V_{Ne}^{\mathrm{el}} = -\frac{Z}{r} + \int \frac{n_e(r') - Z^*\Theta(r'-R)}{|\mathbf{r}-\mathbf{r}'|}\, d^3r'. \tag{13}$$

This result is obtained by minimizing the AA free energy with respect to n_e^0. Paradoxically, this more rigorous approach to determine μ_e^{id} gives anomalously low values of Z^* at low temperatures [18], where it fares worse than models using a local neutrality condition such as Eq. (9). It appears that this more rigorous AA model exposes a detrimental aspect of neglecting the correlations in the surrounding plasma, a feature that must be partly compensated in models that use an ad hoc local neutrality condition.

To summarize, the AA model is a DFT-based, spherically symmetric, self-consistent field atom model embedded in a plasma. The equations are closed with two additional inputs: (1) an external ion distribution function which defines a cavity in a homogeneous (uncorrelated) plasma, and (2) a criterion to determine the non-interacting (field-free) part of the chemical potential of the electrons. The latter cannot be uniquely defined and various sensible choices give different results for the electronic structure of the central ion. The former is a simple and convenient approximation that can be improved upon. In fact, the absence of plasma correlations in the AA model is the most significant piece of physics that is missing in AA models. We now expand the AA model to include interactions within the surrounding plasma and with the central ion while preserving the assumption of spherical symmetry.

3 An Average Atom Model with Plasma Correlations

The importance of the correlations in the surrounding plasma was recognized decades ago and several models have been developed to account for them [26–28, 44]. They affect the AA model shown in Fig. 1 in two ways. First the effective nucleus-electron potential now includes interactions between the electrons and the surrounding ions:

$$\begin{aligned} V_{Ne}^{\mathrm{eff}}(r) = &-\frac{Z}{r} + \int \frac{n_e(r') - Z^* n_I(r')}{|\mathbf{r}-\mathbf{r}'|}\, d^3r' + V_{\mathrm{xc}}^{ee}[n_e(r)] - V_{\mathrm{xc}}^{ee}[n_e^0] \\ &- \frac{1}{\beta}\int \tilde{C}_{Ie}(|\mathbf{r}-\mathbf{r}'|)(n_I(r') - n_I^0)\, d^3r', \end{aligned} \tag{14}$$

where $\tilde{C}_{Ie}$ is the non-Coulombic part of the direct ion-electron correlation function $C_{Ie}(r)$, which is defined by [30]

$$C_{Ie}(r) = \beta Z^*/r + \tilde{C}_{Ie}(r). \tag{15}$$

This new term is highlighted in yellow in Fig. 2. The second modification is that the ion density is described in terms of the ion-ion radial distribution function $g_{II}(r)$:

$$n_I(r) = n_I^0 g_{II}(r). \tag{16}$$

Whereas in the AA model $g_{II}(r)$ is chosen as a simple step function, here it is to be calculated on the basis of the Coulomb interactions in the plasma. As can be seen by comparing Figs. 1 and 2, the structure of the AA model remains unchanged except for the new contribution to $V_{Ne}^{\text{eff}}(r)$. The inputs to the model (Z, n_I^0, β) are the same and a criterion for μ_e^{id} must still be provided. In addition, closure of this modified AA model also requires $\tilde{C}_{Ie}(r)$ and $n_I(r)$ which can be calculated by coupling the AA model with a model of the plasma based on the Ornstein-Zernike integral equations theory of fluids [30]. The calculation of these two quantities increases the complexity of the problem considerably (Fig. 2).

3.1 Ornstein-Zernike Integral Equations Theory of Fluids and the Two-Component Plasma Model

Prior to the advances in computing power that allowed the direct simulation of classical fluids (Monte Carlo and Molecular Dynamics methods), the integral equations theory of fluids [45, 46] offered the most realistic description of an interacting fluid at a reasonable computing cost. It is the theory of choice to develop an AA model with correlations [26–30, 47–49].

The structure and thermodynamics of a fluid system of one type of classical point particles interacting via a pair potential is fully determined by its density n, temperature β and the pair potential $V(r)$. The most widely used integral equations theories of fluids are based on the Ornstein-Zernike equation

$$h(r) = C(r) + n \int C(|\mathbf{r} - \mathbf{r}'|)h(r)\, d^3r'. \tag{17}$$

Expressed in Fourier space, Eq. (17) takes a simple form

$$h(k) = C(k) + nC(k)h(k), \tag{18}$$

where $F(k)$ denotes the Fourier transform of a function $F(r)$. The functions $h(r)$ and $C(r)$ are related to the pair potential by a closure relation

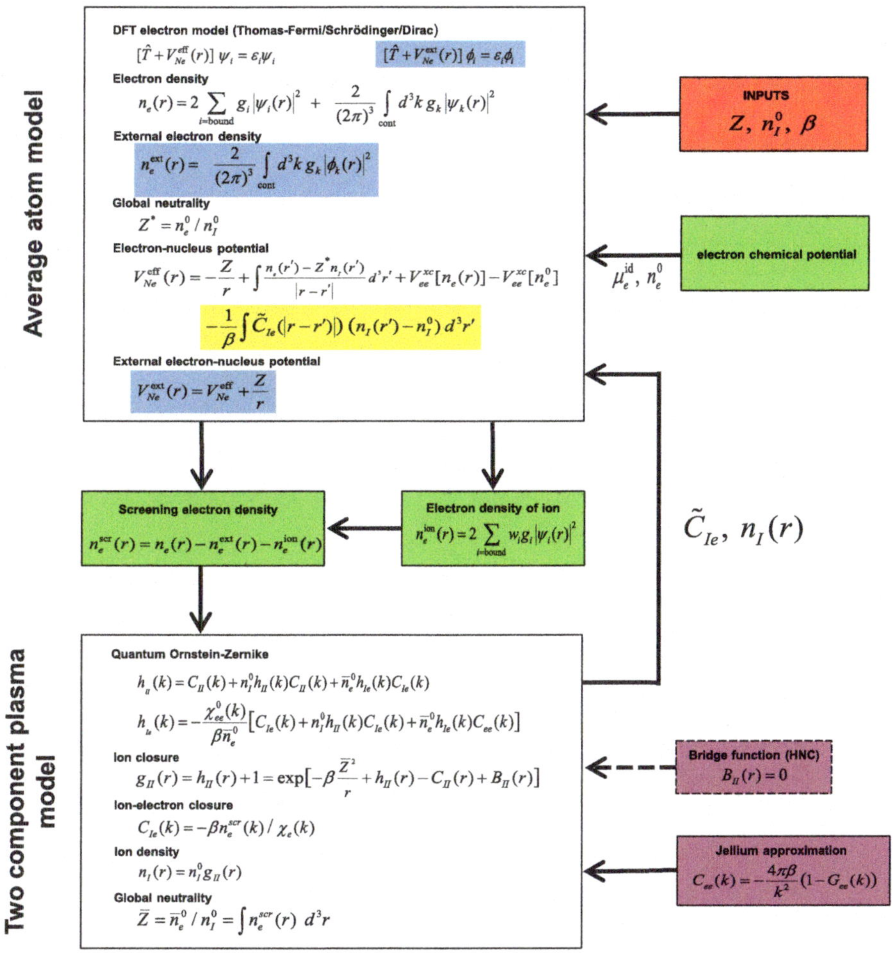

Fig. 2 Structure and key equations of the coupled average atom and two-component plasma model. The *red box* gives the input parameters to the model. *Green boxes* show elements where several choices are possible. Approximations are shown within *purple boxes*, with the *dotted box* indicating an approximation that can readily be improved. *Arrows* show the flow of information between the elements of the model. See text for details

$$g(r) = h(r) + 1 = \exp\left(-\beta V(r) + h(r) - C(r) + B(r)\right). \tag{19}$$

In these equations, $g(r)$ is the pair distribution function, $h(r)$ is the pair correlation function, and $C(r)$ is the direct correlation function which is effectively defined by Eq. (17). Closing these two equations requires the knowledge of the bridge function $B(r)$ which accounts for n-body correlations beyond pair correlations. Various approximations to $B(r)$ are available [46]. Note that these relations are exact if the bridge function is known. Once the structure of the fluid (e.g. $g(r)$) is known, various quantities of interest, such as the thermodynamics can be calculated [45].

The above fluid equations can be readily generalized to mixtures of classical particles given the pair potentials $V_{ij}(r)$ between particles of type i and j [45, 46]. Chihara [50] has shown that the Ornstein-Zernike equation can be further generalized for a plasma of classical ions of density n_I^0 and quantum electrons of density $\bar{n}_e^0$

$$h_{II}(k) = C_{II}(k) + n_I^0 h_{II}(k) C_{II}(k) + \bar{n}_e^0 h_{Ie}(k) C_{Ie}(k) \tag{20}$$

$$h_{Ie}(k) = \frac{-\chi_{ee}^0(k)}{\beta \bar{n}_e^0} \left[C_{Ie}(k) + n_I^0 h_{II}(k) C_{Ie}(k) + \bar{n}_e^0 h_{Ie}(k) C_{ee}(k) \right] \tag{21}$$

where $\chi_{ee}^0(k)$ is the Lindhard response function of the non-interacting quantum electron gas. The quantum nature of the electrons is embodied in the factor $-\chi_{ee}^0(k)/\beta \bar{n}_e^0$. In the limit of a classical electron gas, this factor becomes unity and the classical Ornstein-Zernike equations are recovered. The ion closure relation has the same form as in the one component case

$$g_{II}(r) = h_{II}(r) + 1 = \exp\left(-\beta \frac{\bar{Z}^2}{r} + h_{II}(r) - C_{II}(r) + B_{II}(r) \right) \tag{22}$$

where the pair interaction is a pure Coulomb potential between ions of charge $\bar{Z}$.[3] Global neutrality requires that

$$\bar{Z} = \bar{n}_e^0 / n_I^0 . \tag{23}$$

The last four equations contain seven unknowns: $\bar{Z}$, h_{II}, h_{Ie}, C_{II}, C_{Ie}, C_{ee} and B_{II}. The closing of this system of equations will allow a solution that provides $\tilde{C}_{Ie}$ (from Eq. 15) and $n_I(r) = n_I^0 g_{II}(r)$ that are needed to close the AA model with ion correlations (Fig. 2). The full set of equations will define a model that we will call a "two-component plasma" (TCP) model,

To close the equations of the TCP model, we start by introducing two approximations (purple boxes in Fig. 2). First, the bridge function is assumed to be $B_{II}(r) = 0$. This is the well-known hyper-netted chain (HNC) approximation to the Ornstein-Zernike integral equations theory of fluid (dashed purple box in Fig. 2). A better approximation is to use the bridge function of a similar interacting system that is more easily calculated than that of a general TCP. The bridge functions of the hard sphere fluid [51, 52], of the one-component plasma model [53], or for Yukawa systems [54], or bridge functions optimized with a variational approach [55] have all been used to model plasmas with the integral equations theory of fluids. For simplicity, and with good accuracy for the systems considered here (see Sect. 4), we

[3] Note that this is different from the AA ion charge Z^*, which is further discussed in Sect. 3.2.

will use the HNC approximation. The second approximation is to assume that the direct electron-electron correlation function C_{ee} is given by

$$C_{ee}(k) = \frac{-4\pi\beta}{k^2}\left(1 - G_{ee}(k, \beta, \bar{n}_e^0)\right) \tag{24}$$

where $G_{ee}(k, \beta, \bar{n}_e^0)$ is the local field correction of the jellium model. The jellium approximation (Eq. 24), to which there appears to be no reasonable alternative at this time, implies that the *e–e* correlations are decoupled from the ions. This is reasonable as C_{ee} represent the correlations between unbound electrons. As long as the ion-electron correlations are not too strong, Eq. (24) is a good approximation. On the other hand, it would be a very poor approximation for the bound electrons which are intrinsically very strongly correlated with the ions.

One additional equation is needed to close the TCP model, which is to be provided by the AA model (downward arrows in Fig. 2). This closure will also determine the charge of the ion $\bar{Z}$ and thus the direct ion-ion potential in the TCP model. This closure is not unique, however [27, 28, 30, 49], and is illustrated with the green boxes in Fig. 2. We found that the most satisfactory approach is to use the solution of the AA model to calculate the electron density that screens an individual ion in the plasma. The AA model is naturally suited for this purpose as it is designed to calculate the properties of a single ion in a plasma. This screening electron density is defined in two steps.

In the first step, the electron density profile $n_e(r)$ obtained from the AA model is separated into a contribution that is assigned to a single nucleus (defining a "pseudo-atom" [28, 56, 57]) and an external contribution, due to the electrons in the surrounding plasma. The separation is effected by solving the AA model a second time after removing the central nucleus from the effective electron-nucleus potential $V_{Ne}^{\text{eff}}(r)$, keeping everything else fixed. The resulting electron density is thus computed only from the field of the surrounding plasma $V_{Ne}^{\text{ext}}(r)$ and defines the external electron density $n_e^{\text{ext}}(r)$. The AA equations corresponding to this external system are shown in blue boxes in Fig. 2. Note that in the absence of the central nucleus, the potential $V_{Ne}^{\text{ext}}(r)$ is repulsive almost everywhere and does not support bound states. This pseudo-atom electron density is $n_e^{\text{PA}}(r) = n_e(r) - n_e^{\text{ext}}(r)$.

The second step consists of extracting from $n_e^{\text{PA}}(r)$ the part that will define (along with a central nucleus) an ion, and the remainder which is to be assigned to the electron fluid of the plasma. The most straightforward separation is to assign the bound electrons to the ion and the continuum electrons to the TCP electron fluid. This definition of an ion for the TCP model (green box in Fig. 2) is by no means unique. For instance, the density in the neighborhood of the nucleus associated with continuum resonances of low energy is qualitatively similar to the electron density of a weakly bound state. Thus, it would be reasonable to count those resonant electrons as part of the ion. Such an approach was suggested by Chihara [27] but is not satisfactory in practice as it requires additional arbitrary criteria to determine which resonances should be considered part of the ion and to separate the electron density associated with a resonance from the background continuum electrons.

This simple definition of the ion electron cloud is not without drawbacks. Most flagrant is that the electron density of the bound states is a discontinuous function of n_I^0 when a state crosses into the continuum of positive energies, i.e. when it is pressure-ionized, a phenomenon of great importance in studies of WDM. As a consequence, every quantity appearing in the model becomes discontinuous with density. This can be avoided by applying occupation factors $0 \leq w_i(r) \leq 1$ in the sum

$$n_e^{\text{ion}}(r) = 2 \sum_{i=\text{bound}} w_i(r) g_i |\psi_i(r)|^2 \tag{25}$$

which are calculated from a simple model for the broadening of the bound energy levels [30]. As a weakly bound, broadened level starts to cross into the continuum of positive energies, $w_i(r)$ decreases below unity and vanishes once $\epsilon_i = 0$. Furthermore, $w_i(r)$ includes a smooth radial cutoff that damps the tails of weakly bound states that extend far from the central nucleus [28, 30]. Physically, this accounts for the fact that the bound electron density that is located far from the nucleus cannot be assigned to the central ion in the presence of the other neighboring ions. This cutoff also resolves a numerical difficulty with the spatial integration of long-ranged functions. With these elements it is now possible to define the screening electron density

$$n_e^{\text{scr}}(r) = n_e(r) - n_e^{\text{ext}}(r) - n_e^{\text{ion}}(r). \tag{26}$$

This is the density of electrons that are associated with the central nucleus but are "free", i.e. not part of the central ion. These electrons respond to the attractive field of the ion, screening the ion-ion interaction. From the quantum OZ equations, it can be shown [27, 30, 47] that the response of the electron fluid gives the screening density in terms of direct correlation functions

$$n_e^{\text{scr}}(k) = -\chi_{ee}(k) C_{Ie}(k)/\beta \tag{27}$$

where

$$\chi_{ee}(k) = \frac{\chi_{ee}^0(k)}{1 + \chi_{ee}^0(k) C_{ee}(k)/\beta} \tag{28}$$

is the response function of the correlated quantum electron fluid [58]. Equation (27) has the form of a linear response of the electron fluid to an external pseudo-potential $-C_{Ie}/\beta$. However, in this case the response is highly non-linear since $n_e^{\text{scr}}(r)$ is calculated from the solution of the Schrödinger equation rather than linear response theory. Inverting Eq. (27) gives

$$C_{Ie}(k) = -\beta n_e^{\text{scr}}(k)/\chi_e(k) \tag{29}$$

which is an exact relation within the TCP model. Furthermore, global neutrality of the TCP plasma requires that the ion charge $\bar{Z}$ be related to the screening density by

$$\bar{Z} = \bar{n}_e^0/n_I^0 = \int n_e^{\rm scr}(r)\, d^3r. \tag{30}$$

By taking $n_e^{\rm scr}$ defined from the AA model (Eq. 26), we now have the final equation required to close the TCP model, which provides $\tilde{C}_{Ie}$ (Eq. 15) and $n_I(r)$ that in turn close the AA model (Fig. 2). The two sets of equations are solved iteratively between the AA and TCP models.

Thus, the coupling between the AA and the TCP models is provided by defining a screening density from a solution of the Schrödinger equation for the Z electrons belonging to the central nucleus in the AA model. This particular choice is not unique (hence the green box in Fig. 2). In the same spirit, the Quantum Hypernetted-Chain (QHNC) model of Chihara [27] and Anta and Louis [47], couples the AA and TCP models in terms of the pair correlation function

$$n_e^0 h_{Ie}(r) = n_e^{\rm f}(r) - n_e^0 \tag{31}$$

where $n_e^{\rm f}(r)$ is the density of continuum electrons (second term in Eq. 4). Like Eq. (29), this choice for the coupling relates a quantity associated with the density of free electrons obtained in the AA model ($n_e^{\rm f}$ instead of $n_e^{\rm scr}$) to a correlation function in the TCP model (h_{Ie} instead of C_{Ie}). The QHNC closure has the drawback that $n_e^0 = \bar{n}_e^0$, i.e. $\bar{Z} = Z^*$, which overly constrains the model, as we will see below.

The electron chemical potential (green box in Fig. 2) is given as in the AA model (Eq. 9), with a step function $g_{II}(r)$. This is inconsistent with the pair distribution function obtained in the TCP model but is numerically advantageous and, given the ambiguous nature of the local neutrality condition, is very reasonable. Other choices include imposing charge neutrality as in Eq. (9) but over a correlation sphere of radius $R_c > R$ [48], or $Z = N_b + n_e^0/n_I^0$ as in several AA models.

This completes the formulation of the self-consistent model for an average atom embedded in a plasma of correlated ions. This AA+TCP model assumes spherical symmetry and consists of an AA model coupled to a TCP model based on the Ornstein-Zernike integral equations theory of interacting fluids. A screening electron density is calculated within the AA model to obtain an ion-electron correlation function for the TCP model, which in turns provide correlation functions for the AA atom model. The model has no adjustable parameters and only needs the nuclear charge, ion density and temperature as inputs. It formally recovers the classical one-component plasma model (OCP) [59] and the screened one-component plasma model (SOCP) [60]. The solution of the closed system of equations gives all the pair correlation functions between ions and electrons, the effective ion-ion and nucleus-electron pair potentials, the bound and continuum wave functions and eigenvalues, the ion charge, and the electron densities. These quantities, some of which are directly amenable to experimental measurements,

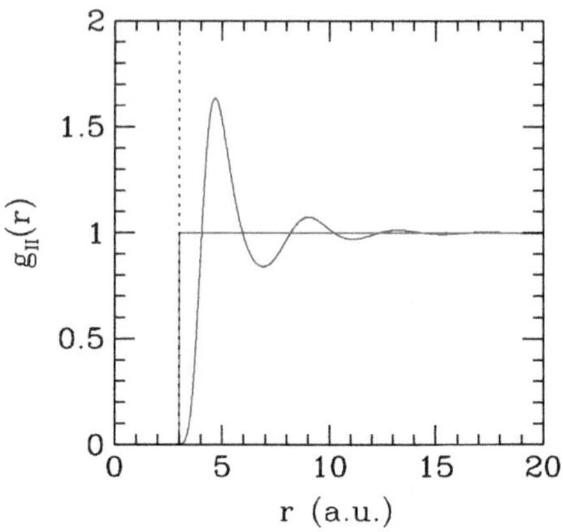

Fig. 3 Pair distribution functions of the JVM model (*red*) and the AA+TCP model (*blue*) for Al at $T = 1\,\text{eV}$ and $\rho = 2.7\,\text{g/cm}^3$. At this density, the ion sphere radius is $R = 2.99\,\text{a.u.}$, which is indicated by a *dotted line*

form the basis upon which WDM properties of interest, such as the equation of state, conductivities, and opacities can be calculated.

3.2 The Role of Plasma Correlations

The role of ion-ion correlations in the AA+TCP model is revealed by comparing some key model quantities calculated with an imposed, step function pair distribution function $g_{II}(r)$ (Eq. 8) as in a typical AA model and with the full self-consistent $g_{II}(r)$ from the AA+TCP model. For clarity, we will call the former the jellium vacancy model (JVM).[4] These results are shown in Figs. 3–6 for the case of Al at $T = 1\,\text{eV}$ (0.0367 Ha) and $\rho = 2.7\,\text{g/cm}^3$ (solid density). Under these conditions, the ions are strongly correlated and $g_{II}(r)$ shows much structure which is only crudely approximated by the step function ion distribution of the JVM (Fig. 3).

The effect of the ion correlations on the effective nucleus-electron potential $V_{Ne}^{\text{eff}}(r)$ that enters the Schrödinger equation of the AA model is shown in Fig. 4. At this density, the $1s^2$, $2s^2$ and $2p^6$ electrons of Al are tightly bound (our model predicts binding energies of -54.6, -3.4 and -2.0 Ha, respectively, measured from the bottom of the valence band) and confined to $r \lesssim 2\,\text{a.u.}$ (see below). Thus the core electrons are not sensitive to the difference in $g_{II}(r)$ and the two potentials are identical in the core. Both show a small positive maximum at $r \sim R$ as the central "ion hole" (Fig. 3) amounts to a repulsive contribution to the effective nucleus-electron potential. Differences, which can be as large as 0.02 Ha (0.5 eV), are mainly confined to $R \lesssim r \lesssim 3R$, as is expected from the difference in the ion-ion distribution function (Fig. 3). At this relatively low temperature, $V_{Ne}^{\text{eff}}(r)$ has Friedel

[4]This is the same model as the JVM of [18] and similar to the JVM of [27].

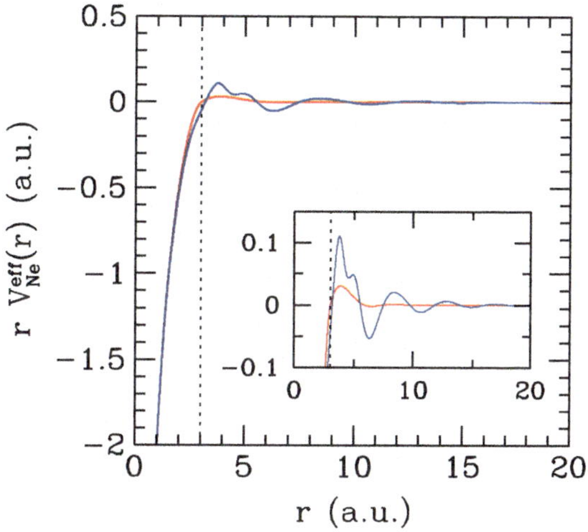

Fig. 4 Effective nucleus-electron potential for the JVM (*red*) and the AA+TCP model (*blue*), from Eqs. (3) and (14), respectively, for the same conditions as in Fig. 3. The inset magnifies the behavior at larger r. The product $rV_{Ne}^{\mathrm{eff}}(r)$ is plotted for clarity and the *dotted line* indicates the ion sphere radius $R = 2.99$ a.u.

oscillations at large r superimposed on the effect of decaying oscillations in $g_{II}(r)$. Both go away at higher T [30, 61].

In the TCP model, an effective ion-ion pair potential that takes into account the electron screening can be defined [27, 30, 47]

$$\beta V(k) = \beta \frac{4\pi \bar{Z}^2}{k^2} - n_e^{\mathrm{scr}}(k) C_{Ie}(k). \tag{32}$$

The first term is the repulsive Coulomb potential between two point ions of charge $\bar{Z}$ and the second term is the screening (mainly attractive) potential from the surrounding electrons that ensures that $V(r)$ is short-ranged. This effective ion-ion potential is plotted in Fig. 5 as $rV(r)$. The ion charge is $\bar{Z} = 3$ and a pure Coulomb potential would appear as a flat line at $rV(r) = 9$. The ion-ion potential reaches this limit as $r \to 0$. Strong screening of the ion charge is evident as $rV(r)$ decreases rapidly to very small values within $\lesssim 2R$. The potential shows two small bumps at $r \sim 0.07$ and 0.85 a.u. that are associated with minima in $n_e^{\mathrm{scr}}(r)$ (see below). This is consistent with Poisson's equation for the electrostatic contribution of n_e^{scr} to the ion-ion potential. Small oscillations outside the ion sphere (inset) are caused by the structure in $g_{II}(r)$.

The various electron densities that appear in the model, $n_e(r)$, $n_e^{\mathrm{ext}}(r)$, $n_e^{\mathrm{PA}}(r)$, $n_e^{\mathrm{ion}}(r)$ and $n_e^{\mathrm{scr}}(r)$ are shown in Fig. 6. At the scale of this figure, these electron densities are nearly identical when computed with either the JVM or the AA+TCP model. Only the latter are shown for clarity. The total electron density $n_e(r)$ (Eq. 4, blue curve) shows two peaks within 1 a.u. that correspond to the $1s^2$ and the $2s^2 2p^6$ bound electrons. The continuum electrons that are attracted to the nucleus and its bound electrons cause an excess of density near the ion sphere radius. At large r, $n_e(r) \to n_e^0$ and $4\pi r^2 n_e(r)$ diverges as r^2. The electron density associated with the plasma external to the central nucleus, $n_e^{\mathrm{ext}}(r)$ (black curve) is similar to $n_e(r)$

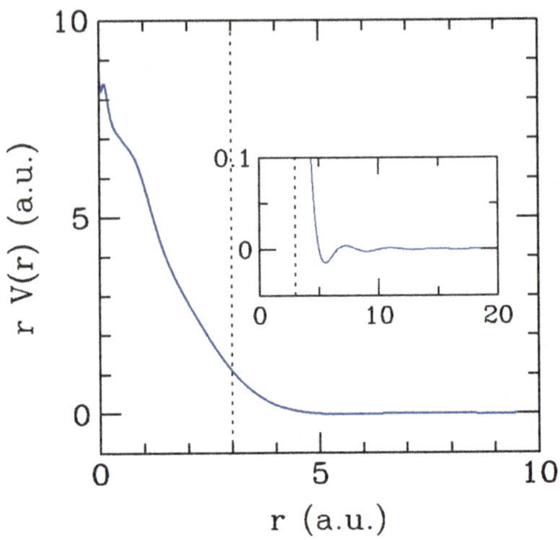

Fig. 5 Effective ion-ion potential in the AA+TCP model, given by Eq. (32) for the same case as in Fig. 3. The inset magnifies the behavior at larger r. The ion sphere radius $R = 2.99$ a.u. is shown by the *dotted line*. The product $rV(r)$ is plotted for clarity

except that it does not support bound states and has no build up of screening charge. The difference between these two electron densities defines the electron cloud associated with the central nucleus, i.e. a pseudo-atom $n_e^{\mathrm{PA}}(r) = n_e(r) - n_e^{\mathrm{ext}}(r)$ (green curve), which is essentially confined within $\sim 2R$. The ion electron cloud ($n_e^{\mathrm{ion}}(r)$) is defined as the density that arises from the bound states (Eq. 25; red curve) which shows peaks associated with the K and L shells. Finally, the electron density that couples the AA and TCP model (Fig. 2 and Eq. 29) is the screening density $n_e^{\mathrm{scr}}(r) = n_e^{\mathrm{PA}}(r) - n_e^{\mathrm{ion}}(r)$ (cyan curve). It forms a broad peak outside of the core states that decays rapidly beyond the ion sphere radius. This results from the attractive force of the net positive charge of the ion (nucleus + bound states), the repulsion from the bound electrons and the orthogonality of the continuum and bound states. Details of $n_e^{\mathrm{scr}}(r)$ are shown in the lower panel of Fig. 6.

It turns out that by running the electron densities obtained with the JVM through a single pass of the TCP model (without requiring the self-consistency, see Fig. 2) gives distribution functions that are good approximations to the fully converged solutions [30]. This realization has been a key element of the successful numerical implementation of the model as it substantially improves the convergence. It also implies that approximate but fairly good $g_{II}(r)$ for the AA+TCP model can be calculated at very little computational cost above that of the JVM model [62].

As pointed out above, the ion charge in the AA part of the model, Z^*, and in the TCP part of the model, $\bar{Z}$ are two distinct quantities, as are the corresponding free electron densities, n_e^0 and $\bar{n}_e^0$. It should be clear from Fig. 2 that we are coupling two different models and that mathematically it is not required that these ion charges be equal. In fact, they are quite different, as can be seen in Fig. 7. As expected, the ion charge increases steadily with temperature. The counterintuitive distinction between $\bar{Z}$ and Z^* can be understood physically as follows. In the AA model, n_e^0 is the density of electrons in the field-free region of space ($V_{Ne}^{\mathrm{eff}} = 0$), far from the central nucleus, which has a clear meaning within a model with a single central

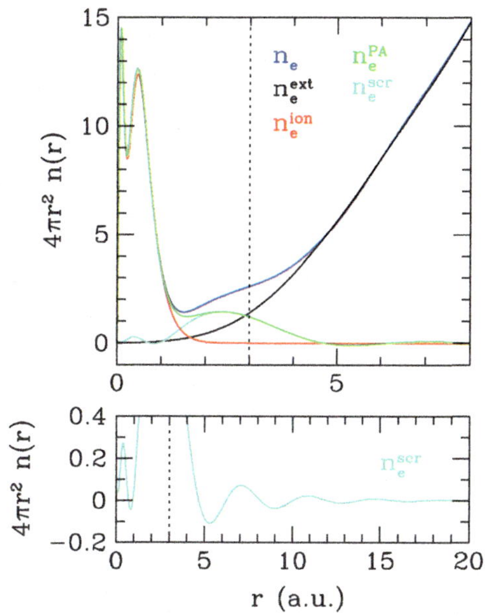

Fig. 6 Electron densities for Al at $T = 1$ eV and $2.7\,\text{g/cm}^3$. Only the electron densities from the AA+TCP model are shown for clarity. The total electron density (Eq. 4) is in *blue*, and the density from the system external to the ion (i.e. without the central nucleus) n_e^{ext} is in *black*. The difference between them is the pseudo-atom density $n_e^{\text{PA}} = n_e - n_e^{\text{ext}}$, which represent the Z electrons, bound and unbound, associated with the central nucleus (*green*). The density of electrons which, along with the central nucleus, define an ion is n_e^{ion} (Eq. 25, *red*). Finally, the screening density $n_e^{\text{scr}} = n_e^{\text{PA}} - n_e^{\text{ion}}$ is shown in *cyan*. The *lower panel*, which is on different scale, shows the smaller structure of n_e^{scr}. In the *upper panel*, the n_e and n_e^{ext} curves diverge as r^2 at large r because both densities become constant and the figure shows the densities multiplied by $4\pi r^2$. The *vertical dotted line* shows the ion sphere radius $R = 2.99$ a.u.

nucleus embedded in a spherically averaged plasma. In the TCP model, on the other hand, $\bar{n}_e^0$ represents the density of the inhomogeneous electron fluid (made up of those electrons that are not included in the ions), averaged over all space. This corresponds more closely to the concept of the electron density in a real dense plasma, where there is no field-free region as an electron always finds itself in the field of some nearby ion(s). Thus, the free electron density in the AA model (n_e^0) is a rather artificial construct that arises from the simplifying concept that the electronic structure around a central nucleus in a smeared out plasma is a good approximation for the "average" electronic structure in a correlated, multi-center plasma. Within the AA+TCP model, the physical electron density (to be used in a conductivity calculation, for example) is $\bar{n}_e^0$, with corresponding ion charge $\bar{Z}$. This is confirmed in Fig. 7 where $\bar{Z} \to 3$ at low T, which is the number of valence electrons of Al under normal conditions. On the other hand, $Z^* \sim 2$ at low T, which would be a poor estimate of the valence electron density in normal aluminum, should it be (wrongly) interpreted as such.

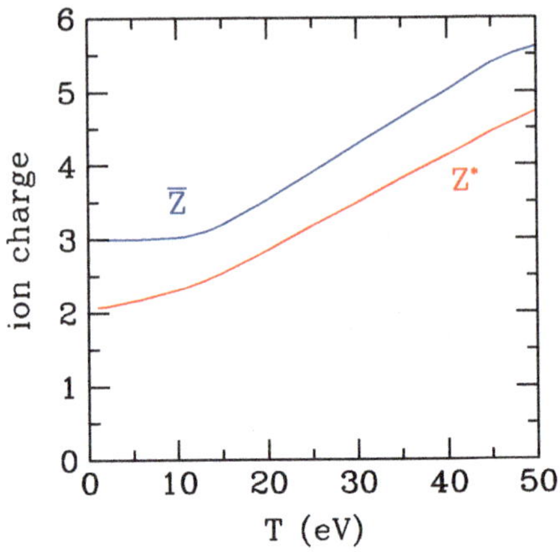

Fig. 7 Ion charge as a function of temperature for solid density Al ($2.7\,\mathrm{g/cm^3}$). The ion charge from the coupled average atom Z^* and the two-component plasma models $\bar{Z}$ are shown. The roles of $\bar{Z}$ and Z^* are illustrated in Fig. 2

3.3 Appraisal of the Model

Given the theoretical development and numerical solutions of the model outlined above, it is possible to ascertain its advantages and limitations at a semi-quantitative level. There are several advantages over computer simulations of WDM. Most immediate is the substantial economy of computer time to converge to a solution. Depending on the temperature and density, the solution of the AA+TCP model is typically 2–3 orders of magnitude faster than a QMD simulation. Computer simulations are inherently subject to statistical noise (fluctuations) due to the finite number of particles considered. The finite size of the simulation box limits spatial sampling of the system, such as the radial extent over which the pair distribution function can be evaluated (Figs. 8 and 9). On the other hand, the AA+TCP model consists of coupled algebraic, differential and integral equations whose solution is smooth within the numerical accuracy of the algorithms. The equations are solved over a comparatively large computational volume of typically 10–20 times the ion sphere radius, which effectively corresponds to an infinite system. It also treats all of the electrons explicitly and on the same footing. There is no pseudo-potential or concern about their transferability. The model gives good solutions over a range of temperatures and densities that is much broader than the typical range of applicability of any one ab initio method. It is also worth noting that a version of the model where the Schrödinger equation is replaced by the semi-classical Thomas-Fermi model of the electrons gives viable results even for strongly coupled systems [30,31], contrary to the conclusions of an earlier effort to develop a Thomas-Fermi AA model with ion-ion correlations [29]. Furthermore, the quantum mechanical version recovers the Thomas-Fermi results in the high-density, high-temperature limit [30]. This is a valuable internal check on the physics and numerical implementation of the model.

While these are significant advantages over ab initio simulations of WDM, the model does not fare as well in other aspects. The substantial savings of computer

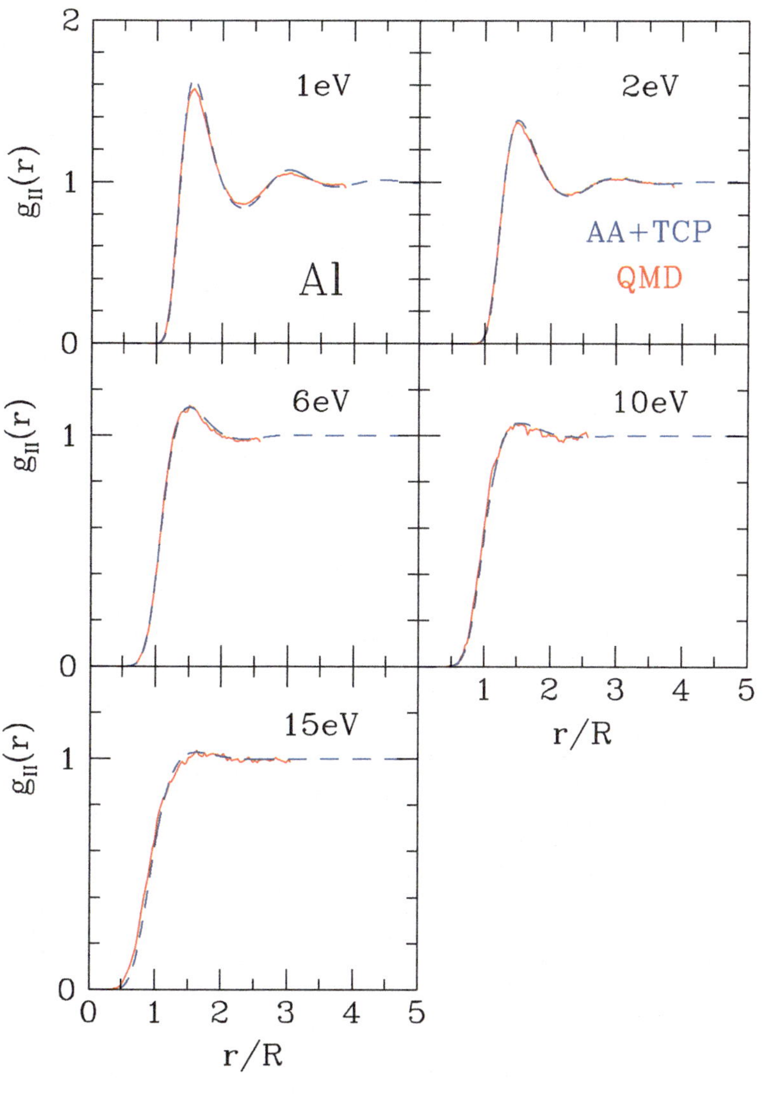

Fig. 8 Pair distribution function of aluminum at solid density (2.7 g/cm^3) and five temperatures ranging from 1 to 15 eV. The *red curves* are quantum molecular dynamics (QMD) simulations [61]. The results of the AA+TCP model are shown in *blue*. *R* is the ion sphere radius (Eq. 1) which is 2.99 a.u. in all cases shown. Note that the last three panels are on a different vertical scale

time come at the cost of more approximate physics. In particular, the model only considers pair interactions and is restricted to spherical symmetry. It cannot account for chemical bonding or angular forces. Its range of validity has been demonstrated to be quite broad in terms of T, ρ and Z, which shows promise for its usefulness to many potential applications, but its limits have not yet been established. It is

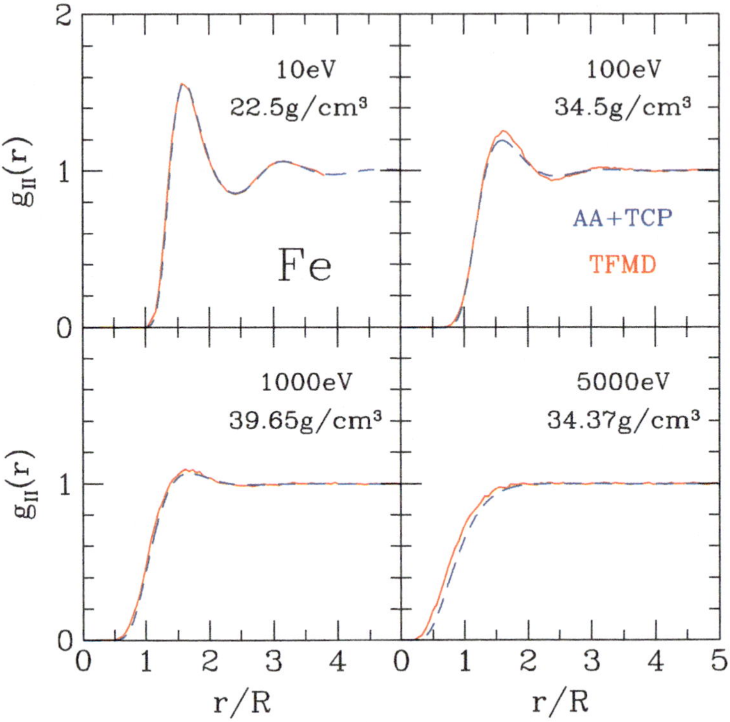

Fig. 9 Pair distribution function of iron for four (T, ρ) points along the principal Hugoniot. The *red curves* are Thomas-Fermi molecular dynamics (TFMD) simulations. The 10 eV calculation includes electron exchange (J.D. Kress, L.A. Collins, Private communication (2012)) while the other three (100–5,000 eV) do not [22]. The AA+TCP calculations (*blue*) were done with the Thomas-Fermi model of the electrons and all include exchange. R is the ion sphere radius (Eq. 1)

worth pointing out that the AA model, like QMD, is based on the finite temperature DFT formalism, and thus both methods share the limitations associated with this approach (such as the well-known underestimation of the electronic band gap, approximate energies for the bound states, the fictitious nature of the unoccupied states, etc).

Perhaps of more concern are the elements of the model that readily admit other possible choices (green boxes in Fig. 2). This introduces a level of arbitrariness in the model whose effect on the solution is not negligible but has not been quantified. Our choices have been guided by earlier work, our understanding of the AA and TCP models, and the pragmatic need for a stable numerical solution of the system of equations. Nonetheless, the choices adopted are reasonable and their validity can be established by assessing the accuracy of the results.

4 Results: Pair Distribution Functions

A converged solution of the system of equations that define the model consists of the electronic wave functions and energies of bound states, the continuum wave functions including resonances, the ion-ion and nucleus-electron effective potentials, the average ion charge, and all the correlation and distribution functions that describe the structure of the two component ion-electron fluid. Of all these quantities, the ion-ion pair distribution function $g_{II}(r)$ is the most readily amenable to a comparison with experimental data and computer simulations. The experimental measurement of the structure of WDM is in its infancy and results are very sparse [5, 6] but a number of ab initio simulations are available. All the calculations shown in this section were performed in the HNC approximation where $B_{II}(r) = 0$. This approximation is expected to become gradually worse as the ion-ion coupling increases such as at lower temperatures or higher ionic densities.

The pair distribution function of aluminum at solid density ($2.7\,\mathrm{g/cm^3}$) and temperatures from 1 to 15 eV is compared to the results of QMD simulations [61] in Fig. 8. The agreement is essentially perfect at $T = 2\,\mathrm{eV}$ and above and remains very good at 1 eV. We emphasize that this excellent agreement is not the result of a fit as there is no adjustable parameter in the AA+TCP model.

For metals around solid density, QMD simulations become computationally impractical when the temperature is comparable to or larger than the Fermi temperature, which typically puts an upper bound of ~10–20 eV on the applicability of the method. However, the method can be extended to higher T by using the semi-classical Thomas-Fermi model for the kinetic energy functional of the electrons rather than a quantum model. Such Thomas-Fermi Molecular Dynamics (TFMD) calculations have been performed along the principal Hugoniot of iron up to 5,000 eV [22] over compression ratios ρ/ρ_0 of 2.9–5.0. For this comparison (Fig. 9) we ran the AA+TCP model with the Thomas-Fermi model of the electrons. At $T = 10\,\mathrm{eV}$ and $22.5\,\mathrm{g/cm^3}$, both calculations include electron exchange and agree perfectly. The agreement remains excellent at the three higher Hugoniot points. For the latter, the TFMD calculations do not include electron exchange; a contribution that should diminish rapidly as T increases. The small shift in the highest temperature point (5,000 eV) for $r/R \lesssim 1$ is somewhat puzzling given the lower ion-ion coupling and the excellent agreement at 1,000 eV. On the other hand, a calculation with the screened one-component plasma model (SOCP [60]), which uses linear response theory to describe the electron screening, is in perfect agreement with our $g_{II}(r)$ at $T = 5{,}000\,\mathrm{eV}$ (not shown, as the curves are indistinguishable on the scale of the figure). This suggests that the departure with the TFMD simulation at this very high temperature may be revealing some statistical inaccuracy in the latter, as suggested by the growing level of noise in the simulations as T increases. The same calculations with the *quantum* AA+TCP give $g_{II}(r)$ that are identical to the Thomas-Fermi results [30], with only small differences appearing at 10 eV. Thus the quantum version of the model recovers the Thomas-Fermi limit at high T and high ρ.

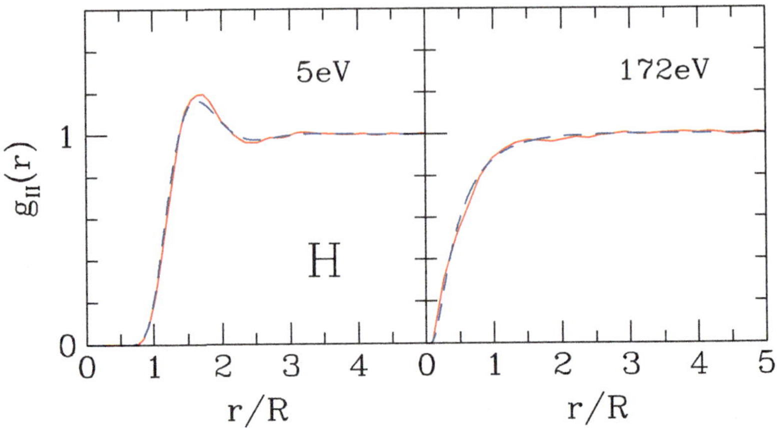

Fig. 10 Pair distribution function of hydrogen at 80 g/cm^3 and $T = 5$ and 172 eV. The *red curves* are quantum molecular dynamics (QMD) simulations [63]. The results of the AA+TCP model are shown in *blue*. $R = 0.323$ a.u. is the ion sphere radius (Eq. 1)

Finally, the $g_{II}(r)$ of very dense hydrogen (80 g/cm^3, or about 800 times the solid density) is found to agree perfectly (within the scatter) with QMD simulations [63] at $T = 172$ eV and to match very well the QMD result at 5 eV (Fig. 10). Again, a calculation with the linear response SOCP model gives an identical $g_{II}(r)$ as with the AA+TCP model (not shown). This demonstrates that in the limit of weak electron-ion coupling (but moderate to strong coupling and electron degeneracy), the quantum AA+TCP model recovers the proper limit of linear-response screening.

While the model must be tested more extensively, several important points can already be made:

1. The numerical solution of the model's equations has been successfully implemented and convergence achieved for a wide range of warm and hot dense matter conditions ($Z = 1 - 26$, $T = 1 - 5,000$ eV, $\rho/\rho_0 = 1 - 800$).
2. Excellent agreement is found in comparisons of the pair distribution function with QMD/TFMD simulations.
3. The AA+TCP model with quantum electrons recovers the results of the AA+TCP with the Thomas-Fermi model of the electrons in the high-T, high-ρ regime.
4. In the limit of weak ion-electron coupling, the quantum AA+TCP model recovers the SOCP model.
5. A Thomas-Fermi AA model with ion correlations can be successfully defined and implemented.

5 Summary and Outlook

We have presented a model of warm dense matter that extends the popular average atom (AA) models by introducing the correlations in the surrounding plasma, which replaces a "one ion" description of WDM by one for the whole plasma. The plasma is a mixture of classical ions and quantum electrons described with the integral equations theory of interacting fluids, which we call a two-component plasma model (TCP). The AA and TCP models are coupled self-consistently, each providing quantities necessary to close the other. The electronic structure surrounding a nucleus takes into account Coulomb interactions with the surrounding plasma, whose ions and the response of the electron fluid (screening) are in turn described by the solution for the central ion. The resulting AA+TCP model has no free parameters and only requires the nuclear charge Z, the temperature T, and the ion density n_I^0 as inputs. The electrons can be treated semi-classically (Thomas-Fermi) or quantum mechanically (Schrödinger or Dirac equation). The form adopted for the coupling between the two models is original and leads to the recognition that the ion charge in the AA model is different from that in the TCP model. The latter is the physically relevant charge related to observables. Earlier AA models with plasma correlations did not recognize this point and were overly constrained.

The introduction of plasma correlations increases considerably the mathematical and numerical complexity of the model, which is mitigated by imposing spherical symmetry on the problem. The model was developed with an emphasis on formal development, internal consistency, and well-defined approximations, which we think, has been achieved to the extent that is possible when combining an AA model with the integral equations theory of fluids. On the other hand, such a coupled model has intrinsic ambiguities that require some ad hoc choices, such as in the criterion for the ideal chemical potential of the electrons and the definition of which electrons are to be counted as part of an ion. These are the less satisfying features of the model. In principle, some of these choices must be better than others, something that can be established by comparison with other theories that are free of such ambiguities (e.g. computer simulations, expansions around a non-interacting plasma state) or with experiments.

So far, the model has been validated by comparing the ion-ion pair distribution function of a wide range of warm and hot dense matter systems with those calculated with ab initio simulations. The agreement is uniformly excellent. For cases where the linear response describes the electron fluid accurately, comparisons with the screened one-component plasma model show perfect agreement. In those cases, deviations with the TFMD ab initio simulations shown suggest that the latter may not be very accurate. The AA+TCP with a Thomas-Fermi model of the electrons is a viable model of hot dense matter. Finally, the AA+TCP model with quantum mechanical electrons recovers the AA+TCP model with the Thomas-Fermi model of the electrons at high densities and temperatures. The goal of developing a computationally efficient model has been achieved: Each of the results shown in Figs. 8–10 runs in about one hour on a single processor workstation.

The realism of the AA+TCP model can be improved in several ways without modifying its structure (Fig. 2) or changing the key assumptions. Foremost is introducing an ion-ion bridge function which will extend the accuracy of the model to low temperatures where the coupling is very strong (e.g. liquid metals). Other possible refinements include the core overlap interaction in the direct ion-ion potential, relativity (Dirac equation), and a more sophisticated exchange and correlation potential, such as a generalized gradient approximation functional.

Computer simulation methods combine theory and sophisticated algorithms to calculate the microscopic properties of dense plasmas. The AA+TCP is a different approach that provides nearly all of the same microscopic properties. Both methods can be thought of as "central engines" around which the calculation of many macroscopic properties of WDM can be built. The AA+TCP model can thus form the basis for the calculation of nearly all the quantities typically obtained with computer simulations such as thermodynamics, conductivity, opacities, diffusion coefficients, and viscosity. It can be applied to the analysis of X-Ray Thomson scattering (XRTS) experiments as well as X-ray absorption near-edge spectroscopy (XANES) experiments. Furthermore, the model can be readily expanded to treat mixtures of ions without any additional approximation or assumption. Here the AA+TCP model offers a distinct advantage over simulations as it can model highly asymmetric and very dilute mixtures.

This new average atom model with plasma correlations has so far shown a very satisfactory degree of physical realism. It is a significant step beyond the more common "atom in a cell" models of warm dense matter. In view of its relatively modest computational cost and its many potential applications, it is a promising approach to produce extensive tabulations of warm dense matter properties.

Acknowledgements We gratefully acknowledge V. Recoules, F. Lambert, J. D. Kress and L. Collins for providing pair distribution functions from their ab initio simulations. This work was performed under the auspices of the United States Department of Energy under contract DE-AC52-06NA25396.

References

1. Report of ReNew workshop, Basic research needs for high energy density laboratory physics (U.S. Department of Energy, 2009). http://science.energy.gov/~/media/fes/pdf/workshop-reports/Hedlp_brn_workshop_report_oct_2010.pdf
2. A.W. DeSilva, J.D. Katsouros, Phys. Rev. E **57**, 5945 (1998)
3. A. Mančić et al., Phys. Rev. Lett. **104**, 035002 (2010)
4. J. Eggert, S. Brygoo, P. Loubeyre, R.S. McWilliams, P.M. Celliers, D.G. Hicks, T.R. Boehly, R. Jeanloz, G.W. Collins, Phys. Rev. Lett. **100**, 124503 (2008)
5. E.G. Saiz et al., Nat. Phys. **4**, 940 (2008)
6. A.L. Kritcher et al., Science **322**, 69 (2008)
7. O. Ciricosta et al., Phys. Rev. Lett. **109**, 065002 (2012)
8. A. Benuzzi-Mounaix, F. Dorchies, V. Recoules, F. Festa, O. Peyrusse, A. Lévy, A. Ravasio, T. Hall, M. Koenig, N. Amadou, E. Brambrink, S. Mazevet, Phys. Rev. Lett. **107**, 165006 (2011)

9. F. Dorchies, A. Lévy, C. Goyon, P. Combis, D. Descamps, C. Fourment, M. Harmand, S. Hulin, P.M. Leguay, S. Petit, O. Peyrusse, J.J. Santos, Phys. Rev. Lett. **107**, 245006 (2011)
10. T. Ma, T. Döppner, R.W. Falcone, L. Fletcher, C. Fortmann, D.O. Gericke, O.L. Landen, H.J. Lee, A. Pak, J. Vorberger, K. Wünsch, S.H. Glenzer, Phys. Rev. Lett. **110**, 065001 (2013)
11. T. Guillot, D.J. Stevenson, W.B. Hubbard, D. Saumon, in *Jupiter – The Planet, Satellites and Magnetosphere* (Chap. 3) F. Bagenal, T. Dowling and W. McKinnon, Eds. (Cambridge University Press, New York, 2004)
12. R. Redmer, T.R. Mattsson, N. Nettelmann, M. French, Icarus **211**, 798 (2011)
13. G. Fontaine, H.M.V. Horn, Astrophys. J. Suppl. **31**, 467 (1976)
14. J.J. Fortney, S.H. Glenzer, M. Koenig, B. Militzer, D. Saumon, D. Valencia, Phys. Plasmas **16**, 1003 (2009)
15. D.A. Liberman, J. Quant. Spectrosc. Ratiat. Transf. **27**, 335 (1982)
16. B. Wilson, V. Sonnad, P. Sterne, W. Isaacs, J. Quant. Spectrosc. Ratiat. Transf. **99**, 658 (2006)
17. P. Sterne, S. Hansen, B. Wilson, W. Isaacs, High Energy Density Phys. **3**, 278 (2007)
18. R. Piron, T. Blenski, Phys. Rev. E **83**, 026403 (2011)
19. S. Mazevet, M.P. Desjarlais, L.A. Collins, J.D. Kress, N.H. Magee, Phys. Rev. E **71**, 016409 (2005)
20. M.P. Desjarlais, J.D. Kress, L.A. Collins, Phys. Rev. E **66**, 025401(R) (2002)
21. G. Zérah, J. Clérouin, E.L. Pollock, Phys. Rev. Lett. **69**, 446 (1992)
22. F. Lambert, J. Clerouin, G. Zerah, Phys. Rev. E **73**, 016403 (2006)
23. D.M. Ceperley, Rev. Mod. Phys. **67**, 279 (1995)
24. K.P. Driver, B. Militzer, Phys. Rev. Lett. **108**, 115502 (2012)
25. B. Holst, R. Redmer, Phys. Rev. B **77**, 184201 (2008)
26. H. Xu, J.P. Hansen, Phys. Rev. E **57**, 211 (1998)
27. J. Chihara, J. Phys. Condens. Matter **3**, 8715 (1991)
28. F. Perrot, Phys. Rev. E **47**, 570 (1993)
29. D. Ofer, E. Nardi, Y. Rosenfeld, Phys. Rev. A **38**, 5801 (1988)
30. C.E. Starrett, D. Saumon, Phys. Rev. E **87**, 013104 (2013)
31. C.E. Starrett, D. Saumon, Phys. Rev. E **88**, 059901(E) (2013)
32. W. Kohn, L.J. Sham, Phys. Rev. **140**, A1133 (1965)
33. N.D. Mermin, Phys. Rev. **137**, A1441 (1965)
34. R. Parr, W. Yang, *Density Functional Theory of Atoms and Molecules* (Oxford University Press, New York, 1989)
35. L.H. Thomas, Proc. Camb. Philos. Soc. **23**, 542 (1927)
36. E. Fermi, Z. Phys. **48**, 73 (1928)
37. R.P. Feynman, N. Metroplis, E. Teller, Phys. Rev. **75**, 1561 (1949)
38. T. Blenski, B. Cichocki, Phys. Rev. E **75**, 056402 (2007)
39. D.A. Liberman, Phys. Rev. B **20**, 4981 (1979)
40. T. Blenski, K. Ishikawa, Phys. Rev. E **51**, 4869 (1995)
41. E. Wigner, F. Seitz, Phys. Rev. **43**, 804 (1933)
42. B.J.B. Crowley, J.W. Harris, J. Quant. Spectros. Ratiat. Transf. **71**, 257 (2001)
43. B.F. Rozsnyai, Phys. Rev. A **5**, 1137 (1972)
44. G. Faussurier, C. Blancard, P. Cossé, P. Renaudin, Phys. Plasma **17**, 052707 (2010)
45. J.P. Hansen, I. McDonald, *Theory of Simple Liquids*, 3rd edn. (Academic, London/Burlington, 2006)
46. C. Caccamo, Phys. Rep. **274**, 1 (1996)
47. J.A. Anta, A.A. Louis, Phys. Rev. B **61**, 11400 (2000)
48. O. Peyrusse, J. Phys. Condens. Matter **20**, 195211 (2008)
49. C.E. Starrett, D. Saumon, Phys. Rev. E **85**, 026403 (2012)
50. J. Chihara, J. Phys. C Solid State Phys. **18**, 3103 (1985)
51. Y. Rosenfeld, N.W. Ashcroft, Phys. Rev. A **20**, 1209 (1979)
52. A. Malijevský, S. Labík, Mol. Phys. **60**, 663 (1987)
53. H. Iyetomi, S. Ogata, S. Ichimaru, Phys. Rev. A **46**, 1051 (1992)
54. W. Daughton, M. Murillo, L. Thode, Phys. Rev. E **61**, 2129 (2000)

55. Y. Rosenfeld, J. Stat. Phys. **42**, 437 (1986)
56. J.M. Ziman, Proc. Phys. Soc. **91**, 701 (1967)
57. L. Dagens, J. Phys. C **5**, 2333 (1972)
58. N.W. Ashcroft, D. Stroud, Solid State Phys. **33**, 1 (1978)
59. S. Ichimaru, Rev. Mod. Phys. **54**, 1017 (1982)
60. G. Chabrier, J. Phys. Fr. **51**, 1607 (1990)
61. D. Saumon, C.E. Starrett, J.D. Kress, J. Clerouin, High Energy Density Phys. **8**, 150 (2012)
62. C.E. Starrett, D. Saumon, High Energy Density Phys. **10**, 35 (2014)
63. V. Recoules, F. Lambert, A. Decoster, B. Canaud, J. Clérouin, Phys. Rev. Lett. **102**, 075002 (2009)

Dynamical Structure Factor in High Energy Density Plasmas and Application to X-Ray Thomson Scattering

Carsten Fortmann

Abstract The Dynamical Structure Factor (DSF) is a key observable in high energy density science that is directly measured e.g. using inelastic x-ray scattering techniques. The DSF delivers a wealth of information about macroscopic equation of state data (temperature of electrons and of ion species, average ionization states, electron density), collective and single-particle dynamics, static correlations, electron-ion collisions, and electronic structure. In a strongly coupled, partly degenerate many-body system, such as warm dense matter generated by intense laser-matter interaction and shock wave compression, calculations of the DSF are a formidable task.

In this talk, I present the state of the art theoretical framework for DSF calculations in the context of x-ray scattering plasma diagnostics The theoretical foundation is given by the many-body quantum field theory and linear response theory.

The DSF or two-particle correlation function is expressed via many-body Green functions and calculated by aid of Feynman diagrams and partial summations to achieve a self-consistent treatment of field propagators and self-energies.

The resulting expressions are identified with the components of the Chihara separation of the DSF into free-free, bound-bound, and bound-free processes and suitable approximations for each process can be given.

The predictive and analytical strength of this approach will be demonstrated using recently obtained experimental data which led to novel insights in high energy density science.

C. Fortmann (✉)
QuantumWise A/S, Lersø Parkalle 107, 2100 Copenhagen, Denmark
e-mail: carsten.fortmann@quantumwise.com

F. Graziani et al. (eds.), *Frontiers and Challenges in Warm Dense Matter*, Lecture Notes in Computational Science and Engineering 96, DOI 10.1007/978-3-319-04912-0_7,

1 Introduction

How can we measure the macroscopic and microscopic physical state variables in a system as complex as high energy density matter? A widely used approach to HEDM diagnostics is to observe the reaction of the system to external perturbations and to scan the variations in the observed quantities as function of variation in the external parameters. These perturbations can be of various type, such as electromagnetic or static fields, particle currents, or pressure. In all cases this approach requires precise ab-initio knowledge about the response of the target, in order to connect the measured response coefficients (e.g. the ratio of applied voltage to measured current, or the transmission or reflectivity at a certain wavelength) to the internal state of the system under investigation. For example, only with a-priori knowledge about the temperature dependence of the electrical conductivity can we measure the temperature through a conductivity measurement. We have to rely on independent experimental data, and, in a complementary sense, on theoretical predictions.

In the present chapter we will focus on one particular method for diagnostics of high energy density matter, namely x-ray Thomson scattering (XRTS) [1]. Within the last decade [2], this technique has become a major player among various x-ray based diagnostic methods and the limits of its applicability are still to be explored. While optical Thomson scattering had been used since long for dilute plasmas, see e.g. Ref. [3], x-rays are required as a scattering probe to investigate solid density plasmas or even plasmas above solid density, such as shock-compressed matter. Only x-rays are capable to penetrate deeply into the target due to their relatively short wavelength. A rough estimate of the wavelength required to penetrate a target can be given by demanding that the x-ray wavelength be shorter than the target's critical wavelength $\lambda_{\text{crit}} = 2\pi c\sqrt{\epsilon_0 m_e/n_e e^2}$ which is a function of the free electron density n_e. Above λ_{crit}, the radiation is reflected at the target surface. To steer clear from resonant absorption near the plasma frequency $\omega_{\text{pl}} = 2\pi c/\lambda_{\text{crit}} = \sqrt{n_\text{e}e^2/\epsilon_0 m_\text{e}}$, one often uses the more conservative criterion $\lambda_{\text{source}} \leq 0.01\lambda_{\text{crit}}$. Based on this constraint, Fig. 1 presents a selection of commonly used x-ray sources and their wavelengths and shows the maximum electron density that can be penetrated. Hard x-ray wavelengths of the order of 1 nm are required to diagnose matter at or above typical solid matter electron densities, i.e. $n_\text{e} \gtrsim 10^{23}\,\text{cm}^{-3}$. Optical wavelengths and their lowest order harmonics are clearly too long. The shown plasma based x-ray sources are short enough, some even to penetrate matter up to $10^{25}\,\text{cm}^{-3}$, as well as x-ray Free Electron Lasers (XFEL) such as the LCLS at Stanford [4] and the future European XFEL [5] in Hamburg. Photons delivered by the FLASH facility [6] penetrate sub-solid density matter of the order $10^{22}\,\text{cm}^{-3}$. This analysis furthermore demonstrates the importance of having a precise and general theoretical understanding of the physics of high energy density matter and its interaction with electromagnetic radiation if one wants to apply x-ray based techniques for plasma diagnostics. It is the objective of this chapter to present the theoretical foundations of x-ray based diagnostic techniques for HEDM research

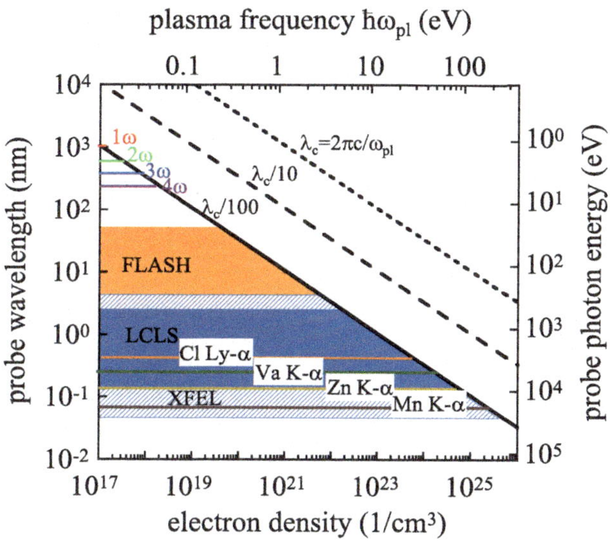

Fig. 1 Critical wavelength (photon energy) as function of the electron density (plasma frequency) for typical electron densities in the high energy density regime. Also shown are typical optical, VUV, and x-ray wavelengths

in general and for XRTS in particular. Starting from a general theory of light and matter, namely Quantum Electrodynamics, we will derive the dynamical structure factor, which directly gives the differential scattering cross section for XRTS. We will discuss how to include non-ideality contributions within this approach, such as collisions and local field effects. An application to experimental data will conclude the chapter and demonstrate the predictive power and analytical capabilities of the theory.

Spectrally resolved XRTS has been applied to many different areas in HEDM research, such as electron density and temperature measurements in solid density plasmas [7], planar, coalescing, and colliding shocks [8–10], strong electron correlations [11], structure resolution [12], and fusion capsule implosions [13]. A rather new domain is XRTS using ultrashort x-ray free electron laser radiation with the potential to resolve non-equilibrium processes in high energy density matter [14]. XRTS provides very precise measurements for electron density and ionization states in addition to the electron and ion temperatures. Furthermore, x-ray scattering infers the electric conductivity, as well as the ionic and electronic structure. One distinguishes collective scattering, sensitive to long range correlations and collective behavior, and non-collective scattering, sensitive to single-particle properties, i.e. the electronic momentum distribution function, ionization potentials, and ionic charge distribution. Collective excitations in the correlated plasma, such as the electron plasma waves (plasmons) and ion acoustic waves are of special interest. For example the plasmon position, amplitude, and spectral width yield thermodynamical and transport properties provided a consistent theoretical framework exists to interpret the experimental data.

2 Theory

2.1 Many-Body QED

The basis for the theoretical treatment of correlated systems of charged particles with electromagnetic fields is provided by combining Many-Body Theory with Quantum Electrodynamics [15]. Whereas QED predicts the elementary interaction processes of electric charges with each other and with the gauge field, many-body theory is successful in describing the macroscopic properties of systems containing a large number of individual particles. Correlations among the systems constituents, their reaction and response to radiation is formulated within a single framework based on the QED Lagrangian

$$\mathscr{L}(x) = \sum_c \bar{\psi}_c(x)\left(i\hbar c\gamma^\mu\partial_\mu - m_c c^2\right)\psi_c(x) - \frac{\epsilon_0 c^2}{4}F_{\mu\nu}(x)F^{\mu\nu}(x) + \sum_c Z_c e j_{\mu c}(x)A^\mu(x)\,, \quad (1)$$

with $\psi_c(x)$ the Dirac spinor for fermions of species c, $F^{\mu\nu}$ is the field tensor of the electromagnetic field $A(x)$, $F^{\mu\nu}(x) = \partial^\mu A^\nu(x) - \partial^\nu A^\mu(x)$, and $j_c(x)$ the charged current; $x = (ct, \mathbf{x})$ is the space-time four vector. The first term in Eq. (1) describes the unperturbed dynamics of free Fermions, we recognize the structure of the Dirac equation. The second term governs the evolution of the gauge field, i.e. the source-free Maxwell equations. Finally, the last term contains the interaction between charged particles, described by the charged 4-current $j_c(x)$ with the electromagnetic field $A(x)$. This term contains e.g. the electrostatic interaction via Coulomb forces and radiative effects like bremsstrahlung and Compton scattering.

The Lagrangian density (1) is gauge invariant and Lorentz-invariant. However, to obtain the actual equations of motion for the fields, one has to perform a Legendre transform with respect to the dynamical variables $\psi_c(x)$ and $A(x)$. In this way, we obtain the Hamiltonian density, which we can quantize by imposing (anti)commutation rules on the creation and annihilation operators of the fields. Hence, the hamiltonian formulation explicitly breaks the Lorentz invariance, because commutation rules can only be defined in a specific coordinate system. Within the non-relativistic treatment, that we are going to pursue in the following, this is not a problem. For relativistic treatments of the many-body problem we refer to the excellent monograph by Lindgren [16], which contains numerous further references.

2.2 Hamiltonian and Green Functions

In the non-relativistic limit, the quantization method that was outlined above leads to the well-known Hamiltonian for a system of charged particles exposed to an external electromagnetic field,

$$
\begin{aligned}
H = \Omega_0 \sum_c \int \frac{d^3p}{(2\pi)^3} \varepsilon_{\mathbf{p},c} \mathrm{a}_c^\dagger(\mathbf{p}) \mathrm{a}_c(\mathbf{p}) + \Omega_0 \sum_{\lambda=1}^{2} \hbar\omega_{\mathbf{k}} \mathrm{b}_\lambda^\dagger(\mathbf{k}) \mathrm{b}_\lambda(-\mathbf{k}) \\
- \Omega_0 \sum_{\lambda=1}^{2} \hat{\mathbf{j}}(\mathbf{k}) \cdot \boldsymbol{\varepsilon}_{\mathbf{k}\lambda} \left(\frac{\hbar\Omega_0}{2\epsilon_0\omega_{\mathbf{k}}} \right)^{1/2} \left(\mathrm{b}_\lambda^\dagger(-\mathbf{k}) + \mathrm{b}_\lambda(\mathbf{k}) \right) \\
+ \frac{1}{2}\Omega_0^3 \sum_{cc'} \int \frac{d^3p\, d^3p'\, d^3k}{(2\pi)^9} V_{cc'}(\mathbf{k}) \mathrm{a}_c^\dagger(\mathbf{p}-\mathbf{k}) \mathrm{a}_{c'}^\dagger(\mathbf{p}'+\mathbf{k}) \mathrm{a}_{c'}(\mathbf{p}') \mathrm{a}_c(\mathbf{p}) \,. \quad (2)
\end{aligned}
$$

$a_{\mathbf{p},c}$ (b_λ) and $a_{\mathbf{p},c}^\dagger$ (b_λ) are annihilation and creation operators for Fermions (Bosons), respectively. The non-relativistic single-particle energy $\varepsilon_{\mathbf{p},c} = \hbar^2\mathbf{p}^2/2m_c$ was introduced. λ is the photon polarization, Ω_0 is a normalization volume.

The state of the many-body system is uniquely defined through the density operator ρ, i.e. quantum-statistical ensemble averages are calculated as a trace over the operator representing the observable multiplied by the density operator,

$$
\langle O \rangle = \mathrm{Tr}\{\rho O\} \,. \quad (3)
$$

The density operator obeys the Liouville-von Neumann equation

$$
-i\hbar\dot{\rho}(t) + [H, \rho(t)] = 0. \quad (4)
$$

In thermal equilibrium, which we will treat from now on,[1] the density operator is time independent. The Liouville-von Neumann equation is then formally solved as

$$
\rho_{\mathrm{eq}} = \rho = \frac{1}{Z} \mathrm{e}^{-\beta\mathscr{H}} \,, \quad Z = -\frac{1}{\beta} \mathrm{Tr}\{\mathrm{e}^{-\beta\mathscr{H}}\} \,, \quad (5)
$$

with $\mathscr{H} = H - \sum_{c,\mathbf{p}} \mu_c a_{\mathbf{p},c}^\dagger a_{\mathbf{p},c}$ and μ_c the chemical potential for species c.[2] For an interacting quantum system, the operator $\exp(-\beta\mathscr{H}) = \exp(-\beta(\mathscr{H}_0 + H'))$ has to be evaluated, which is a formidable task due to the non-vanishing

[1] Indeed, we constrain ourselves to the regime of linear response, i.e. the response functions are calculated as a perturbation expansion of equilibrium correlation functions.

[2] We treat the grand canonical ensemble.

commutator $[H, H']$. This expression is the starting point for the Quantum MonteCarlo method [17]. Here, we will use a different approach based on a systematic perturbation expansion of the Hamiltonian and resummation of infinite perturbation series.

The perturbation series is formulated in terms of thermodynamic Green functions and self-energies. Instead of presenting a formal derivation here, we introduce the single particle Green function through the following consideration and analogies: The single particle momentum distribution for a system of non-interacting Fermions in thermal equilibrium is given by the Fermi distribution

$$n_{\mathrm{F}}(p) = \left[\exp((\epsilon_p - \mu)/k_{\mathrm{B}}T) + 1\right]^{-1} . \tag{6}$$

The Fermi distribution can be expanded in a Fourier series over imaginary frequencies,

$$n_{\mathrm{F}}(p) = \frac{1}{2} - \sum_{\nu} \frac{1}{z_\nu - \varepsilon_p + \mu} , \quad z_\nu = i\pi \frac{2\nu + 1}{\hbar k_{\mathrm{B}} T} , \nu = 0, \pm 1, \pm 2, \ldots \tag{7}$$

The term under the sum,

$$G_c^{(0)}(\mathbf{p}, z_\nu) = \frac{1}{z_\nu - \varepsilon_p + \mu}, \tag{8}$$

has precisely the structure of a free Green function as known from ordinary quantum mechanics [18], $G^0(E) = [H - E + i\eta]^{-1}, \eta \to 0^+$, where H is the Hamiltonian and E is an eigenvalue. By analogy, we therefore introduce the thermodynamic or quantum-statistical Green function

$$G_c(\mathbf{p}, z_\nu) = \frac{1}{z_\nu - \varepsilon_{p,c} + \mu_c - \Sigma_c(\mathbf{p}, z_\nu)} . \tag{9}$$

$\Sigma_c(\mathbf{p}, z_\nu)$ is the self-energy that describes many-body correlations not contained in the free Green function (8). The self-energy itself obeys another Dyson equation, which, besides the Green function also involves the Green function for the Maxwell field, as well as the effective coupling between particles and fields, expressed through the vertex function. The photon Green function and its respective self-energy, the polarization function, have their own Dyson equation as well. Summarizing, the full particle and field dynamics can be cast into a formally closed set of non-linear integro-differential equations, the Hedin equations [19]. In relativistic quantum field theory, these equations are known as Dyson-Schwinger equations [20, 21], see also [22, 23]. They are conveniently given in diagrammatical form:

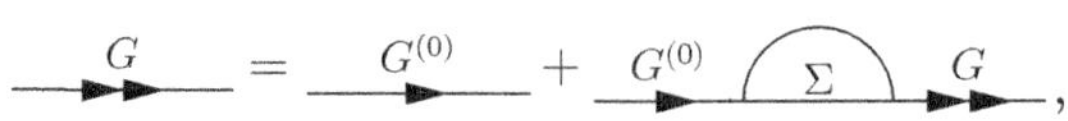

(10)

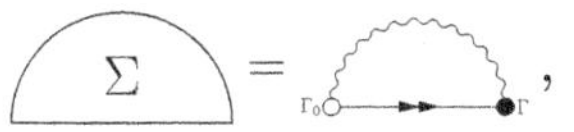

(11)

(12)

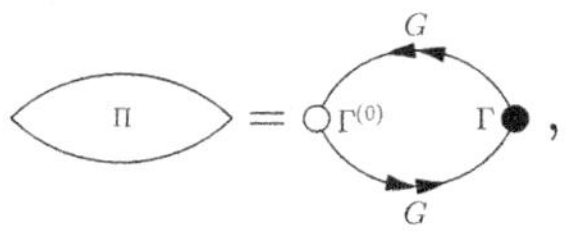

(13)

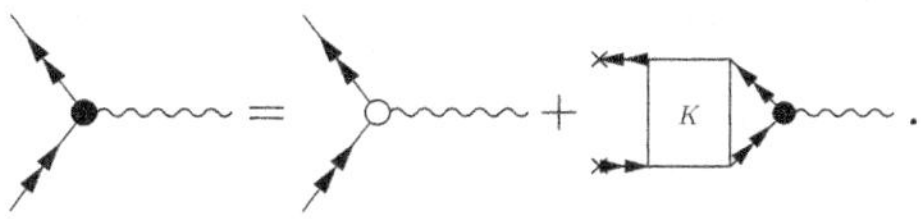

(14)

The last equation is the vertex equation. The vertex is represented by a solid circle while it's non-interacting counterpart, $\Gamma^{(0)} = Ze$ is represented through an empty circle. The practical importance of the Hedin equations lies in the Fredholm type structure with the term on the left hand side of the equation (the "full" propagator or self-energy) also appearing on the right hand side. Such equations can typically be solved iteratively. In practice, one does not operate on the discrete Matsubara frequencies z_ν, ω_μ (although this is possible, see e.g. [24]) but rather performs an analytical continuation $z_\nu \to \omega + i\eta, \eta \to 0$ to the real frequency axis. For example instead of the Green function, one iterates the single-particle spectral function

$$A_c(\mathbf{p}, \omega) = -2\mathrm{Im}\, G(\mathbf{p}, \omega + i0^+) \tag{15}$$

$$= \frac{-2\mathrm{Im}\, \Sigma_c(\mathbf{p}, \omega + i0^+)}{\left[\omega - \varepsilon_p^c - \mathrm{Re}\, \Sigma_c(\mathbf{p}, \omega + i0^+)\right]^2 + \left[\mathrm{Im}\, \Sigma_c(\mathbf{p}, \omega + i0^+)\right]^2}. \tag{16}$$

A schematic of the single-particle spectral function is illustrated in Fig. 2. In the limiting case of non-interacting particles, the self-energy vanishes $\Sigma \equiv 0$ and the spectral function becomes an on-shell delta distribution (black vertical line), located at the free-particle dispersion, $A^{(0)}(\mathbf{p}, \omega) = 2\pi\, \delta(\hbar\tilde{\omega} - \varepsilon_{\mathbf{p}}), \tilde{\omega} = \omega + \mu/\hbar$. The first non-trivial approximation to the self-energy is a constant real self-energy (Hartree-Fock approximation), $\Sigma^{\mathrm{HF}} = \Sigma(p)$. This describes a mean-field shift of the particle energy away from the free particle dispersion, $A^{\mathrm{HF}}(\mathbf{p}, \omega) = 2\pi\, \delta(\hbar\tilde{\omega} - E^{\mathrm{HF}}(p)), E^{\mathrm{HF}}(p) = \varepsilon_{\mathbf{p}} + \Sigma(p)$. The Hartree-Fock spectral function is again a delta distribution describing quasi-particle states with infinite life time. Approximations beyond the Hartree-Fock mean field theory take into account correlations. This leads to non-diagonal matrix elements in the Hamiltonian in the free particle basis, i.e. to a finite life time of the free particle states. The spectral function becomes a distribution with a finite width. The red curve in Fig. 2 represents such a finite-width spectral function, but the real spectral function might also exhibit more complex structures e.g. satellite resonances, describing collective excitations of the system.

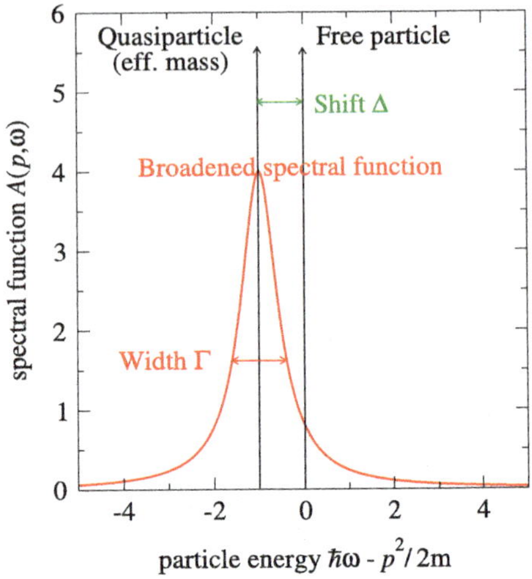

Fig. 2 Schematic illustration of the single-particle spectral function

The Hedin equations are not exactly solvable. Suitable approximations have to be found that reflect the basic physics of the system at hand. The nearest-to-full approximation is the $GW\Gamma$ approximation, where one starts from a suitable functional form for the vertex function Γ, which is kept throughout the calculation. The construction of Γ takes into account the Ward-Takahashi identities [25], a generalized continuity equation which connects the vertex function to the gradient of the self-energy. The remaining equations (10)–(13) are then solved and in each iteration the vertex function is updated [24].

Neglecting vertex corrections beyond the zero-order term $\Gamma^{(0)} = Ze$ defines the GW approximation. The GW self-energy is the convolution of the full Green function G and the screened interaction (full photon propagator) W [26]. One further simplifies the equations and only iterates the particle self-energy and Green function, keeping the screened interaction on the level of the self-consistent one-loop approximation or RPA [27], giving rise to the $GW^{(0)}$ approximation.

Figures 3 and 4 show the $GW^{(0)}$ spectral function for a sample hydrogen plasma at temperature $k_BT = 1\,\text{keV}$ and two different electron densities, $n_e = 7\times10^{21}\,\text{cm}^{-3}$ and $n_e = 1\times10^{26}\,\text{cm}^{-3}$, respectively. The spectral function in the first dilute case is a relatively narrow distribution with maximum near the free-particle dispersion (indicated as black line on the bottom). The width of the spectral function decreases with increasing momentum p, as the particle becomes less sensitive to collisions. In contrast, the second, dense plasma spectral function exhibits a much more complex structure with satellite resonances in the small p region. These satellites describe coherent electron-plasmon scattering events, leading to so-called plasmaron quasi-particles.

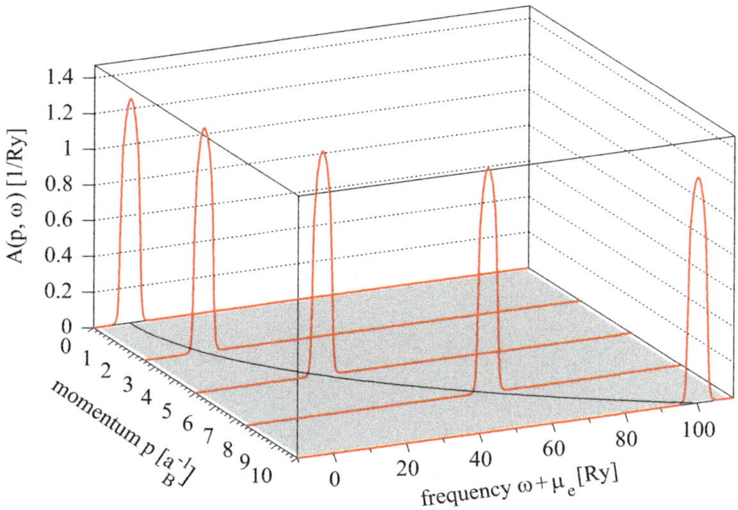

Fig. 3 Single-particle spectral function for a model hydrogen plasma with $n_e = 7 \times 10^{21}\,\mathrm{cm}^{-3}$ and $T = 1\,\mathrm{keV}$

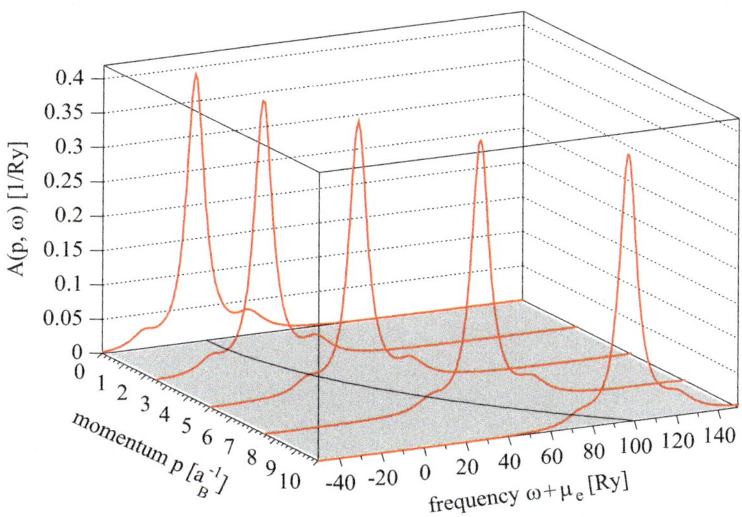

Fig. 4 Single-particle spectral function for a model hydrogen plasma with $n_e = 1 \times 10^{26}\,\mathrm{cm}^{-3}$ and $T = 1\,\mathrm{keV}$

Having calculated the spectral function, we have now access to a wealth of macroscopic plasma observables, such as the equation of state [28],

$$n_a(\mu_a, T) = \sum_\sigma \int_{-\infty}^{\infty} \frac{\mathrm{d}\omega}{2\pi} \int \frac{\mathrm{d}^3 p}{(2\pi)^3} f_a(\omega) A_{a,\sigma}(\mathbf{p}, \omega) \,, \tag{17}$$

the particle dispersion (bandstructure) by solving the dispersion equation

$$E_p = \hbar^2 p^2/2m - \mu + \mathrm{Re}\, \Sigma(\mathbf{p}, E_p), \tag{18}$$

and optical properties via the polarization function [29]

$$\Pi_{\mathrm{e}}(\mathbf{k}, \omega_\mu) = \sum_{z_\nu, \mathbf{p}} \int_{-\infty}^{\infty} \frac{\mathrm{d}\omega}{2\pi} \frac{\mathrm{d}\omega'}{2\pi} \Gamma_0 \frac{A_e(\mathbf{p}+\mathbf{k})}{\omega - z_\nu - \omega_\mu} \frac{A_e(\mathbf{p}, \omega')}{\omega' - z_\nu} \Gamma(\omega_\mu + z_\nu, z_\nu; \mathbf{p}+\mathbf{k}, \mathbf{p}). \tag{19}$$

In particular, the last equation brings us back to our original task, the calculation of x-ray scattering spectra.

2.3 Dynamical Structure Factor

The polarization function (19) describes the response of the system to external fields which couple with the charge density fluctuations, e.g. electrostatic and electromagnetic fields. It can be shown that the dynamical structure factor (DSF), i.e. the correlation function of density fluctuations,

$$S(\mathbf{k}, \omega) = \int_{-\infty}^{\infty} dt\, \exp(i\omega t) \langle \delta\rho^*(\mathbf{k}, t) \delta\rho(\mathbf{k}, 0) \rangle, \tag{20}$$

can be expressed via the polarization function as

$$S(\mathbf{k}, \omega) = -\frac{\hbar}{\pi n_{\mathrm{e}}} \mathrm{Im}\, \Pi^{-1}(\mathbf{k}, \omega) \,, \tag{21}$$

Eq. (21) is called the fluctuation-dissipation theorem. It provides the link between the microscopic theory of particle and field dynamics and the macroscopic response to external fields.

As such, the scattered power P_s is given by the dynamical structure factor

$$P_s(\mathbf{R}, \omega) d\Omega d\omega = \frac{P_i r_0^2 d\Omega}{2\pi A} \left| \hat{\mathbf{k}}_f \times (\hat{\mathbf{k}}_f \times \hat{\mathbf{E}}_{0i}) \right|^2 NS(\mathbf{k}, \omega) d\omega, \tag{22}$$

with incoming power P_i, irradiated area A, frequency and wavevector shift ω and $\mathbf{k}$, respectively, as well as detector solid angle $d\Omega$ and frequency binning $d\omega$

Hence, we are now in the position to calculate the system response, measurable by the scattered power per frequency interval and per solid angle, from the microscopic equations of motion of the many-body system. All required quantities are given, once the single-particle spectral function is computed.

2.3.1 The Chihara Formula

The usual starting point for dynamical structure factor calculations for XRTS applications [30, 31] is a formula first presented by Chihara [32]. Here, we will touch it only briefly, because it will serve for a comparison to a result that we will derive within our Green function approach.

In a partially ionized plasma, it is assumed that the electronic charge density separates into free electrons and bound electrons, $\rho_e(\mathbf{k}, t) = \rho_{\text{free}}(\mathbf{k}, t) + \rho_{\text{bound}}(\mathbf{k}, t)$. Consequently, the density autocorrelation (20) yields three components, free-free, bound-bound, and the bound-free correlations, respectively. Each correlation relates to one interaction channel with the electromagnetic field: free-free absorption and scattering (inverse bremsstrahlung, resonance absorption, and Compton (plasmon) scattering), bound-bound transitions (line emission/absorption and Raman scattering), and bound-free transitions (photo-ionization and Compton scattering). This approach allows to calculate each component of the DSF using a well-chosen approximation for each one. The free-free component is dominated by the dynamics of the freely moving particles, being disturbed by collisions with the heavier particles (ions) and electronic correlations. The electronic structure of ions does not enter this term (at least in lowest order approximations), whereas it is the dominating term in the bound-bound contribution. This "chemical picture"[3] is expressed by the Chihara formula,

$$S(k, \omega) = Z_{\text{f}} S_{\text{ee}}^{(0)}(k, \omega) + |f(k) + q(k)|^2 \, S_{\text{ii}}(k, \omega) + Z_{\text{core}} S_{\text{core}}(k, \omega) \qquad (23)$$

$f(k)$ is the ionic form factor, i.e. the electron density of bound electrons, $q(k)$ is the so-called "screening cloud", i.e. the density distribution of free electrons, $S_{ii}(k, \omega)$ is the ion-ion structure factor, and $S_{ee}^0(k, \omega)$ is the electron one-component structure factor that describes the free electron response, i.e. in the absence of bound states and ions. Finally, $S_{\text{core}}(k, \omega)$ is the bound-free scattering term [31, 33].

2.3.2 The Two-Component Random Phase Approximation

We will now continue in our task to derive usable expressions for the DSF within the Green function formalism. To this end, we choose a rather simple model system

---

[3]As opposed to the "physical picture", where electrons are neither bound nor free but have to be described as a dynamical superposition of bound and scattering states.

using the lowest order non-trivial approximation, the RPA. We assume only two species, free electrons and completely stripped ions, e.g. a completely ionized hydrogen plasma. This defines the two component plasma RPA (TCP-RPA) model.

For the spectral function, we assume a free particle spectral function $A_c(\mathbf{p}, \omega) = 2\pi\delta(\hbar\omega - \varepsilon_{\mathbf{p},c} + \mu)$. Then, the polarization function (19) becomes the free particle polarization[4]

$$\Pi_c^{(0)}(\mathbf{k}, \omega) =$$  $$= \frac{1}{\Omega_0} \sum_{\mathbf{p}} \frac{f^c_{\mathbf{p}+\mathbf{k}/2} - f^c_{\mathbf{p}-\mathbf{k}/2}}{E^c_{\mathbf{p}+\mathbf{k}/2} - E^c_{\mathbf{p}-\mathbf{k}/2} - \hbar\omega} \,. \tag{24}$$

It is convenient to calculate directly $\chi_{cc'} = \Pi_{cc'} / [1 - V_{cc'}\Pi_{cc'}]$, the pair correlation function, instead of the polarization function. From the Hedin equations we obtain for $W_{cc'}$ and $\chi_{cc'}$.

$$\chi_{cc'}(\mathbf{k}, \omega) = \chi_c^{(0)}(\mathbf{k}, \omega)\delta_{cc'} + \sum_{d=e,i} \chi_c^{(0)}(\mathbf{k}, \omega) W_{cd}(\mathbf{k}, \omega) \chi_{dc'}(\mathbf{k}, \omega), \tag{25}$$

$$W_{cc'}(\mathbf{k}, \omega) = V_{cc'}(\mathbf{k}) + \sum_{d=e,i} V_{cd}(\mathbf{k}) \chi_d^{(0)}(\mathbf{k}, \omega) W_{dc'}(\mathbf{k}, \omega). \tag{26}$$

Here, indices c, d indicate particle species, i.e. electrons and ions and $\chi_c^{(0)}(\mathbf{k}, \omega) = \Pi_c^{(0)}$. $V_{cc'}(\mathbf{k})$ is the unscreened two-particle interaction potential. In the considered fully ionized plasma, it is the Coulomb potential

$$V_{cc'}(\mathbf{k}) = \frac{q_c q_{c'}}{\epsilon_0 k^2} \,, \tag{27}$$

but can be replaced by an effective potential to account for core electrons in partially ionized plasmas.

In the self-consistent mean field approximation, we truncate the Dyson series (25) and (26) after the first non-trivial term,

$$\chi_{cc'}^{RPA}(\mathbf{k}, \omega) = \chi_c^{(0)}(\mathbf{k}, \omega)\delta_{cc'} + \chi_c^{(0)}(\mathbf{k}, \omega) W_{cc'}(\mathbf{k}, \omega) \chi_{c'}^{(0)}(\mathbf{k}, \omega), \tag{28}$$

which can be solved by algebraic matrix inversion [35].

$$\chi_{ee}^{RPA}(\mathbf{k}, \omega) = \frac{\chi_e^{(0)} - \chi_e^{(0)} V_{ii} \chi_i^{(0)}}{1 - V_{ee}\chi_e^{(0)} - V_{ii}\chi_i^{(0)}} \qquad \chi_{ei}^{RPA}(\mathbf{k}, \omega) = \frac{2\chi_e^{(0)} V_{ei} \chi_i^{(0)}}{1 - V_{ee}\chi_e^{(0)} - V_{ii}\chi_i^{(0)}} \tag{29}$$

$$\chi_{ie}^{RPA}(\mathbf{k}, \omega) = \chi_{ei}^{RPA}(\mathbf{k}, \omega) \qquad \chi_{ii}^{RPA}(\mathbf{k}, \omega) = \frac{\chi_i^{(0)} - \chi_i^{(0)} V_{ee} \chi_e^{(0)}}{1 - V_{ee}\chi_e^{(0)} - V_{ii}\chi_i^{(0)}} . \tag{30}$$

[4]Calculations of optical properties using interacting spectral functions within the $GW^{(0)}$ approximation, see Ref. [34].

Next, we calculate the DSF (21) by taking the imaginary part of $\chi_{ee}(\mathbf{k},\omega)$,

$$\operatorname{Im}\chi_{ee} = \operatorname{Im}\chi_{ee}^{(0)} + V_{ei}^2\Big\{\Big[\big(\operatorname{Re}\chi_{ee}^0\big)^2 - \big(\operatorname{Im}\chi_{ee}^0\big)^2\Big]\operatorname{Im}\chi_{ii} + \\ + 2\operatorname{Re}\chi_{ee}^0\operatorname{Im}\chi_{ee}^0\operatorname{Re}\chi_{ii}\Big\}, \quad \chi_{ee}^0 = \frac{\chi_e^0}{1 - V_{ee}\chi_e^0}. \tag{31}$$

This expression can be further simplified. The ion-ion response is only significant in a very narrow frequency interval of the order of the ion acoustic frequency $\omega = c_s k$, with the sound speed $c_s = \sqrt{k_B T/m_i}$. At such low frequencies, the Im $\chi_{ee}^{(0)}$ term is essentially 0. Hence, we neglect all terms of the order Im $\chi_{ee}^0\,\chi_{ii}$ and obtain

$$\operatorname{Im}\chi_{ee}(k,\omega) \simeq \operatorname{Im}\chi_{ee}^0(k,\omega) + \big[V_{ei}(k)\operatorname{Re}\chi_{ee}^0(k,0)\big]^2\operatorname{Im}\chi_{ii}(k,\omega)\,. \tag{32}$$

This result is also given e.g. in Evans et al. [36].

Let us compare our result to the Chihara formula (23). Our TCP-RPA result, Eq. (31) contains all the terms that are left in the Chihara formula, if bound states are excluded, i.e. $f(k) = Z_{\mathrm{core}} = 0$: We identify the high frequency component $S_{ee}^{(0)}$ and the screening cloud $q(k)$ in our two component plasma model. Moreover, our model contains additional terms, which do not appear in the Chihara formula. There, these terms are missing because all ionic correlation functions are treated as static functions, i.e. independent of frequency arguments, and hence do not have an imaginary part. How important are these additional dynamic terms?

We evaluate both our TCP-RPA model as well as Chihara's model with all structure factors $S_{\alpha\beta}$ replaced by their RPA counterpart given in Eq. (29) and (30). Results for a model hydrogen plasma at $n_e = 10^{21}\,\mathrm{cm}^{-3}$ and $T = 10\,\mathrm{eV}$ are shown in Fig. 5. Panel (a) shows the TCP-RPA result (red) and the Chihara-RPA (green) as a function of the frequency shift on a double-log scale. The additional terms in the TCP-RPA produce a slight modification at the foot of the low frequency component ("ion feature"). The black curve is the electronic one component DSF, i.e. only the first term in Eq. (31). It completely dominates the high frequency behavior. Figure 5b shows only the low frequency signal and demonstrates that both implementations are numerically equal. Under present conditions, the additional terms in the TCP-RPA do not make a significant contribution. This is an important confirmation of Chihara's result, as it's validity has been put into question frequently: The TCP-RPA result, which is numerically equal to Chihara's formula was derived within a rigorous, well defined model starting from the macroscopic equations of motion of all constituent particles.

However, more thorough investigations are needed to answer the question if the additional terms found in the TCP-RPA might become important under different plasma conditions, in particular, if the plasma is more strongly correlated, i.e. at

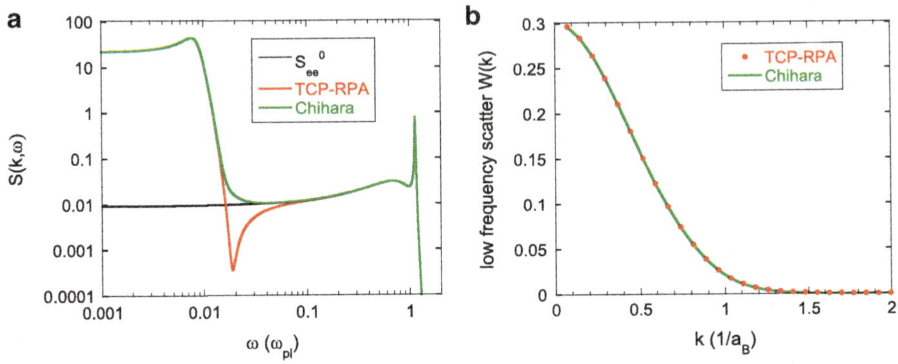

Fig. 5 Ion feature using the two component RPA compared to Chihara formula. Parameters: $n_e = 10^{21}\,\mathrm{cm}^{-3}$, $T = 10\,\mathrm{eV}$. (**a**) Dynamical structure factor for H, $Z_f = 1$, $\alpha = 3$. (**b**) Integrated low frequency signal

higher density and/or lower temperatures. However, under those conditions, the RPA becomes invalid and a comparison to the TCP-RPA model becomes questionable. The following section describes how to incorporate many-body effects in the high frequency component of the DSF, $S_{ee}^{(0)}$ and go beyond the RPA.

2.3.3 Local Field Corrections and Collision Frequency

The established notation to describe many-particle correlations in the DSF, such as electron-ion scattering or electron-electron interaction is the notion of the local field correction (LFC) factor $G_{\alpha\beta}(\mathbf{k},\omega)$, giving rise to a modified Dyson equation for the response function,

$$\chi_{\alpha\beta}(\mathbf{k},\omega) = \chi_\alpha^{(0)}\delta_{\alpha\beta} + \sum_\gamma \chi_\alpha^{(0)} V_{\alpha\gamma}(k)\left(1 - G_{\alpha\gamma}(\mathbf{k},\omega)\right)\chi_{\gamma\beta}(\mathbf{k},\omega)\,, \tag{33}$$

α and β characterize different particle species or spin projections. For example the electron-electron response becomes

$$\chi_{ee}^{\mathrm{OCP}}(\mathbf{k},\omega) = \frac{\chi_e^{(0)}(\mathbf{k},\omega)}{1 - V_{ee}(k)(1 - G_{ee}(\mathbf{k},\omega))\chi_e^{(0)}(\mathbf{k},\omega)}\,, \tag{34}$$

with the Lindhard function $\chi_\alpha^{(0)}$, see Eq. (24). In the limit $G_{\alpha\beta}(\mathbf{k},\omega) = 0$, the RPA is recovered.

The LFC can be expressed through a Feynman diagram as

$$G_{\alpha\beta}(\mathbf{k},\omega) =$$

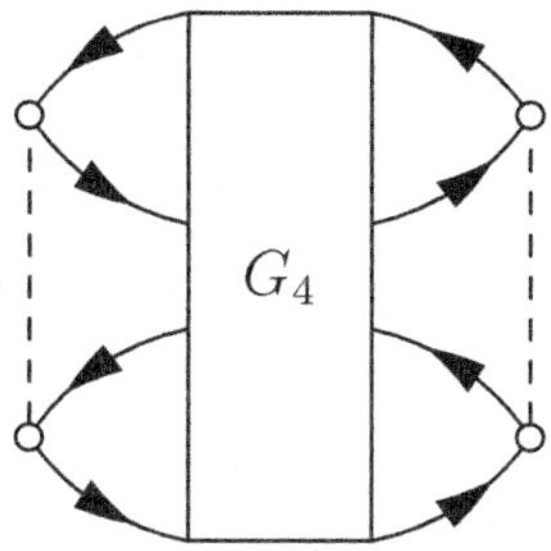

$$. \qquad (35)$$

Here, G_4 is the 4-body correlation function [37] that has to be evaluated using suitable approximations, e.g. the Born approximation, the Lenard-Balescu approximation, using a dynamically screened interaction potential, or the t-matrix approximation to describe strong collisions [38, 39].

Limiting cases, such as the long wavelength limit $k \to 0$ put constraints on the LFC as we shall see in the following. In the long wavelength limit, the dielectric function

$$\epsilon(\mathbf{k},\omega) = 1 - \frac{1}{\epsilon_0 k^2}\Pi(\mathbf{k},\omega) = 1 - \frac{1}{\epsilon_0 k^2}\frac{\Pi^{RPA}(\mathbf{k},\omega)}{1 + \frac{1}{\epsilon_0 k^2}G(\mathbf{k},\omega)\Pi^{RPA}(\mathbf{k},\omega)} . \qquad (36)$$

is expected to coincide with the generalized Drude expression

$$\epsilon(k \to 0,\omega) = \epsilon^{\text{Drude}}(\omega) = 1 - \frac{\omega_{\text{pl}}^2}{\omega(\omega + i\nu(\tilde{\omega}))} \qquad (37)$$

with the dynamical collision frequency $\nu(\omega)$. By comparing Eqs. (34)–(37), one establishes the relation

$$\nu(\omega) = -i\frac{\omega_p^2}{\omega}G(k \to 0,\omega) , \qquad (38)$$

while

$$G(\mathbf{k},\omega) = \epsilon_0 k^2 \left[\frac{1}{\Pi(\mathbf{k},\omega)} - \frac{1}{\Pi^{RPA}(\mathbf{k},\omega)}\right] . \qquad (39)$$

The long wavelength limit of the LFC is the dynamical collision frequency, for which various approximations exist [15]. In this limit, $\nu(\omega)$ only contains electron-ion collisions, whereas electron-electron collisions do not contribute.

For example using the Born approximation in Eq. (35) and in the long wavelength limit, we find for the electron-ion collision frequency

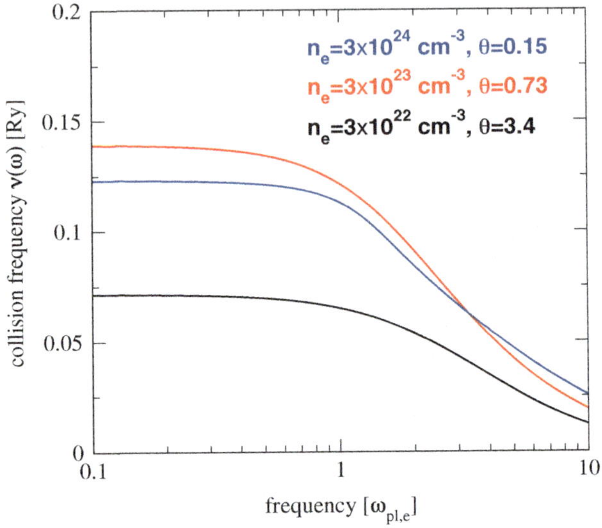

Fig. 6 Imaginary part of the collision frequency for three different electron degeneracy parameters

$$\nu^{\mathrm{Born}}(\omega) = -i\,\frac{Z_{\mathrm{f}}}{6\pi^2 m_e}\int_0^\infty dq\,q^4 V_S^2(q) S_{\mathrm{ii}}^{\mathrm{OCP}}(q)\left[\chi^{(0)}(q,\omega) - \chi^{(0)}(q,0)\right]\,. \tag{40}$$

This expression is evaluated in Fig. 6. The imaginary part of the Born collision frequency is shown as a function of transfer frequency ω for a Be plasma at 13 eV electron temperature and electron densities varying between $3 \times 10^{22}\,\mathrm{cm}^{-3}$ and $3 \times 10^{24}\,\mathrm{cm}^{-3}$. Correspondingly, the electron degeneracy parameter varies from $\theta = 3.4$ (non-degenerate plasma) via 0.73 (partly degenerate) to 0.15 (strongly degenerate). This has an immediate effect on the collisionality of the plasma: The collision frequency increases as the density rises from $3 \times 10^{22}\,\mathrm{cm}^{-3}$ to $3 \times 10^{23}\,\mathrm{cm}^{-3}$. Then, it decreases as the density rises further to $3 \times 10^{24}\,\mathrm{cm}^{-3}$ and the plasma goes into the degenerate regime, because small energy transfers (compared to the Fermi energy), are suppressed due to Pauli blocking. The collision has to reach a final state beyond the Fermi energy to be allowed. Consequently, the collision frequency at larger frequencies still rises as the density increases beyond $3 \times 10^{22}\,\mathrm{cm}^{-3}$.

For applications to XRTS, we need to calculate the DSF (i.e. the response function) at finite transfer wavevector k, while the Drude expression only gives us the low k limit. By replacing $G(k,\omega)$ by $\nu(\omega)$ according to Eq. (38), we find the so-called Mermin [40, 41] expression,

$$\chi_{ee}^{\mathrm{M}}(\mathbf{k},\omega) = \left(1 - \frac{i\omega}{\nu}\right)\left[\frac{1}{\chi_e^{(0)}(\mathbf{k},0)} - \frac{i\omega/\nu}{\chi_e^{(0)}(\mathbf{k},\omega + i\nu)}\right]^{-1}\,. \tag{41}$$

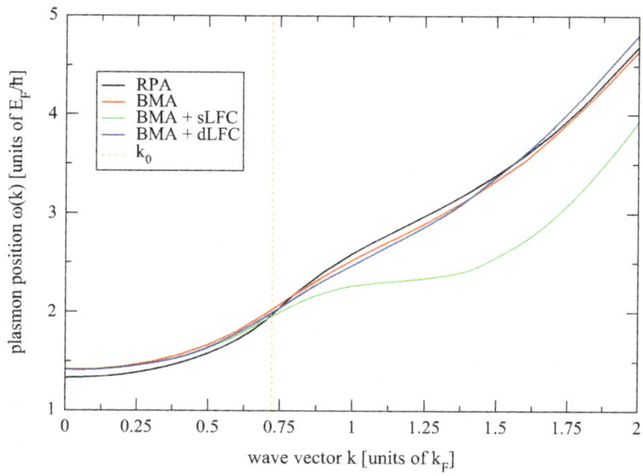

Fig. 7 Plasmon dispersion as function of transfer wavevector for electron density $n_e = 2 \times 10^{23}\,\mathrm{cm}^{-3}$ and $T = 0.01\,\mathrm{eV}$

The Mermin expression allows us to calculate the response function and thereby the dynamical structure factor for a correlated electron-ion plasma at finite wavenumber (k) and frequency ω. Note however, that the quantity that describes correlations, the dynamical collision frequency, is evaluated in the limit $k \to 0$. Effects of spatial correlation (finite-k effects) cannot be described in this way, as well as electron-electron correlations.

As a solution, Wierling [42] proposed an extended Mermin expression, that includes both electron-ion collisions and electron-electron correlations by replacing the RPA response in the Mermin expression Eq. (41) by the electron-electron OCP response function Eq. (34) that contains the purely electronic LFC $G_{ee}(k, \omega)$.

$$\chi^{\mathrm{xM}}(k,\omega) = \left(1 - \frac{i\omega}{\nu}\right)\left[\frac{1}{\chi_{ee}^{\mathrm{OCP}}(\mathbf{k},0)} - \frac{i\omega/\nu}{\chi_{e}^{\mathrm{OCP}}(\mathbf{k},\omega+i\nu)}\right]^{-1}. \tag{42}$$

In this way, one systematically combines the descriptions of two correlated OCP's, each with their respective local field factor or structure factor, to form a coupled TCP, where the interaction between the correlated electronic and the correlated ionic subsystem is contained in the dynamical collision frequency, which, itself depends on the OCP correlation functions.

Figure 7 shows an application of the combined Mermin-LFC model, i.e. the plasmon dispersion of a degenerate plasma ($k_B T_e \ll E_F$) as a function of the transfer wavevector k. Calculations using the RPA (black curve), the Born-Mermin approximation without LFCs (red curve), Born-Mermin plus static LFCs (green curve) using the parametrization by Farid [43] and Born-Mermin plus dynamic LFCs [42] (blue curve) are shown. k_0 marks the onset of Landau damping. Above k_0, the static LFC results in a horizontal plasmon dispersion, whereas all other models predict a quadratic dispersion. We will see that the BMA combined with static LFC is successful in describing experimental plasmon scattering data in this regime of plasma parameters. This is in marked contract to the fact that the strong corrections seen in the static LFC are almost completely taken back in the dynamical LFC calculation and points out that the interplay between static and dynamic correlations is all but well understood.

2.3.4 Bound States in the DSF

The influence of bound states is an important issue for XRTS. In high Z targets, Compton scattering from bound states can become dominant over free-free scattering. How to treat this contribution as an extension to the RPA dielectric function has been outlined in [44]. This can provide a straightforward extension to the TCP-RPA, see Sect. 2.3.2. However, state-of-the-art 'physical' approaches will be able to handle this problem properly in the future.

Within the Chihara approach, bound states are treated by the third term in Eq. (23). The core electron structure factor $S_{\mathrm{core}}(k,\omega)$ is related to the bound-free transition matrix element

$$S_{\mathrm{core}}(k,\omega) \propto |\langle\psi_{\mathrm{bound}}|\exp(i\mathbf{k}\cdot\mathbf{r})|\,\psi_{\mathrm{free}}\rangle|^2 \;. \tag{43}$$

Various approximations with respect to the choice of the bound state wavefunctions ψ_{bound} and the scattering states ψ_{free} exist in the literature, for a detailed discussion we refer to Ref. [33]. A more recent account is also given in Ref. [45] using real space Green functions. Besides the choice of the wavefunctions, the account for plasma shielding effects and resulting depletion of the ionization potential is a matter of intensive debate. Traditionally [30], the Stewart-Pyatt shift [46] was employed to calculate the effective ionization potential as a function of the free electron density and average charge state of the system. Recent XFEL near edge absorption spectroscopy measurements [47] however support the Ecker-Kroll model [48], which was historically dismissed due to thermodynamic inconsistencies [46].

A rigorous account for electron-ion interaction amounts to treat electrons not in a chemical picture, as done within the Chihara framework or in the TCP-RPA, but in the physical picture, where the electron wavefunction is a superposition of bound states and scattering states. This goes beyond the scope of this work and we refer to Ref. [49] as a recently achieved milestone in this direction using the DFT-MD method.

3 Application: Angle and Energy Resolved XRTS in Shock Compressed Boron

Let us conclude this chapter with an application of our developed theoretical framework to an experiment that was performed at the Jupiter Laser Facility at the Lawrence Livermore National Laboratory [11]. In the experiment, samples of solid boron (mass density 2.45 g/cm^3) have been dynamically compressed by interaction

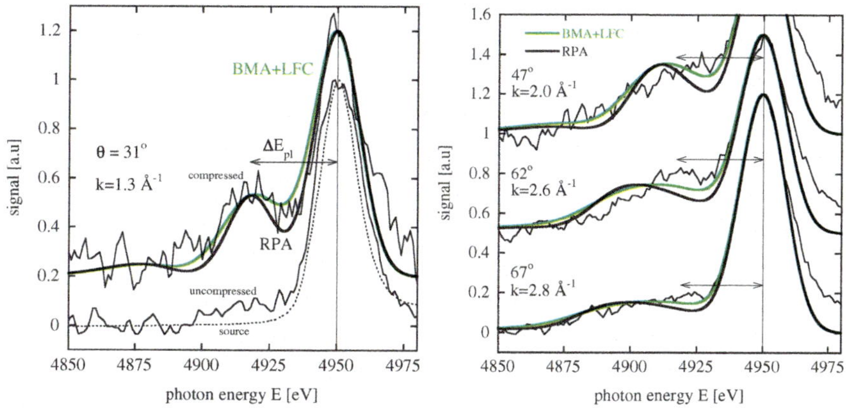

(a) 31° scattering in compressed boron

(b) 47°, 62°, and 67° scattering in compressed boron

Fig. 8 (**a**) X-ray scattering spectra from compressed targets at 31° scattering angle compared to fit calculations using BMA+LFC (*green*) and the RPA (*black*). for three different densities. Also shown are the V probe spectrum (*dotted*) and scattering from uncompressed targets. (**b**) Scattering spectra for $\theta = 47°, 62°$, and 67° and fit calculations using BMA+LFC (*green*) and RPA (*black*). *Arrows* indicate the experimental plasmon shift $\Delta E_{\rm pl}(k)$

with high energy laser pulses. A 2ω (527 nm) drive laser with energy of 400 J compresses the boron samples in a 10 ns-long flat top pulse. Subsequently, ultrashort pulse laser produced K-α radiation probes the compressed boron plasma by x-ray Thomson scattering. Energies of 250 J are delivered at a pulse duration of 10 ps and at 1053 nm wavelength, yielding intensities of $\approx 10^{17}$ W/cm^2. Vanadium and copper foils of 10 μm thickness provide K-α photon energies of 4.95 and 8.05 keV respectively, with k-vectors between $k = 1.3\,\text{Å}^{-1}$ and $k = 4.3\,\text{Å}^{-1}$ controlled through the variation of the scattering angle. The scattered X-rays disperse in a cylindrically bent large graphite crystal (highly-oriented pyrolytic graphite) and are focused onto an imaging plate detector.

Figure 8 shows the experimental scattering data compared to theoretical calculations using the Chihara formula with the free electron contribution calculated within the BMA+LFC model (solid curves) and the RPA (dashed curves) for various electron densities. The large k data at $k = 4.3\,\text{Å}$ is not shown here. At small momentum transfer ($k = 1.3/\,\text{Å}$, panel (a)), a strong plasmon resonance is observed, downshifted by $\Delta E_{\rm pl} = (32 \pm 3)$ eV with respect to the photon energy of the V K-α probe, $E_0 = 4.95$ keV. This feature is one order of magnitude stronger compared to scatter from uncompressed targets, indicating the presence of a large fraction of free electrons due to shock induced ionization. Strong ionization has also been observed in earlier boron compression experiments using the diamond anvil-cell technique [50]. As the plasmon shift at small k is independent of the approximation for the DSF, (Fig. 7), we obtain a model independent measurement of the electron density $n_{\rm e} = (4.0 \pm 0.7) \times 10^{23}\,\text{cm}^{-3}$. For the ionization we find $Z_f = 2.3$ by comparing the integrated inelastic scatter to the elastic part in the non-collective signal at $4.3\,\text{Å}^{-1}$ [12]. Finally, from n_e and Z_f we deduce the mass density $\rho = 3.2\,\text{g/cm}^3$, corresponding to a compression ratio $\rho/\rho_0 = 1.3$. The electron temperature was estimated by radiation hydrodynamic simulations as $T_e \simeq 0.2\,\text{eV}$ [11].

The theoretical fit calculations for the small k spectrum also demonstrate, that only the BMA+LFC theory, with electron-ion collisions is capable to describe the experimental data. Indeed, in such nearly Fermi degenerate systems $\theta = k_B T/E_F \simeq 0.01$, only electron-ion collisions determine the plasmon width for small momentum transfers $\mathbf{k}$, where non-collisional (Landau) damping is suppressed, [51], and the plasmon width is a direct measure of the optical conductivity. This is different from dilute plasmas, e.g. space or magnetic fusion plasmas, where Landau damping dominates over collisional damping. The electron ion collision frequency accounts for differences of 3 eV in plasmon width and the electron-electron LFC results in a 10 eV decreased shift when compared to the RPA. This demonstrates the significance of many-body corrections for applications of x-ray Thomson scattering to plasma diagnostics.

With increasing momentum transfer k the plasmon is increasingly damped, and between $k = 1.7$ and $2.0\,\text{Å}^{-1}$ the plasmon energy is constant within an error bar of 5 eV, see Fig. 8b. This behavior is consistent with the BMA+LFC calculations and indicate increasing local field effects due to strong electron coupling. Since $k_B T_e \simeq 0.2\,\text{eV}$, we deal with a Fermi degenerate plasma, $k_B T_e/E_F = 0.01$, the Fermi energy is $E_F = \hbar^2 k_F^2/2m_e = 19.8\,\text{eV}$. In this regime, the plasmon spectrum is not sensitive to T_e. $G(k)$ scales as k^2 at wavenumbers $k \ll k_F$, $k_F = (3\pi^2 n_e)^{1/3}$ being the Fermi wavenumber. Consequently, for conditions of Fig. 8a (small k) the plasmon energy shift is only weakly influenced by the LFC. In this limit the plasmon energy shift in BMA+LFC agrees with the values obtained in the RPA.

These k-vector resolved measurements of the plasmon resonance energy reflect electron-electron interactions that affect the dispersion in the vicinity of the Fermi momentum. The random phase approximation (RPA) as well as the BMA accounting for electron-ion collisions predict a parabolic (Bohm-Gross) dispersion [52,53]. Plasmon dispersion data deviating from the RPA have been observed in cold metals [54–56]. In shock-compressed matter, the decrease of the plasmon dispersion due to electron-electron local field effects will occur simultaneously with electron-ion collisions affecting plasmon damping.

Figure 9a shows the experimental results for $\Delta E_{\text{pl}}(k)$ compared to the RPA and BMA with and without local field corrections. In the limit $k \to 0$, all three models converge towards the plasma frequency, in agreement with the data. With increasing k the experimental dispersion is constant at $\Delta E_{\text{pl}}(k) \simeq 33\,\text{eV}$. The plasmon group velocity $v_G = d\omega(k)/dk$ vanishes, i.e. plasmons do not propagate at this energy. This behavior is in good agreement with the fully collisional BMA+LFC calculation but contradicting the RPA as well as the usual BMA that neglects the LFC factor [52]. In the strongly non-collective regime, the BMA+LFC calculation quickly converges into the single particle dispersion (Compton shift $\Delta E_C = \hbar^2 k^2/2m_e$), as

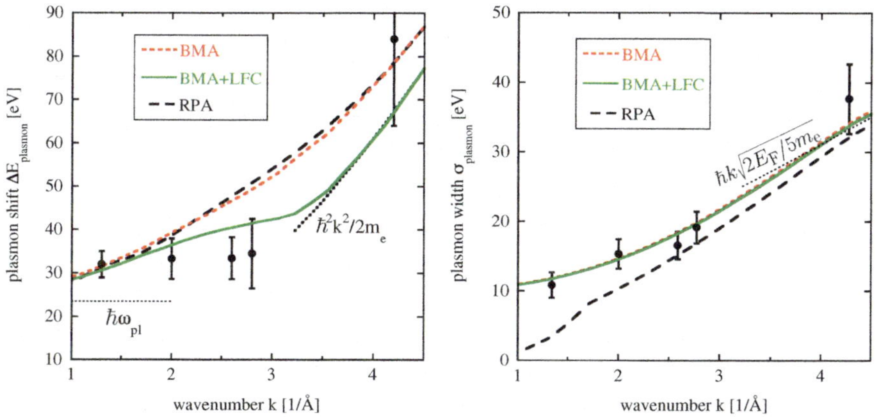

(a) Plasmon dispersion in compressed boron and exact limits at high and low k

(b) Plasmon width in compressed boron and Compton limit.

Fig. 9 Plasmon dispersion (**a**) and rms-width (**b**) as function of the scattering wavevector k. Experimental data are compared to RPA (*black*), BMA with LFC (*green*) and BMA without LFC (*red*) calculations. The *black dotted lines* represent the limits of dispersion and damping at small and large k, respectively

expected and consistent with the RPA and the usual BMA. In this limit correlations become less important since distances smaller than the screening length are probed. Within the error margins, all three models are consistent with the non-collective scattering data at $k = 4.3\,\text{Å}^{-1}$.

Finally, we analyse the plasmon rms width as a function of k, results are shown in Fig. 9b and compared to RPA and BMA with and without local field corrections. The two points at $k = 1.3$ and $2.0\,\text{Å}^{-1}$ corresponding to collective scattering show that plasmon damping is mainly due to electron-ion collisions. Conversely, the RPA underestimates the plasmon width.

At the present density, the critical wavenumber up to which plasmons are a well-defined mode is $k_\text{c} = 1.7\,\text{Å}^{-1}$. Above k_c, the plasmon merges with the single-pair continuum [51] giving rise to Landau damping in the RPA theory. At $k = 1.3\,\text{Å}^{-1}$, we find the plasmon resonance at $\Delta E_\text{pl} = 32\,\text{eV}$, which lies beyond the cut-off energy of the single-pair continuum $\hbar\omega_\text{pair} \leq \hbar^2(k^2+2kk_\text{F})/2m_\text{e} = 29.1\,\text{eV}$. The sharp edge of the pair continuum reflects the steplike Fermi distribution of degenerate electrons. Hence, within the RPA, the plasmon mode cannot couple to the single-pair continuum and therefore is undamped.

At 47° scattering angle, $k = 2.0\,\text{Å}^{-1}$, the plasmon appears at $E_\text{pl} = 33\,\text{eV}$, which is within the single-pair continuum $\hbar\omega_\text{pair} \leq 50.1\,\text{eV}$. Hence, Landau damping is finite, explaining the increased spectral width in the RPA calculation as compared to the 31° case. However, RPA still underestimates the plasmon damping, whereas we obtain good agreement using the BMA+LFC model. In the Compton scattering regime (large k), where the width approaches that of the Compton profile

$\sigma_C = \hbar k \sqrt{2E_F/5m_e}$ both RPA and BMA agree and are consistent with the data. The calculations explain the plasmon broadening due to electron-ion collisions as well as vanishing plasmon dispersion due to strong electron-electron coupling. In the limits of small and large $k-$vectors, we observe the well-known analytical limits for the plasmon dispersion: At small k, the resonance lies at the plasma frequency, at large k, it converges to the Compton energy $E_C = \hbar^2 k^2/2m_e$, while the width is the Compton width $\Delta E_C = \hbar k \sqrt{2E_F/5m_e}$.

4 Conclusions

In this chapter, we have demonstrated how to calculate the response function of a high energy density plasma. We have used a fundamental approach, based on the combination of many-body perturbation theory and QED. The response function yields the dynamical structure factor, i.e. the differential cross section for inelastic x-ray scattering or Thomson scattering. The Born-Mermin approximation, complemented by local field corrections is the state-of-the art analytical model to include electron-ion collisions and electron-electron correlations in a correlated plasma. As an illustration, we discussed x-ray scattering data from shock-compressed boron and demonstrated that our BMA+LFC model successfully describes collisional plasmon damping as well as the non-dispersing plasmon resonance energy.

Certainly, the presented theory has its limitations, some of which we would like to mention briefly before closing this chapter:

- All derivations assume that particle velocities are non-relativistic. A relativistic treatment, as presented for optical Thomson scattering [57, 58] has to be developed also for x-ray Thomson scattering. The framework of QED plus many-body theory as presented here is a natural starting point.
- Not only the free electron response but also the ion-ion structure factor and the bound-free term in the Chihara formula contain valuable information about ionic correlations and electronic structure in HEDM. As discussed only briefly, newer experimental data point to some inconsistencies in existing models for these contributions. It would therefore be desirable to overcome the Chihara formula, i.e. the chemical picture, and to develop means to calculate the complete DSF based on a physical picture. DFT-MD simulations seem to be a promising tool as shown in Ref. [49].
- The calculations that are presented in this chapter all assume thermal equilibrium in the plasma volume where the scattering takes place. This limitation can be overcome by integrating the scattering cross-section over density- and temperature profiles as demonstrated e.g. in Ref. [59] and [60].

Acknowledgements The author gratefully acknowledges continued fruitful collaboration with S. Glenzer, R. Redmer, H. Reinholz, G. Röpke, M. Desjarlais, G. Gregori and many others. Part of the presented work was made possible through support by the Alexander von Humboldt foundation. Special thanks go to F. Graziani.

References

1. S.H. Glenzer, R. Redmer, X-ray Thomson scattering (XRTS) in high energy density plasmas. Rev. Mod. Phys. **81**(4), 1625–1663 (2009)
2. S.H. Glenzer, G. Gregori, R.W. Lee, F.J. Rogers, S.W. Pollaine, O.L. Landen, Demonstration of spectrally resolved x-ray scattering in dense plasmas. Phys. Rev. Lett. **90**(17), 17–20 (2003)
3. S.C. Snyder, G.D. Lassahn, L.D. Reynolds, Direct evidence of departure from local thermodynamic equilibrium in a free-burning arc-discharge plasma. Phys. Rev. E **48**(5), 4124–4127 (1993)
4. http://lcls.slac.stanford.edu
5. www.xfel.eu
6. http://flash.desy.de
7. S.H. Glenzer, X-ray Thomson scattering from dense plasmas. AIP Conf. Proc. **926**, 8 (2007)
8. H.J. Lee, P. Neumayer, J. Castor, T. Döppner, R.W. Falcone, C. Fortmann, B.A. Hammel, A.L. Kritcher, O.L. Landen, R.W. Lee, D.D. Meyerhofer, D.H. Munro, R. Redmer, S.P. Regan, S. Weber, S.H. Glenzer, X-ray Thomson-scattering measurements of density and temperature in shock-compressed beryllium. Phys. Rev. Lett. **102**(11), 115001 (2009)
9. A.L. Kritcher, P. Neumayer, J. Castor, T. Döppner, R.W. Falcone, O.L. Landen, H.J. Lee, R.W. Lee, E.C. Morse, A. Ng, S. Pollaine, D. Price, S.H. Glenzer, Ultrafast x-ray Thomson scattering of shock-compressed matter. Science (New York) **322**(5898), 69–71 (2008)
10. C. Fortmann, H.J. Lee, T. Döppner, R.W. Falcone, A.L. Kritcher, O.L. Landen, S.H. Glenzer, Measurement of the adiabatic index in be compressed by counterpropagating shocks. Phys. Rev. Lett. **108**(17), 175006 (2012)
11. P. Neumayer, C. Fortmann, T. Döppner, P.F. Davis, R.W. Falcone, A.L. Kritcher, O.L. Landen, H.J. Lee, R.W. Lee, C. Niemann, S. Le Pape, S.H. Glenzer, Plasmons in strongly coupled shock-compressed matter. Phys. Rev. Lett. **105**(7), 075003 (2010)
12. S. Le Pape, P. Neumayer, C. Fortmann, T. Döppner, P. Davis, A. Kritcher, O. Landen, S. Glenzer, X-ray radiography and scattering diagnosis of dense shock-compressed matter. Phys. Plasma **17**, 56309 (2010)
13. A. Kritcher, T. Döppner, C. Fortmann, T. Ma, O. Landen, R. Wallace, S. Glenzer, In-flight measurements of capsule shell adiabats in laser-driven implosions. Phys. Rev. Lett. **107**(1), 015002 (2011)
14. R.R. Fäustlin, Th. Bornath, T. Döppner, S. Düsterer, E. Förster, C. Fortmann, S.H. Glenzer, S. Göde, G. Gregori, R. Irsig, T. Laarmann, H.J. Lee, B. Li, K.-H. Meiwes-Broer, J. Mithen, B. Nagler, A. Przystawik, H. Redlin, R. Redmer, H. Reinholz, G. Röpke, F. Tavella, R. Thiele, J. Tiggesbäumker, S. Toleikis, I. Uschmann, S.M. Vinko, T. Whitcher, U. Zastrau, B. Ziaja, Th. Tschentscher, Observation of ultrafast nonequilibrium collective dynamics in warm dense hydrogen. Phys. Rev. Lett. **104**(12), 125002 (2010)
15. H. Reinholz, Dielectric and optical properties of dense plasmas. Annales de Physique **30**(4–5), 1–187 (2005)
16. I. Lindgren, *Relativistic Many-Body Theory: A New Field Theoretical Approach* (Springer, New York, 2011)
17. E. Brown, M.A. Morales, C. Pierleoni, D. Ceperley, Quantum Monte Carlo techniques and applications for warm dense matter, Chapter 5 of this book
18. A. Messiah, *Quantum Mechanics, Two Volumes Bound as One* (Dover, Mineola, 1999)
19. L. Hedin, New method for calculating the one-particle green's function with application to the electron-gas problem. Phys. Rev. **139**(3A), A796–A823 (1965)
20. F.J. Dyson, The S matrix in quantum electrodynamics. Phys. Rev. **75**(11), 1736–1755 (1949)
21. J. Schwinger, On the green's functions of quantized fields: I. Proc. Nat. Acad. Sci. **37**(7), 452–455 (1951)
22. W.D. Kraeft, D. Kremp, W. Ebeling, G. Röpke, *Quantum Statistics of Charged Particle Systems* (Akademie-Verlag, Berlin, 1986)

23. R. Redmer, in *Plasmas in Planetary Interiors*, ed. by A. Dinklage, T. Klinger, G. Marx, L. Schweikhard. Lecture Notes in Physics, vol. 670, pp. 331–348 (Springer, Berlin/New York, 2005)
24. Y. Takada, Inclusion of vertex corrections in the self-consistent calculation of quasiparticles in metals. Phys. Rev. Lett. **87**(22), 226402 (2001)
25. J.C. Ward, An identity in quantum electrodynamics. Phys. Rev. **78**(2), 182 (1950)
26. B. Holm, U. von Barth, Fully self-consistent GW self-energy of the electron gas. Phys. Rev. B **57**(4), 2108–2117 (1998)
27. U. von Barth and B. Holm, Self-consistent GW0 results for the electron gas: fixed screened potential W0 within the random-phase approximation. Phys. Rev. B **54**(12), 8411–8419 (1996)
28. C. Fortmann, Single-particle spectral function for the classical one-component plasma. Phys. Rev. E **79**(1), 16404 (2009)
29. C. Fortmann, T. Döppner, T. Laarmann, A. Przystawik, T. Raitza, G. Röpke, J. Tiggesbäumker, N.X. Truong, On the occurrence of balmer spectra in expanding microplasmas from laser irradiated liquid hydrogen. Arxiv preprint physics/0703177 (2007)
30. G. Gregori, S.H. Glenzer, O.L. Landen, Strong coupling corrections in the analysis of x-ray Thomson scattering measurements. J. Phys. A **36**(22), 5971 (2003)
31. G. Gregori, S.H. Glenzer, H.K. Chung, D.H. Froula, R.W. Lee, N.B. Meezan, J.D. Moody, C. Niemann, O.L. Landen, B. Holst, and Others, Measurement of carbon ionization balance in high-temperature plasma mixtures by temporally resolved x-ray scattering. J. Quant. Spectrosc. Radiat. Transf. **99**(1–3), 225–237 (2006)
32. J. Chihara, Difference in x-ray scattering between metallic and non-metallic liquids due to conduction electrons. J. Phys. F Met. Phys. **17**, 295 (1987)
33. M. Schumacher, F. Smend, I. Borchert, Incoherent scattering of gamma rays by inner-shell electrons. J. Phys. B At. Mol. Phys. **8**(9), 1428–1439 (1975)
34. C. Fortmann, G. Röpke, A. Wierling, Optical properties and one-particle spectral function in non-ideal plasmas. Contrib. Plasma Phys. **47**(4–5), 297–308 (2007)
35. C. Fortmann, R. Redmer, H. Reinholz, G. Röpke, A. Wierling, W. Rozmus, Bremsstrahlung vs. Thomson scattering in vuv-fel plasma experiments. High Energy Density Phys. **2**(3–4), 57–69 (2006)
36. D.E. Evans, J. Katzenstein, Laser light scattering in laboratory plasmas. Rep. Prog. Phys. **7**(1), 79–271 (1969)
37. F. Hoehne, R. Redmer, G. Röpke, H. Wegener, Linear response theory for thermoelectric transport coefficients of a partially ionized plasma. Phys. A Stat. Theor. Phys. **128**(3), 643–675 (1984)
38. H. Reinholz, I. Morozov, G. Röpke, T. Millat, Internal versus external conductivity of a dense plasma: many-particle theory and simulations. Phys. Rev. E **69**(6), 66412 (2004)
39. R. Thiele, R. Redmer, H. Reinholz, G. Röpke, Using the Gould–DeWitt scheme to approximate the dynamic collision frequency in a dense electron gas. J. Phys. A Math. Gen. **39**(17), 4365–4368 (2006)
40. N.D. Mermin, Lindhard dielectric function in the relaxation-time approximation. Phys. Rev. B **1**(5), 2362–2363 (1970)
41. A. Selchow, G. Röpke, A. Wierling, H. Reinholz, T. Pschiwul, G. Zwicknagel, Dynamic structure factor for a two-component model plasma. Phys. Rev. E **64**(5), 56410 (2001)
42. A. Wierling, Dynamic local field corrections for two-component plasmas at intermediate coupling. J. Phys. A Math. Theor. **42**(21), 214051 (2009)
43. B. Farid, V. Heine, G.E. Engel, I.J. Robertson, Extremal properties of the Harris-Foulkes functional and an improved screening calculation for the electron gas. Phys. Rev. B **48**(16), 11602–11621 (1993)
44. G. Röpke and R. Der, The influence of two-particle states (excitons) on the dielectric function of the electron—hole plasma. Phys. Status Solidi (b) **92**(2), 501–510 (1979)
45. B.A. Mattern, G.T. Seidler, J.J. Kas, J.I. Pacold, J.J. Rehr, Real-space green's function calculations of compton profiles. Phys. Rev. B **85**(11), 115135 (2012)

46. J.C. Stewart, K.D. Pyatt Jr., Lowering of ionization potentials in plasmas. Astrophys. J. **144**, 1203 (1966)
47. S.M. Vinko, O. Ciricosta, B.I. Cho, K. Engelhorn, H.-K. Chung, C.R.D. Brown, T. Burian, J. Chalupsky, R.W. Falcone, C. Graves, V. Hajkova, A. Higginbotham, L. Juha, J. Krzywinski, H.J. Lee, M. Messerschmidt, C.D. Murphy, Y. Ping, A. Scherz, W. Schlotter, S. Toleikis, J.J. Turner, L. Vysin, T. Wang, B. Wu, U. Zastrau, D. Zhu, R.W. Lee, P.A. Heimann, B. Nagler, J.S. Wark, Creation and diagnosis of a solid-density plasma with an x-ray free-electron laser. Nature **482**(7383), 59–U75 (2012)
48. G. Ecker, W. Kröll, Lowering of the ionization energy for a plasma in thermodynamic equilibrium. Phys. Fluids **6**(1), 62–69 (1963)
49. K.-U. Plagemann, P. Sperling, R. Thiele, M.P. Desjarlais, C. Fortmann, T. Döppner, H.J. Lee, S.H. Glenzer, R. Redmer, Dynamic structure factor in warm dense beryllium. New J. Phys. **14**(5), 055020 (2012)
50. M.I. Eremets, V.V. Struzhkin, H. Mao, R.J. Hemley, Superconductivity in boron. Science (New York) **293**(5528), 272–4 (2001)
51. D. Pines, P. Nozières, *The Theory of Quantum Liquids*, vol. 1 (Benjamin, New York, 1966)
52. R. Thiele, T. Bornath, C. Fortmann, A. Höll, R. Redmer, H. Reinholz, G. Röpke, A. Wierling, S. Glenzer, G. Gregori, Plasmon resonance in warm dense matter. Phys. Rev. E **78**(2), 1–8 (2008)
53. D. Bohm, E.P. Gross, Theory of plasma oscillations. A. Origin of medium-like behavior. Phys. Rev. **75**(12), 1851–1864 (1949)
54. W. Schülke, H. Nagasawa, S. Mourikis, P. Lanzki, Dynamic structure of electrons in li metal: inelastic synchrotron x-ray scattering results and interpretation beyond the random-phase approximation. Phys. Rev. B **33**(10), 6744–6757 (1986)
55. W. Schülke, K. Höppner, A. Kaprolat, Semiempirical local-field correction function for electrons in li metal. Phys. Rev. B **54**(24), 17464–17468 (1996)
56. B.C. Larson, J.Z. Tischler, E.D. Isaacs, P. Zschack, A. Fleszar, A.G. Eguiluz, Inelastic x-ray scattering as a probe of the many-body local-field factor in metals. Phys. Rev. Lett. **77**(7), 1346–1349 (1996)
57. J.P. Palastro, J.S. Ross, B. Pollock, L. Divol, D.H. Froula, S.H. Glenzer, Fully relativistic form factor for Thomson scattering. Phys. Rev. E **81**, 036411 (2010)
58. J.S. Ross, S.H. Glenzer, J.P. Palastro, B.B. Pollock, D. Price, L. Divol, G.R. Tynan, D.H. Froula, Observation of relativistic effects in collective Thomson scattering. Phys. Rev. Lett. **104**(10), 1–4 (2010)
59. C. Fortmann, R. Thiele, R.R. Fäustlin, Th. Bornath, B. Holst, W.-D. Kraeft, V. Schwarz, S. Toleikis, Th. Tschentscher, R. Redmer, Thomson scattering in dense plasmas with density and temperature gradients. High Energy Density Phys. **5**(3), 208–211 (2009)
60. P. Sperling, R. Thiele, B. Holst, C. Fortmann, S.H. Glenzer, S. Toleikis, Th. Tschentscher, R. Redmer, Two-color Thomson scattering at flash. High Energy Density Phys. **7**(3), 145–149 (2011)

Progress in Warm Dense Matter and Planetary Physics

Winfried Lorenzen, Andreas Becker, and Ronald Redmer

Abstract We give an introduction into the method of quantum molecular dynamics simulations which combines density functional theory with classical molecular dynamics. This method has demonstrated its predictive power in determining the thermophysical properties of matter under extreme conditions as found, e.g., in astrophysical objects like giant planets and brown dwarfs. Such extreme states of matter can also be probed by state-of-the-art shock wave experiments in the laboratory. We give exemplary ab initio results for the behavior of the simplest and most abundant elements hydrogen and helium under extreme conditions. In addition, we also show results for more complex molecular systems such as water. The light elements H and He, the hydrides of C (CH_4), N (NH_3), and O (H_2O) and, in particular, mixtures of these systems have rich high-pressure phase diagrams which are important for the structure, evolution and magnetic field of gas giant planets like Jupiter and ice giant planets like Neptune. Finally, we describe the impact of these results on the design of advanced interior, evolution, and dynamo models and give exemplary results for solar and extrasolar giant planets.

1 Introduction

1.1 Warm Dense Matter

The properties of warm dense matter (WDM) are of lively interest [1–3]. As an intermediate state between cold condensed matter and hot plasmas, WDM is

W. Lorenzen • A. Becker (✉) • R. Redmer
Institute of Physics, University of Rostock, 18051 Rostock, Germany
e-mail: winfried.lorenzen@uni-rostock.de; andreas.becker@uni-rostock.de; ronald.redmer@uni-rostock.de

F. Graziani et al. (eds.), *Frontiers and Challenges in Warm Dense Matter*, Lecture Notes in Computational Science and Engineering 96, DOI 10.1007/978-3-319-04912-0_8,

characterized by high densities as typical for condensed matter and temperatures of several eV (1 eV$\hat{=}$11,604.5 K). Partial ionization, strong correlations and quantum effects are important under such conditions so that WDM is a perfect test case for new concepts of many-particle theory. On the other hand, the implementation of new experimental high-pressure platforms that launch strong shock waves by using, e.g., high-power lasers, gas guns, pulsed power or chemical explosions, and the availability of intense x-ray sources, in particular of free electron lasers (FEL), allows to probe and diagnose WDM states with better and better accuracy [4].

The strongest and perhaps most sustaining influence on WDM research originates from problems in astrophysics [5, 6]. For instance, the interior of giant planets (GPs) such as Jupiter and Saturn is in WDM states [7, 8]. The detection of a great number of extrasolar planets (exoplanets) since 1995 has boosted the interest in WDM as well [9–12]. The formation, evolution, interior structure, and magnetic field configuration of planets is intimately connected with the equation of state (EOS), the phase diagram, the transport and optical properties of planetary materials at extreme conditions of pressure and temperature, i.e. in WDM states. Interesting phenomena such as demixing [13, 14] and nonmetal-to-metal transitions [15, 16] are of fundamental interest in this context but also new and exotic phases such as proton conductors might be important in the deep interior of, e.g., Neptune-like planets [17–19].

An adequate theoretical treatment of WDM faces enormous computational challenges which are caused by inherent strong correlation and quantum effects. This prevents the application of methods of perturbation theory using effective pair potentials or of corresponding cross sections. Therefore, and also due to the huge progress in computational power, ab initio simulation techniques have been developed. Both Path Integral Monte Carlo (PIMC) simulations and Quantum Molecular Dynamics simulations, based on density functional theory (DFT-MD), have demonstrated their ability to determine thermophysical properties of WDM accurately.

The key elements of PIMC simulations are the representation of the density matrix via a path-integral and the evaluation of the corresponding integrals via Monte Carlo methods; for details, see [20]. In DFT-MD simulations, the electronic structure is determined via finite-temperature density functional theory at every timestep of a classical molecular dynamics simulation for the ions; see e.g. [21, 22].

In this paper we give a brief introduction into the DFT-MD method and present exemplary results for the most abundant planetary materials, especially hydrogen and helium, covering equation of state (EOS) data, the phase diagram, and the electrical conductivity. We discuss the impact of these ab initio data on state-of-the-art models of planetary interiors, evolution scenarios, and magnetic fields configurations.

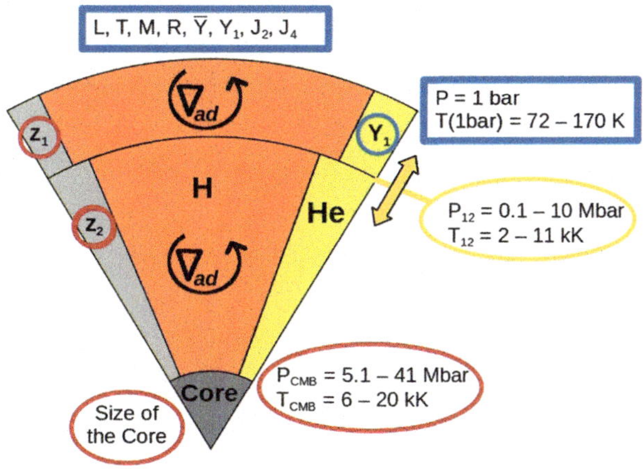

Fig. 1 Standard three-layer structure model assumed for the solar GPs with observational constraints (*blue*), variable parameters (*yellow*) and results from the modeling procedure (*red*)

1.2 Giant Planets in the Solar System

Our solar system consists of four rocky planets (Mercury, Venus, Earth, and Mars) and four GPs (Jupiter, Saturn, Uranus, and Neptune). In this paper we focus on the behavior of the lightest elements hydrogen and helium, their mixtures, and of hydrides such as water under conditions as relevant for the interior of GPs. Figure 1 gives a schematic representation of the interiors of the solar GPs within a standard three-layer model [7]: a central rocky core is surrounded by two fluid envelopes which are assumed to be fully convective so that the pressure-temperature profile follows an adiabat. The mass fractions of hydrogen (X), helium (Y), and of all heavier elements (i.e. *metals*, Z) can be different in each layer. From observations we know for the solar GPs the mass M, radius R, surface pressure $P(R)$ and temperature $T(R)$, the mean helium mass fraction $\overline{Y}$ and the atmospheric helium mass fraction Y_1, the atmospheric metallicity Z_1, the period of rotation, the lowest-order gravitational moments J_{2i}, the age of the planet and its luminosity L. These *observational constraints* are used to infer details about the interior structure, see Sect. 4.1. The interior of the solar GPs is in WDM states, see Fig. 1: the boundary between the fluid envelopes is at about $P_{12} \sim (0.1 - 10)$ Mbar and $T_{12} \sim (2{,}000 - 11{,}000)$ K, while the core-mantle boundary is predicted at about $P_{\text{CMB}} \sim (5 - 40)$ Mbar and $T_{\text{CMB}} \sim (6{,}000 - 20{,}000)$ K. Therefore, the thermophysical properties of WDM are an essential ingredient in planetary physics studies. Some fundamental problems in this context are

- The number and composition of the layers (e.g. less or more than three?),
- The physical origin of the layer boundaries (e.g. due to *metallization* of hydrogen or due to *demixing* of helium from hydrogen?),
- Their stability and location inside the planet (e.g. does vertical mass transport and/or core erosion occur?),

- The size and thermodynamic state of the core (e.g. is there a core at all and, if yes, is it solid or liquid?),
- The derivation of mass-radius relations for GPs dependent on their composition,
- The origin of the planetary dynamo and the shape of the magnetic field (i.e. dipolar field or higher multipole moments, axisymmetric or tilted?),
- The understanding of the formation process of GPs (i.e. *core accretion* or *disk instability*?),
- The presence of internal heat sources such as helium settling, and the heat transport mechanisms (i.e. convection only or semi-convection?).

1.3 Extrasolar Giant Planets

The number of extrasolar GPs is rapidly increasing, with already hundreds of detected planets around other stars [9–12]. Therefore, planetary physics has been revolutionized since the discovery of the first exoplanet around a solar-like star in 1995 (51 Pegasi [23]). The most successful observational techniques are the radial velocity (RV) and the transit method. The first one investigates variations of spectral lines emitted by the star via the Doppler effect. These change periodically due to the star's motion around the common center-of-mass if the star has one or more planetary companions. The RV method gives information on the minimum mass of the accompanying planet. The transit method yields the mean radius of the planet eclipsing its host star since the decrease of the apparent stellar flux is proportional to the size of the planet, i.e. $\sim R_P^2$. Most important in this context is that the CoRoT [24] and Kepler [25] missions were designed in order to detect a large number of transiting planets, among them hopefully the first Earth-like planet in the habitable zone of its parent star. Today, the sample size of exoplanets with known masses and radii for which their bulk density can be derived is exceeding 200 [26, 27]. Based on this data, new models for their composition and interior structure, their formation and evolution can be made which are in many aspects very different from the assumptions that have been proposed for the solar GPs until 1995. Other methods to detect exoplanets are pulsar timing, direct imaging, and microlensing, see [9, 12]. Further substantial progress in planetary physics is expected since the number of exoplanets and, therefore, the sample size, is rapidly increasing.

All known exoplanets detected by these methods (color coded) are shown in Fig. 2 together with the solar system planets. We display their mass with respect to the orbital distance to their host stars (semi-mayor axis). Since transiting probability and frequency decrease with increasing distance from the star, most of the detected transiting planets are very close to their host stars. On the other hand, large planetary masses lead to stronger Doppler shifts in the star spectrum and, thus, to a better detectability so that most of the detected RV planets so far have Jupiter-like masses.

The two big arrows in Fig. 2 indicate that the closer a planet orbits its host star the hotter is its interior. Simultaneously, the density in the interior of a planet increases

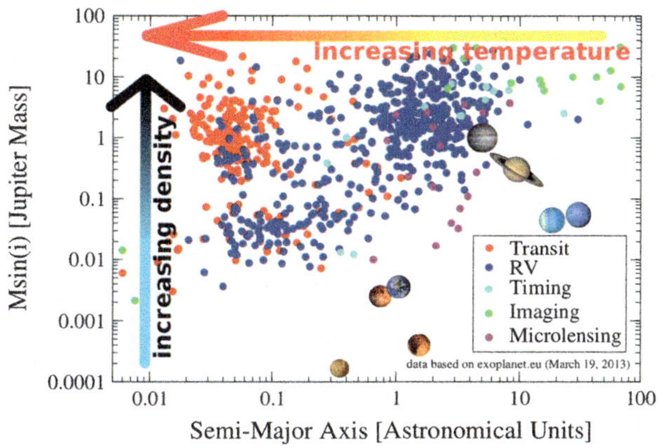

Fig. 2 The mass of exoplanets versus their semi-major axis (data taken from [28]). The color code represents the detection method: *dark blue* – RV method, *red* – transit method, *light blue* – pulsar timing, *green* – imaging, *magenta* – microlensing. The solar system planets are shown for comparison (size not true to scale)

with the total mass of the planet. Therefore, the solar GPs and the rapidly growing sample of exoplanets represent a perfect laboratory to scan the density-temperature plane and, therefore, to study correlation and quantum effects as inherent in WDM.

As of April 2013, there are about 850 confirmed exoplanet candidates. A further 2,000 candidates were detected by the Kepler mission [25, 29]. Their statistics indicates that most exoplanets are of Neptune-size or smaller which leads to a classification depending on their mass and radius (and hence their density) in Earth-like, Neptune-like, Jupiter-like planets and Super-Earths. The latter planetary class has no representative in the solar system, an example is GJ 1214b [30, 31]. Another interesting planet is Kepler 22 with a radius of 2.4 R_E ($1\,R_E = 6{,}378$ km) and a mass of 10–35 M_E ($1\,M_E = 5.974 \cdot 10^{24}$ kg) – the first detected exoplanet in the habitable zone of its host star [32]. The Kepler 11 system [33] consists of at least six transiting planets, the inner five have radii of few Earth radii and masses between that of Earth and Neptune, i.e. (1–18) M_E. Their mean densities imply that planets b and c are Super-Earths and d-f represent perhaps small gaseous planets, i.e. planets with no analog in our solar system. The region of detected exoplanets has been extended recently to objects even smaller than Mercury: the Kepler 37 system [34] consists of three transiting planets, the innermost having a radius of only $\sim$0.3 R_E and thus its size is similar to that of the Moon. The first exoplanet observed by direct imaging is Formalhaut b, a Jupiter-like planet orbiting its host star in an eccentric orbit with a mean distance of 115 AU within a dust ring [35]. A catalogue of confirmed exoplanets can be found at [28]. The classification of exoplanets is shown in Fig. 3 by means of a mass-radius relation.

Some points are important for the future development of this field. First, the rapidly growing number of exoplanets will have a strong impact on astrophysics and, simultaneously, will initiate further interest in WDM research. Second, new high-pressure platforms such as NIF and diagnostic methods implemented at FELs will lead to breakthroughs in dense plasma experimental research. Third, the ongoing increase of computing power will promote the development and application

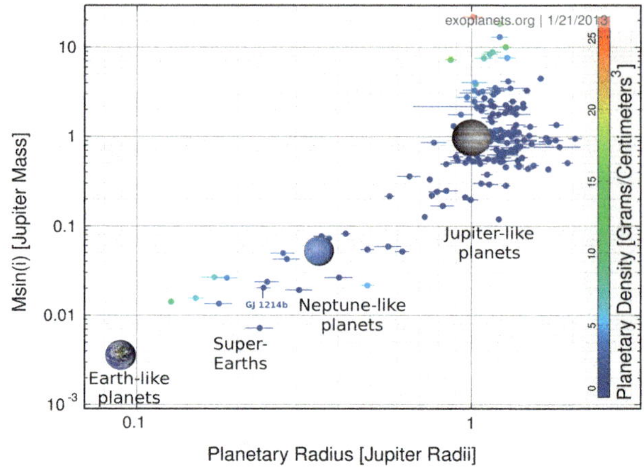

Fig. 3 Classification of exoplanets with respect to Earth-like, Neptune-like and Jupiter-like planets and Super-Earths (e.g. GJ 1214b) [36]

of ab initio methods in WDM research. These trends will advance a much stronger overlap and interplay between plasma, planetary, and computational physics in the future.

2 DFT-MD Simulations

As outlined in the introduction, DFT-MD simulations combine quantum mechanical DFT calculations for the electrons with a classical MD simulation for the ions. The basic workflow is shown in Fig. 4, and the constituents, especially the DFT part, are explained in the following sections. This method has first been applied to plasma physics and WDM in the 1990s [40, 41]. Other chapters in this book deal with special aspects of this method. New developments for higher accuracy and predictability (e.g. deriving better XC functionals) and more efficient methods to solve the Kohn-Sham equations (see Sect. 2.1, e.g. using plane-waves [22] or localized orbitals [42] as basis sets) will trigger future progress in this area.

2.1 Kohn-Sham Equations

The solution of the many-particle Schrödinger equation requires an enormous amount of dimensions which is not feasible for more than a few (<10) electrons. This problem can be circumvented by density functional theory, which deals with the quantum mechanical description of electrons in an external potential. It is based on the theorems of Hohenberg and Kohn [43] which state:

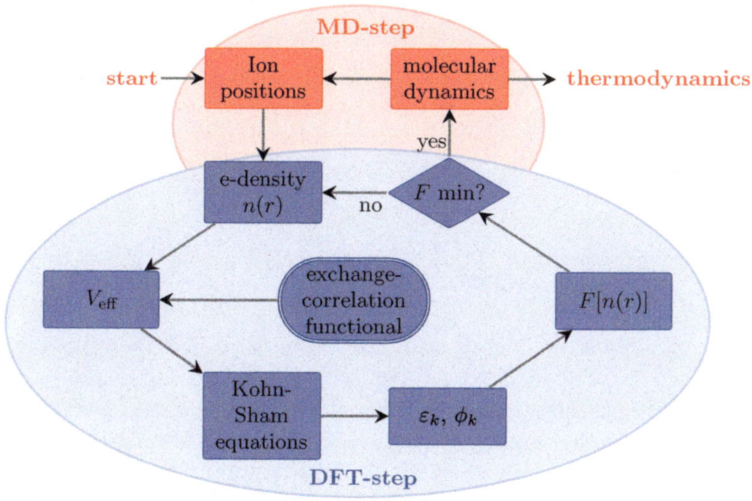

Fig. 4 DFT-MD workflow as implemented in VASP [22, 37–39]. The central part of the method is the DFT calculation (*blue*) which is performed for each MD step (*red*). Basic input for the DFT calculation is the XC functional

1. If two (electron) systems with an external potential $v_1(\mathbf{r})$ and $v_2(\mathbf{r})$ have the same ground state density $n(\mathbf{r})$, then the potentials can only differ by a constant.
2. The density functional $E[n(\mathbf{r})]$ has its minimum at the ground state density.

This was generalized to finite temperatures by Mermin [44] where a functional $\Omega[n(\mathbf{r})]$, corresponding to the grand potential or, more relevant for DFT-MD simulations, a free energy functional $F[n(\mathbf{r})]$ is used instead of $E[n(\mathbf{r})]$. These theorems have a huge advantage over the usual formulation of quantum mechanics, because they basically say that we do not need to solve the many-particle Schrödinger equation, we "only" need to find the electron density which yields the minimum energy. The difficult part is, how to find the "correct" energy functional $E[n(\mathbf{r})]$ and how to calculate from this the ground state energy and density.

The basis of DFT is the energy functional $E[n(\mathbf{r})]$. While Hohenberg and Kohn and Kohn and Sham proposed functionals for constant and slowly varying density, the search for accurate functionals is still ongoing and the results of the calculations can depend heavily on the used functional [45]. Formally, the energy functional can be written in the form

$$E[n(\mathbf{r})] = V_{\text{ext}}[n(\mathbf{r})] + U_{\text{H}}[n(\mathbf{r})] + T_{\text{s}}[n(\mathbf{r})] + E_{\text{XC}}, \tag{1}$$

where

$$V_{\text{ext}} = \int v_{\text{ext}}(\mathbf{r})n(\mathbf{r})d^3r \tag{2}$$

is the energy due to the external potential $v_{\text{ext}}(\mathbf{r})$,

$$U_{\text{H}}[n(\mathbf{r})] = \frac{1}{2}\int\int \frac{n(\mathbf{r})n(\mathbf{r}')}{|\mathbf{r}-\mathbf{r}'|} d^3r d^3r' \tag{3}$$

is the Hartree energy,

$$T_{\text{s}}[n(\mathbf{r})] = \sum_i^N \int \phi_i^*(\mathbf{r}) \left(-\frac{1}{2}\nabla^2\right) \phi_i(\mathbf{r}) d^3r \tag{4}$$

is the Kohn-Sham kinetic energy, and E_{XC} is the so-called exchange-correlation (XC) functional. Note that Hartree atomic units are used throughout this chapter. In this XC functional all unknown properties of the interacting system are gathered, and this functional is the central ingredient for DFT calculations.

The simplest approximation is the local density approximation (LDA) [46–48] where the XC contribution is calculated for a homogeneous electron gas at the local density $n(\mathbf{r})$. This treatment is correct for slowly varying electron densities and yields the correct limiting case for very high densities. Generalized gradient approximations (GGA) are a better approach, where the XC functional depends not only on the electron density, but also on its gradient. The GGA by Perdew, Burke, and Ernzerhof (PBE) [49] is used for most results shown in Sect. 3, except for few calculations with the hybrid functional proposed by Heyd, Scuseria, and Ernzerhof (HSE) [50, 51]. Even better XC functionals are computationally much more demanding, and up to now not feasible for large-scale simulations.

Since the beginning of DFT in 1965 up till now more than 50 XC functionals have been developed, partly based on empirical data and partly based on ab initio theory. Several works are dedicated to benchmarking different functionals for e.g. lattice constants [45, 52–56], bulk moduli [52, 53], bond lengths [57–59], and band gaps [53, 54]. In general, there is no XC functional which performs equally well for all properties and all elements.

The PBE functional yields often not the most accurate results, but for many applications it performs quite well compared to its computational demands. Additionally, it is an ab initio functional in the sense that it is nonempirical. One of the main problems of the PBE functional (and other LDA and GGA functionals as well) is the self-interaction error, i.e. they are not self-interaction free, leading to too small band gaps [60, 61]. In principle, this can be improved by better functionals like HSE [53, 61] or EXX-LDA [62], however, with much higher computational costs.

The Hohenberg-Kohn theorems were used by [46] to derive a set of equations

$$\left(-\frac{1}{2}\nabla^2 + v_{\text{eff}}(\mathbf{r})\right) \phi_i(\mathbf{r}) = \varepsilon_i \phi_i(\mathbf{r}), \tag{5}$$

$$n(\mathbf{r}) = \sum_i^N |\phi_i(\mathbf{r})|^2, \tag{6}$$

$$v_{\text{eff}} = v_{\text{ext}}(\mathbf{r}) + \int \frac{n(\mathbf{r}')}{|\mathbf{r} - \mathbf{r}'|} d^3 r' + \frac{\delta E_{xc}}{\delta n(\mathbf{r})}, \tag{7}$$

which have to be solved self-consistently, thus reformulating the many-body Schrödinger equation into a set of effective one-particle Schrödinger equations. Hereby, an iterative recipe to solve the problem is available, by starting from a guessed ground state density $n(\mathbf{r})$, constructing the effective potential v_{eff} (Eq. 7), solving Eq. 5, and calculating the new density $n(\mathbf{r})$ (Eq. 6). This iteration has to be performed until the energy is converged, see also Fig. 4.

2.2 Numerics and Convergence

For the evaluation of the Kohn-Sham equations, one has to represent the wavefunctions in a convenient way. Many available codes use an expansion into plane waves, which have useful features for calculations, e.g. the implementation of periodic boundary conditions is straightforward and Fourier transformations can easily be done. One problem of plane waves is the representation of the wavefunctions near the ions, since the Coulomb potential leads to strong oscillations. To represent these oscillations many plane waves are needed, which make calculations very demanding. To circumvent this problem, pseudopotentials were introduced. The general idea is to replace the wavefunctions inside a sphere around the ions with smooth functions, without changing the physical outcome of the calculations. The first pseudopotentials introduced where the so-called norm-conserving pseudopotentials [63, 64], which conserved the norm of the wavefunctions at the sphere boundary. But still, for some elements, e.g. transition-metal elements or elements with d or f electrons, very "hard" pseudopotentials, i.e. small sphere cutoff radii, are needed, which still leads to high plane wave cutoffs. This could be solved by ultrasoft pseudopotentials [65] which relaxed the norm-conserving condition at the cost of a more involved pseudopotential generation. However, these generated pseudopotentials have to be tested extensively [66]. A better approach is the projector augmented wave (PAW) method [66, 67] which is based on a linear transformation between the all-electron and pseudo-wavefunctions. In principle, all these techniques are only of technical nature, i.e. they should only reduce the computational costs without changing the physical outcome. This has to be checked carefully in convergence tests with respect to plane wave cutoff (number of plane waves), and the radial augmentation cutoff.

Due to the periodic boundary conditions it is necessary to perform some of the calculations in reciprocal space by integrating over the Brillouin zone. However, for DFT-MD simulations it is not possible (and usually not needed) to sample the Brillouin zone with a very fine grid, so that the integration is replaced by a summation over some special $\mathbf{k}$ points. In fact, often only one special $\mathbf{k}$ point is needed, but obviously this calls for convergence tests with respect to the $\mathbf{k}$-point sampling. There are several possibilities to sample the Brillouin zone. One very

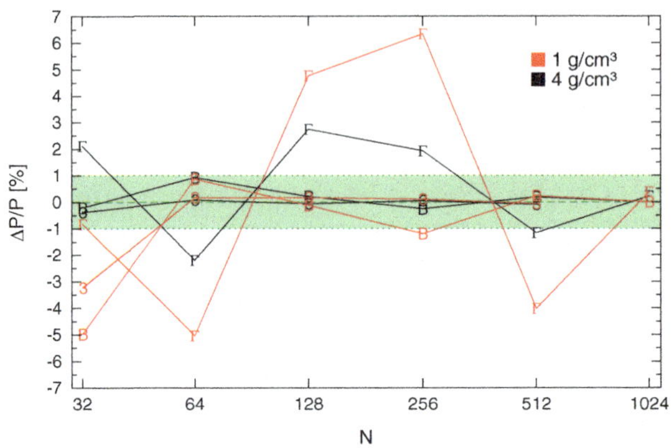

Fig. 5 Convergence of the pressure with respect to the particle number for the Γ point (Γ), the BMVP (B), and a $3 \times 3 \times 3$ Monkhorst-Pack grid (3). Shown are results for liquid hydrogen at a density of 1 g/cm^3 (*red*) and 4 g/cm^3 (*black*) and a temperature of 1,000 K

common way for DFT-MD simulations is to use only the Γ point. Choosing this point has (apart from being only one point) the advantage that the wavefunctions have real values and the code does not need to deal with complex numbers, which can speed up the calculations. However, one has to check very carefully if the results are converged, which is sometimes only the case for higher particle numbers compared to other choices.

A common method for choosing more **k** points is the method of Monkhorst and Pack [68] where a discrete grid of **k** points is generated. Another special point that can be used instead is the so called "Baldereschi mean-value point" (BMVP) [69]. It has the advantage that many quantities converge very fast with respect to higher **k**-point sets. An example for convergence tests with respect to **k**-point sampling and particle number is shown in Fig. 5, where the good convergence of the BMVP for hydrogen simulations can be seen.

By using the Born-Oppenheimer approximation, one can perform classical MD simulations for the ions, and use DFT for the electrons. The basic principle is illustrated in Fig. 4. Like the classical MD simulations it starts with an ion configuration inside a simulation box. These ions provide the external potential for the DFT calculations for the electron system. With a (guessed) initial electron density $n(r)$ the Kohn-Sham equations are solved, and from the Kohn-Sham eigenvalues and wavefunctions a new electron density and the resulting energy functional can be calculated. This procedure can be repeated with the new electron density, until the energy functional reaches its minimum. Afterwards the forces on the ions are calculated from the Hellmann-Feynman theorem [70] and, as in the classical MD simulations, the ions are moved in a finite timestep. In this way the forces are calculated ab initio for each timestep and the fundamental problem of classical MD simulations (knowing the potential) can be solved. This workflow is implemented in several codes such as VASP [22, 37–39] and abinit [71, 72].

2.3 Electrical Conductivity

Within DFT the dynamic electrical conductivity can be calculated with the Kubo-Greenwood formula [73–78]

$$\sigma(\omega) = \frac{2\pi}{3V\omega}\sum_{\mathbf{k}} w_{\mathbf{k}} \sum_{j=1}^{N_b}\sum_{i=1}^{N_b}\left[f_{j,\mathbf{k}} - f_{i,\mathbf{k}}\right] \times \left|\langle\phi_{j,\mathbf{k}}|\hat{\mathbf{p}}|\phi_{i,\mathbf{k}}\rangle\right|^2 \delta\left(\varepsilon_{i,\mathbf{k}} - \varepsilon_{j,\mathbf{k}} - \omega\right), \tag{8}$$

where V is the volume of the simulation box and ω the frequency. The summation over the matrix elements of the Kohn-Sham orbitals with the momentum operator $\hat{\mathbf{p}}$, weighted with the difference of the Fermi occupation numbers $f_{i,\mathbf{k}}$, is performed over all N_b bands. This is, however, very time consuming and needs much disk space. Therefore, it cannot be calculated for each timestep of the simulation, but only for snapshots of the simulation. Convergence tests have to be done with respect to the **k**-point sum [79]. Additionally, it is possible within the Kubo theory to derive the thermal conductivity and the thermopower [78]. While it is possible to calculate optical properties like the reflectivity from the dynamic conductivity [78, 80], only results for the static limit ($\omega \to 0$) are shown in Sect. 3.

Since the Kubo-Greenwood formula evaluates transitions between discrete energy eigenvalues obeying energy conservation via $\delta\left(E_{i,\mathbf{k}} - E_{j,\mathbf{k}} - \omega\right)$, the electrical conductivity is zero most of the time and has only finite values when the frequency ω is equal to an exact energy difference between two bands. However, since the simulation can only be performed for a given number of electrons in a finite box, the bands form a discrete spectrum while they lie energetically very close for a real system. To circumvent this (unphysical) problem, the δ-function is replaced by a Gaussian with a finite width. Depending on this width the dynamical conductivity will be more or less smooth.

But still, the conductivity would drop down for small ω, approximately when ω is smaller than the mean band distance. In general this effect can be reduced with higher particle numbers, since then the mean band distance gets smaller. Nevertheless, the direct limit of $\omega = 0$ cannot be reached. Two approaches are possible for obtaining this limit. In the first approach, the width of the Gaussian can be increased, until smooth functions up to $\omega = 0$ are obtained. But especially for small particle numbers the resulting static conductivity would depend very much on the chosen width and the results would be to some extent arbitrary. A better approach is to perform a regression (either a linear or an exponential function) in the area of small frequencies which are still higher than the mean band distance.

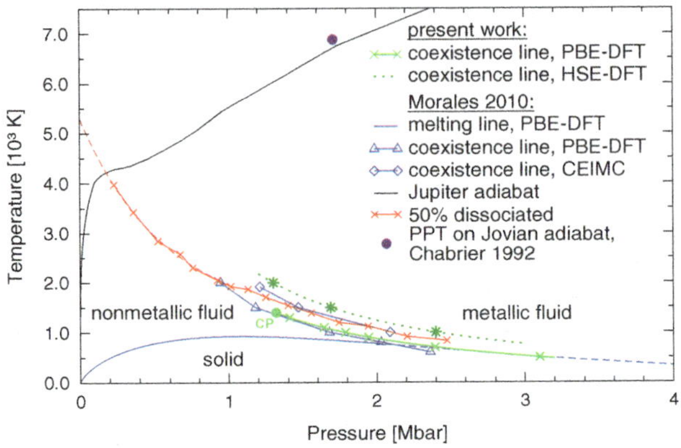

Fig. 6 High pressure phase diagram of hydrogen. Present work (*green*) within PBE-DFT [87] and HSE-DFT (W. Lorenzen. unpublished) is compared with that of Morales et al. (*blue*) [86] using PBE-DFT and CEIMC. Also shown are their melting line (*blue solid*), the Jupiter adiabat (*black line*) [89], and a *red line* characterizing 50 % dissociation [90]

By this approach the unphysical drop in the conductivity due to the finite system size is ignored, but a reasonable limit can be obtained.

Note that this Kubo-Greenwood formalism applies only to the electronic contribution to the conductivity. In principle it is possible to calculate the ionic conductivity from DFT-MD simulations [81], but the effective charges carried by each ion have to be known, which is very involved for complex mixtures [82].

3 Ab Initio Results for WDM

3.1 *Hydrogen*

Hydrogen as the simplest element is an ideal test case for WDM research. At the same time it is the most abundant element in the universe and the largest constituent of GPs. Especially for modeling GPs, the EOS for hydrogen is needed, which was therefore the aim of many studies, see e.g. [78, 83–85]. However, the main uncertainty and the main discrepancy between chemical models and the DFT-MD simulations was the existence of a first-order liquid-liquid phase transition from a nonmetallic molecular fluid to a dissociated metallic fluid (PPT). With the recent progress of ab initio simulations the existence of this phase transition seems to be clear [86, 87] and the problem is now reduced to determining the exact location of the transition. However, the experimental validation is still missing. An anomaly in the heating curve of liquid hydrogen was obtained recently [88] using pulsed lasers at static pressures in the Mbar region; this feature is closely correlated with our theoretical predictions for the PPT.

In Fig. 6 the high pressure phase diagram of hydrogen is shown. At temperatures below 1,000 K solid hydrogen occurs in various phases which are also of great interest, see [84, 85]. At higher temperatures, but still below 2,000 K, the above mentioned first-order phase transition is found. This liquid-liquid transition is

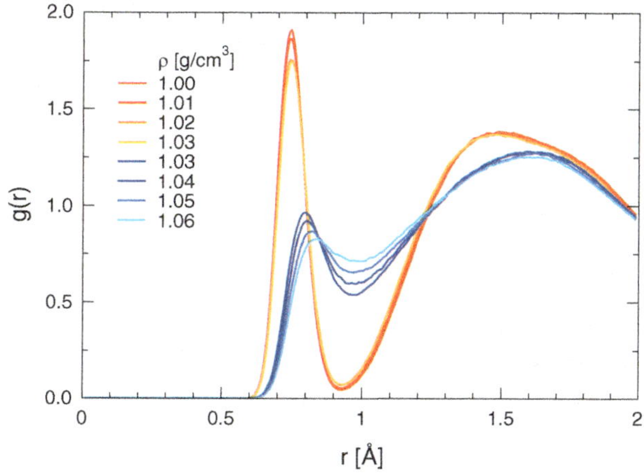

Fig. 7 Proton-proton pair correlation functions $g(r)$ at 700 K for densities near the phase transition [87]. The molecular peak vanishes abruptly at 1.03 g/cm^3 indicating that the transition is driven by dissociation of H_2 molecules

characterized by pressure dissociation as derived from the abrupt vanishing of the molecular peak, see Fig. 7. Simultaneously, a nonmetal-to-metal transition occurs which is reflected in the electrical conductivity data, see Fig. 8. The strong increase in the molecular phase (semiconducting fluid) is followed by a jump over 1–2 orders of magnitude when crossing the coexistence line. An almost temperature and density independent behavior is observed in the dense metallic fluid. The jump disappears above the critical point and the increase is steep but continuous there.

However, as can be seen from Fig. 6, the XC functional used in the DFT-MD simulations has an influence on the location of the liquid-liquid phase transition since the underlying nonmetal-to-metal transition is induced by bandgap closure. The PBE XC functional is known to yield too small bandgaps so that the transition pressure is underestimated (green line with crosses and blue line with triangles). More reliable results are expected using the HSE nonlocal XC functional (preliminary results are shown by the green dotted line with stars) and CEIMC simulations (blue line with diamonds). The impact of nuclear quantum effects and of van der Waals corrections in addition to that of a nonlocal XC functional as shown above has been studied recently [91].

3.2 *Hydrogen-Helium Mixtures*

A prominent example for a mixture in WDM is hydrogen and helium. This mixture is especially important for astrophysical objects, since hydrogen is in principle always accompanied by helium. Most of the modeling so far applies mixing rules to describe the properties of the mixture, especially for the EOS. However, more and more studies are devoted to the real mixture, see e.g. [92, 93]. An important effect which cannot be explained with mixing rules is the demixing of hydrogen and helium at high pressures. In particular, this process has been predicted to occur

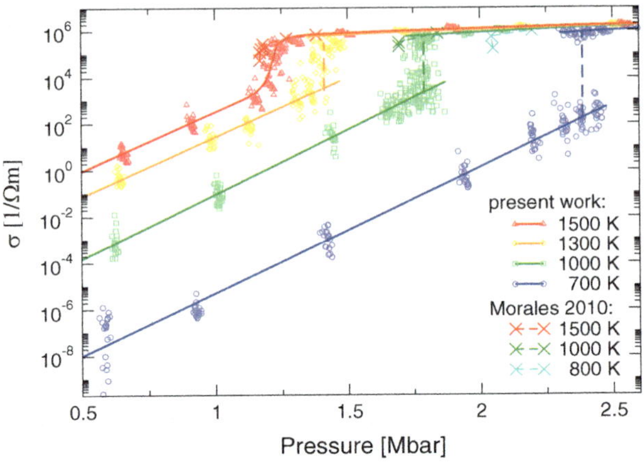

Fig. 8 Electrical conductivity for various isotherms [87]. The molecular phase is a semiconducting fluid. A jump occurs whenever crossing the coexistence line. An almost temperature and density independent slope is characteristic of the metallic fluid

in Saturn which would influence the thermal evolution of this planet strongly, see Sect. 4.3. To calculate this effect, one has to evaluate the Gibbs free energy of mixing

$$\Delta G(x) = \Delta U(x) + P\Delta V(x) - T\Delta S(x)\,, \tag{9}$$

where G is the Gibbs free energy, U the internal energy, P the pressure, V the volume, T the temperature, and S the entropy. The Δ denotes the difference between the mixed state and a linear mixing of the pure systems, e.g. $\Delta U(x) = U(x) - xU(1) - (1-x)U(0)$. Whenever the system can minimize its Gibbs free energy by demixing into two phases with different helium fractions x, demixing occurs. The only problem in DFT-MD calculations is the difficulty to access the entropy of mixing. The simplest approach is to use the ideal entropy of mixing

$$\Delta S_{\mathrm{id}}(x) = -k_B\left[x \ln x + (1-x)\ln(1-x)\right]\,, \tag{10}$$

while in principle better results can be obtained with thermodynamic integration techniques [94]. Two recent results for conditions in the interior of Jupiter and Saturn are shown in Fig. 9, in comparison to the phase diagrams of pure hydrogen and helium. The isentropes of both Jupiter and Saturn enter the demixing region at about 1 Mbar. While the jovian isentrope leaves the demixing region again at about 2.5 Mbar, and therefore has only a small fraction of its interior inside this region, the whole interior of Saturn is in the demixing region. This implies only a small effect on the structure and evolution of Jupiter, but has drastic effects on the evolution of Saturn, see Sect. 4.3. New results for lower pressures, i.e. in the molecular phase of hydrogen are in preparation [104]. All recent ab initio results reduced the uncertainty in the demixing temperatures drastically compared with previous predictions [13, 105, 106].

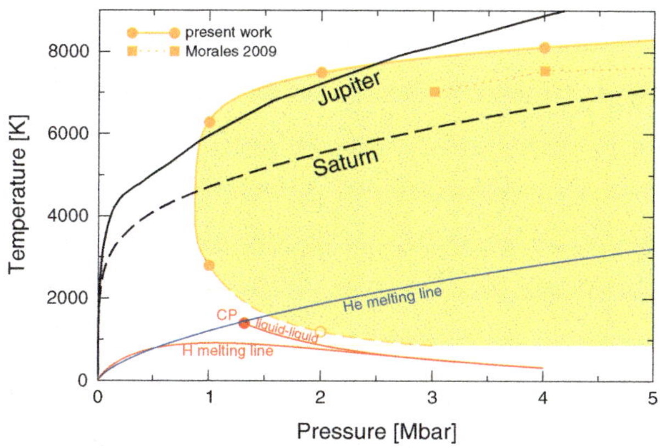

Fig. 9 Demixing region (*yellow area*) for a helium fraction of $x = 0.086$, i.e. the mean helium fraction in the planetary interiors, in comparison with the isentropes of Jupiter [95] (*solid black*) and Saturn (*dashed black*), see [96]. The *filled circles* are the calculated data from the Gibbs free energy, the *open circle* at 2 Mbar is extracted from conductivity data. The line is a fit to these points. The results of Morales et al. [94] for the demixing line are also shown (*squares*). We compare with the phase diagram of pure hydrogen (*red*). The melting line is taken from Ref. [86] and the coexistence line of the liquid-liquid phase transition with its critical point from Ref. [87]. For helium we show the melting line (*blue*; Kechin-type fit [97] to experimental data [98–102]) and our recent high-pressure prediction [103]

3.3 *Equation of State and Shock Wave Experiments for Hydrogen*

Probing the EOS in the WDM regime is a challenging task. Sophisticated experiments have to be performed to reach such extreme conditions. Pressures of several megabar can be generated using static diamond anvil cells or dynamic compression methods. Drivers for strong shock waves are, e.g., high-power lasers [107, 108], gas guns [109, 110], pulsed power [111, 112], or chemical explosions [113, 114]. For instance, all final states of single shock experiments performed with deuterium are located on the principal Hugoniot curve for a given initial state in the cryogenic liquid at 20 K and 0.17 g/cm^3. Recent experiments indicate a maximum compression of about 4.25–4.5 [108, 111, 115, 116]. States off the principal Hugoniot can be reached by varying the initial condition, e.g. by using a precompressed sample [117], by applying reverberation techniques [109, 111, 116], or by generating quasi-isentropic compression paths [118, 119]. All techniques but the last one yield the thermal EOS $P(\rho, T)$ or at least $P(\rho)$. Up to now, quasi-isentropic experiments only measure the compression ratio and determine pressures and temperatures via an EOS that is used in a hydrocode in order to reproduce the same compression.

Ab initio EOS data are benchmarked by shock wave experiments measuring $P(\rho, T)$ or $P(\rho)$. In Fig. 10 experimental data for the principal Hugoniot (orange

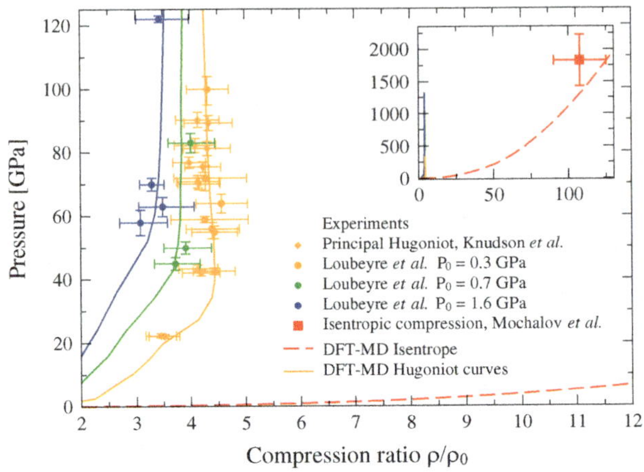

Fig. 10 Isentropic compression of deuterium (*red dashed*) starting at 0.04 g/cm^3 and 283 K reaches 108-fold initial density (inset) [119]. Hugoniot experiments with precompressed targets (*circles*) [117] start at 300 K and 0.087 g/cm^3 (*orange*), 0.12 g/cm^3 (*green*) and 0.16 g/cm^3 (*blue*). The *orange curve* is almost identical with the principal Hugoniot starting at 20 K and 0.0855 g/cm^3 for which we show data (*diamonds*) [111]. *Solid* and *broken lines* are DFT-MD results [120]

circles [117] and diamonds [111]) and Hugoniot curves for precompressed states of deuterium (blue and green circles [117]) are shown as well as a point representing a 108-fold compression of gaseous deuterium [119]. The DFT-MD Hugoniot curves for different initial conditions fixed by the experiment are derived from our ab initio EOS data [120] which coincide with another ab initio data set [84]. Both reproduce the experimental pressures very well while they underestimate the temperature onset of dissociation slightly [117]. This is again due to the bandgap problem when using the GGA-PBE functional. Note that higher precompression leads to smaller compression ratios. This has been shown earlier for hydrogen [84, 121] and helium [122].

Quasi-isentropic compression paths can be reconstructed evaluating Eq. (12) in Sect. 4.1 with a given EOS. We calculate the compression paths of deuterium starting from an initial temperature of 283 K and a density of 0.04 g/cm^3 [119] using our DFT-MD data [120]. For the measured compression ratio we expect a final pressure of ~13 Mbar at 1,500 K, significantly lower than the predicted 18 Mbar at 3,500 K as derived from a semi-empirical EOS used in the hydrocode. However, the data point is located on a quasi-isentrope, see Fig. 10, illustrating impressively that high pressures and low temperatures as relevant for states deep in planetary interiors can be reached in the laboratory today. More accurate experiments should be performed to discriminate between competing interior models and, thus, competing EOS data; see next Sect. 3.4.

3.4 Water

After hydrogen and helium, carbon, nitrogen and oxygen (and neon) are the next abundant elements. Thus, the high-pressure behavior of their hydrides CH_4, NH_3 and H_2O, and that of their mixtures, is of great interest in planetary physics, especially for Uranus and Neptune in the solar system [123]. This is further highlighted by the fact that Neptune-like exoplanets might be very common, see Fig. 3. We review here some new results for *warm dense water* as a prototypical example for molecular systems. A better knowledge of their high-pressure behavior is a prerequisite for an advanced understanding of the structure and evolution of such exoplanets.

Extensive DFT-MD simulations were performed to determine the EOS and the transport properties of warm dense water [21, 81, 124, 125]. We show in Fig. 11 the high-pressure phase diagram as predicted in Ref. [17]. Increasing the temperature in the liquid phase transforms water continuously from a molecular through a dissociated (ionic) to a plasma region (with *free* electrons). At high pressures we find various solid (ice) phases and, most interestingly, also an exotic superionic phase as predicted earlier [21, 124]. This phase consists of an oxygen lattice but mobile protons which would represent a proton conductor. The type the oxygen lattice under these conditions is still under debate, whether or not it retains the bcc structure of the high-pressure ice VII and ice X phases, or transforms to a fcc lattice with a slightly higher melting line [19].

The total electrical conductivity of warm dense water is shown in Fig. 12. The electronic contribution is calculated via the Kubo-Greenwood formula (8) using the HSE XC functional in order to obtain correct bandgaps [81]. The ionic contribution is determined by assigning an effective charge to the protons, see [21]. This analysis based on the diffusion coefficients can be improved by applying, e.g., polarization theory within each MD timestep in order to determine the effective charge of the proton rigorously which is, however, numerically very demanding [82]. It turns out that both methods agree well for warm dense water but not for the superionic phase where cross correlation terms (which are neglected in the diffusion analysis) have to be considered.

The curves show a systematic increase of the total conductivity with density and temperature. In the fluid phase below 4,000 K, the conductivity is mostly due to protons while the electronic contribution dominates at higher temperatures. Interestingly, the conductivity drops slightly (factor two) along the plasma-to-superionic phase transition (dotted lines). This is due to an increased localization of the electrons which stabilizes the oxygen lattice. At higher densities the conductivity increases again so that a nonmetal-to-metal transition is predicted in the superionic phase, simultaneously with a proton rearrangement. Protons occupy ice X sites more frequently at lower densities but favor octahedral sites in the high-density regime.

Shock-wave experiments with a so far unprecedented accuracy were performed recently at the Sandia Z machine that allow us to benchmark the high-pressure water EOS. We show the corresponding Hugoniot data for water (red symbols)

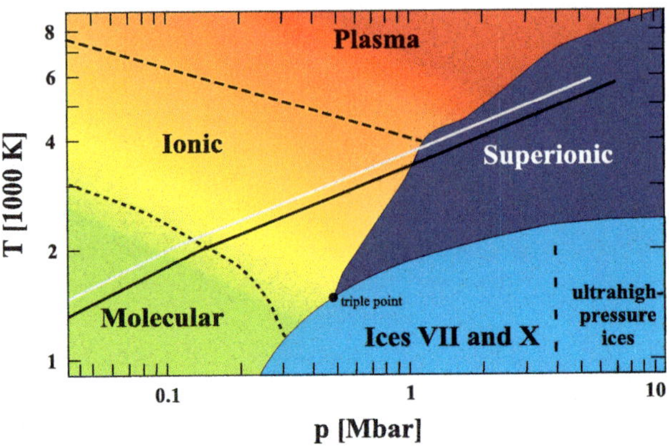

Fig. 11 High-pressure phase diagram of water [17]. The liquid phase includes a molecular, a dissociated (ionic) and a plasma region at higher temperatures. Besides the various solid phases (shown are ice VII and ice X), an exotic superionic phase is predicted. see also [19, 21, 124]. For comparison the theoretical isentropes of the water-rich planets Uranus (*grey solid*) and Neptune (*black solid*) are shown as well

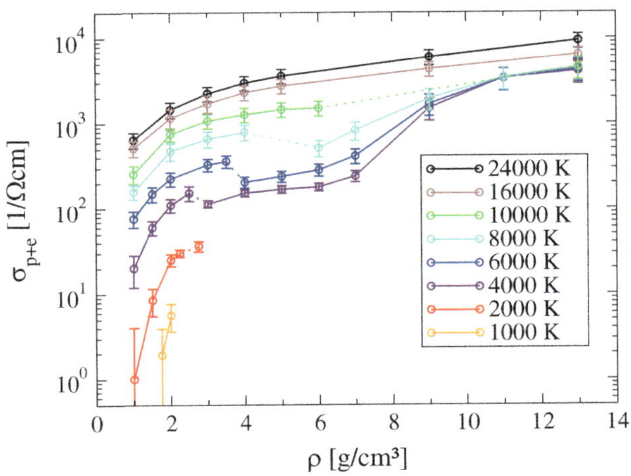

Fig. 12 Total electrical conductivity of water, i.e. electron plus proton contribution [81]. The *dotted lines* indicate the location of the phase transition between the plasma and the superionic phase, see Fig. 11

in Fig. 13 [126]. The Z data are in good agreement with gas gun data [128] and explosively driven shock data [129] but indicate a significantly lower compressibility than laser-driven shock data [130]. The vastly reduced uncertainty in the new Z data (roughly an order of magnitude), strongly suggests that water is much less compressible than the standard Sesame model predicts, and that water is instead very accurately described by ab initio EOS data [125, 131]. Furthermore, the reanalyzed ultra-high pressure point [132] is also in very good agreement with the ab initio

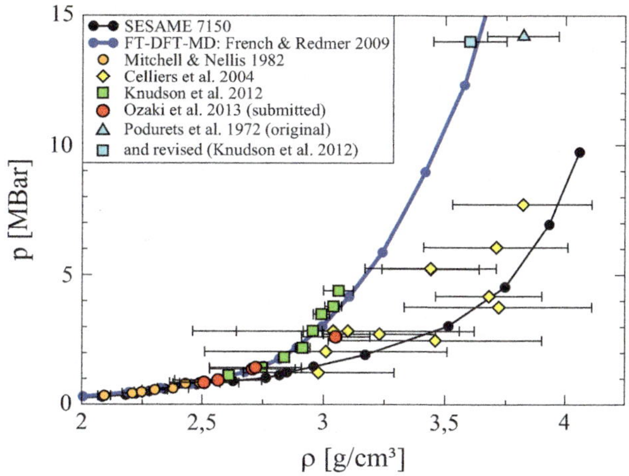

Fig. 13 Hugoniot curve for water. Z data (*green boxes* [126]) are compared with laser data (*yellow diamonds* [130] and *red circles* [127]) and gas gun data (*orange circles* [128]). Models include SESAME (*black line*) and DFT-MD (*blue line* [131]). The high-pressure point of Podurets et al. (*light blue triangle* [132]) has been reanalyzed (*light blue box*)

EOS. Thus, with the exception of the laser data, the EOS based on DFT-MD for water matches all experimental Hugoniot data up to 1.4 TPa. Furthermore, very good agreement is also achieved for second shock data and the reflectivity along the principal Hugoniot curve, see [126] for details. We conclude that this ab initio EOS can now be used to construct more reliable interior models for Uranus and Neptune [8, 17, 133], see also Sect. 4.2 and Fig. 17, but also for water-rich exoplanets, e.g., for GJ 436b [31] and the Super-Earths and Sub-Neptunes in the Kepler 11 system [18].

Similar experimental and theoretical studies are performed for ammonia [124, 134–141] and methane [134, 142–144] which have a rich phase diagram as well. Of central importance in this context is the treatment of multi-component C-N-O-H-He mixtures. Corresponding EOS data would serve as a new and more realistic data base for interior, evolution, and dynamo models of Neptune-like planets. These studies are closely connected with *high-pressure chemistry* since chemical equilibria are shifted at extreme conditions and new reaction channels might occur which are not important at ambient conditions. Interesting effects such as demixing and polymerization might occur as well. First promising results were obtained for a *synthetic* Uranus composition of H:O:C:N = 28:7:4:1 [145] and mixtures of CH_4 and H_2O [146].

3.5 Other Materials

The detection of Super-Earths has enhanced the interest in geophysics and initiated studies of the high-pressure behavior of typical mantle materials like MgO, SiO_2, $MgSiO_3$, $CaSiO_3$ etc. on a large scale; for reviews, see [147–150]. For instance, new predictions for the melting line based on ab initio simulations [151, 152] and a first-order liquid-liquid phase transition in silicate melts derived from laser-driven shock compression experiments were given [153] which could be important for stabilizing convective layering and compositional stratification in Super-Earths. These data are needed to predict, e.g., accurate mass-radius relationships for exoplanets [154] and the interior structure of Super-Earths [155].

Furthermore, extended data sets are generated for warm dense Fe, Ni, and Fe-Ni alloys which are expected to be the dominant core materials of rocky, ice-giant and gas-giant planets, including exoplanets. Of central importance in this context is the phase diagram of iron. Various phases transformations are predicted in the solid up to 1 Gbar (bcc-hcp-fcc-bcc) [156]. The slope of the high-pressure melting line is important for deciding whether or not the inner core of Super-Earths is solid (as in the Earth) or liquid [157, 158]. Solidification of iron at the boundary between the inner (solid) and outer (liquid) core is critical for the planetary magnetic field because most of the lighter elements are excluded from the inner core upon freezing into the corresponding lattice structure which releases most of the energy that drives convection and, thus, the planetary dynamo. The strength of the planetary magnetic field also depends on the electrical conductivity of solid and liquid iron under such conditions [159]. The influence of lighter elements on the electrical conductivity of, e.g., liquid Fe-Si and Fe-O alloys is also crucial [160]. For instance, the slope of the high-pressure melting line of iron decides whether or not Super-Earths are just up-scaled Earths (with an outer liquid core above an inner solid core) or solid bodies [161–163]. This is of paramount importance for the generation of possible magnetic fields and, thus, for the existence of even simplest life-forms on exoplanets.

4 WDM and Giant Planets

In this section we outline how giant planets serve as astrophysical laboratories for WDM calculations since typical parameters for their deep interior cover pressures of several ten Mbar and moderate temperatures of few eV, see Sect. 1. In particular, we discuss the influence of the plasma phase transition (PPT) in hydrogen (Sect. 3.1) and of demixing in hydrogen-helium systems (Sect. 3.2) on the interior and evolution of GPs, especially of Jupiter and Saturn.

4.1 Interior Models for Jupiter and Saturn

Interior structure models for Jupiter and Saturn have been calculated for decades, see e.g. [7, 89, 95, 164–169]. Those models have to reproduce all observational constraints, see Sect. 1.2, especially the shape of the gravity field. Caused by the rather rapid rotation of the planet (for Jupiter and Saturn, the rotational period is $\sim$10 h, for Uranus and Neptune $\sim$17 h) its gravitational field departs from a purely spherical function. To describe these deviations the field is expanded into Legendre polynomials $P_{2i}(\cos\Theta)$ with the polar angle Θ,

$$V(r,\Theta) = -\frac{GM}{r}\left(1 - \sum_{i=1}^{\infty}\left(\frac{R_{eq}}{r}\right)^{2i} J_{2i}\, P_{2i}(\cos\Theta)\right), \tag{11}$$

with the gravitational constant G and the equatorial radius R_{eq}. The gravitational moments J_{2i} can be measured through the analysis of the orbits of planetary satellites and of the trajectories of spacecraft flybys. These *constraints* have to be matched by any theoretical interior model that uses EOS data for the relevant planetary materials at WDM conditions. This is realized in the standard three-layer model by redistributing the amount of heavy elements within both adiabatic layers, see [170]. For a corresponding multi-layer approach including double-diffusive convection and compositional gradients, see [171].

A central goal of these efforts is the determination of the size of the planetary cores which would give valuable information on the formation process of planets, i.e. either via core accretion [172, 173] or via disk instability [174, 175]. Even for Jupiter, the prototypical and best studied GP in the solar system, the predictions vary from light cores (8.3 M_E [165], 5 M_E [166], (3–3.5) M_E [167], (0–14) M_E [168], (0–8) M_E [89]) to massive ones (14–18) M_E [169], see Fig. 14. Obviously there is still a big uncertainty whether Jupiter has a core or not. In contrast Saturn's core mass can be as high as 20 M_E [176, 177].

While interior models based on the EOS for hydrogen and helium within a chemical picture proposed by Saumon, Chabrier, and van Horn [178] (SCvH-EOS) agree in predicting a rather small core for Jupiter, the first Jupiter models that used an ab initio EOS differed significantly in their core mass prognosis, as can be seen in Fig. 14. The underlying planetary models are a three-layer model in case of Nettelmann et al. [89], using LM-REOS and fitting the gravity data by choosing corresponding metallicities in the two envelopes, and a two-layer model in the approach of Militzer et al. [169] who also use a DFT-MD based EOS and assume deep zonal winds to fit J_4. This discrepancy was resolved recently by Militzer [182] who showed that linear mixing of separate ab initio EOS for hydrogen and helium, as performed by Nettelmann et al., leads to the same results for an adiabat under Jovian conditions as derived from a real mixture of hydrogen and helium via thermodynamic integration.

A three-layer interior model (see Fig. 1) of Jupiter that fulfills all observational constraints is shown in Fig. 15. The model shown here has a transition pressure of 8 Mbar between the two adiabatic layers at 0.629 of Jupiter's radius, a core of 3.56 M_E, a metallicity of 0.038 ($\sim$2.5 Z_{solar}) in the outer layer, and of 0.128 in the inner layer. The core-mantle boundary is characterized by a pressure of 42 Mbar at moderate temperatures of 19,500 K. Former models based on the SCvH-EOS [166] argue that the interior of Jupiter is influenced by the PPT in hydrogen with a critical point at T_c = 15,300 K, P_c = 0.614 Mbar and ρ_c = 0.35 g/cm^3 leading to a jump in entropy at 1.17 Mbar and 6,880 K. Their prediction for the critical point was shown

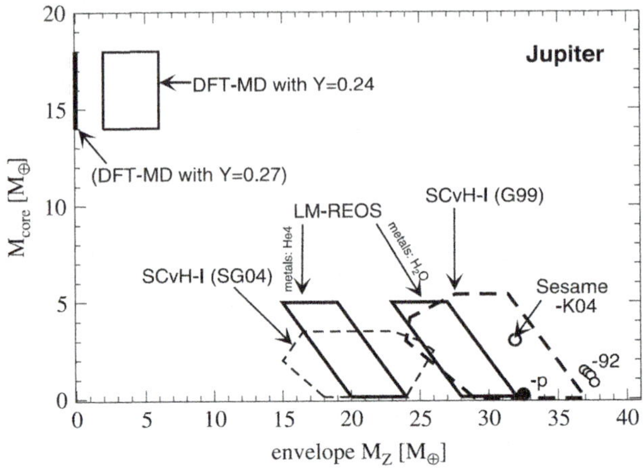

Fig. 14 Core mass vs. metallicity for Jupiter using different EOS data and interior models [170]. Shown are results using the H- and He-Sesame-92 EOS [179] and a modified version of Kerley (K04) [180]. LM-REOS results use a scaled He-EOS (left quad) and the H_2O-REOS (right quad) as a representative for metals. The other DFT-MD EOS [169] assumes $Y_1 = Y$ throughout the envelope. The SCvH-I models taken from [181] (SG04) assume $Z_1 = Z_2$ and those from [168] (G99) $Z_1 \neq Z_2$

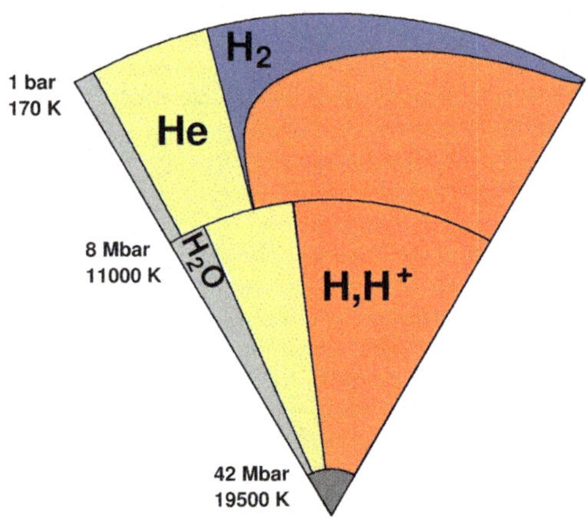

Fig. 15 Jupiter model based on LM-REOS and a three-layer structure with a transition pressure $P_{12} = 8$ Mbar at 0.629 of the Jovian radius [89, 186]. At the core-mantle boundary we find a temperature of 19,500 K and a pressure of 42 Mbar. The mass of the core amounts in this model 3.56 M_E. The fraction of metals in the fluid layers is $Z_1 = 0.038$ and $Z_2 = 0.128$

in Fig. 6 (phase diagram of hydrogen) together with the Jovian isentrope according to Fig. 15. The isentrope, however, is not influenced by the PPT as we conclude from the available ab initio calculations, but by the continuous transition from the molecular to the conducting fluid which still occurs above the critical point. The red line taken from Tamblyn and Bonev [90] indicates 50 % dissociation in hydrogen

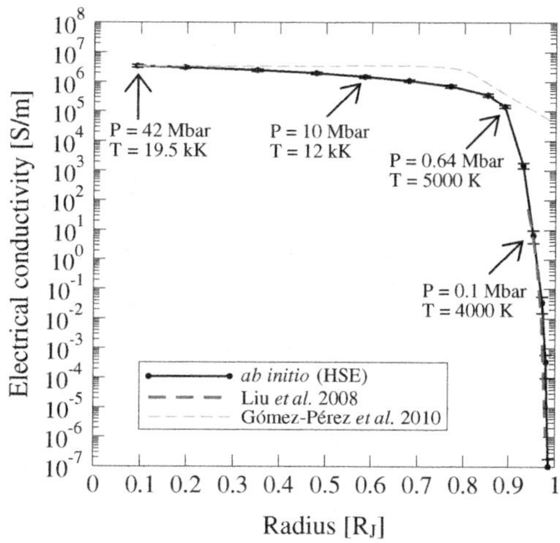

Fig. 16 Electrical conductivity along the Jovian adiabat, see [186]. The *solid line* represents an evaluation of the Kubo-Greenwood formula 8 based on ab initio simulations using the HSE functional. It is in perfect agreement with the semiconductor model of Liu et al. [183] (*red dashed line*) based on experimental data. Also shown is a model of Gomez-Perez et al. [187] used in dynamo simulations of Jupiter's magnetic field

for a given P-T value. It intersects the isentrope just in the region where it flattens. The isentrope obeys the relation

$$\left(\frac{\partial T}{\partial \rho}\right)_S = \frac{T}{\rho^2}\frac{\left(\frac{\partial P}{\partial T}\right)_\rho}{\left(\frac{\partial u}{\partial T}\right)_\rho} = \frac{T}{\rho}\gamma \tag{12}$$

with the Grüneisen parameter γ. The change of temperature T with density ρ at constant entropy S is governed by the absolute values of T and ρ and the derivatives of the pressure P and the specific internal energy $u = U/m$ with respect to T at constant ρ. The flattening of the isentrope is driven by the additional energy required to break the hydrogen bonds leading to an increasing denominator, namely c_V. This dissociation is accompanied by a sharp rise in electrical conductivity, see Fig. 16. It is calculated applying the Kubo-Greenwood formula (8) on VASP simulations using the HSE functional, see Sects. 2.3 and 3.1. The results are in very good agreement with a semiconductor model of Liu et al. [183] for the conductivity in the outer regions of Jupiter. It fits a linear decrease of the band gap with density according to shock wave data [109, 184, 185]. The conductivity and further material properties along the Jovian adiabat were calculated by French et al. [186]. These data will serve as a more realistic input in future dynamo simulations for planetary magnetic fields [187] that so far assume a superexponential behavior of the electrical conductivity (orange line in Fig. 16).

4.2 Interior Models for Uranus and Neptune

Jupiter and Saturn, which are mainly composed of hydrogen and helium have a mean density of 1.33 g/cm^3 and 0.69 g/cm^3, respectively. However, Uranus and

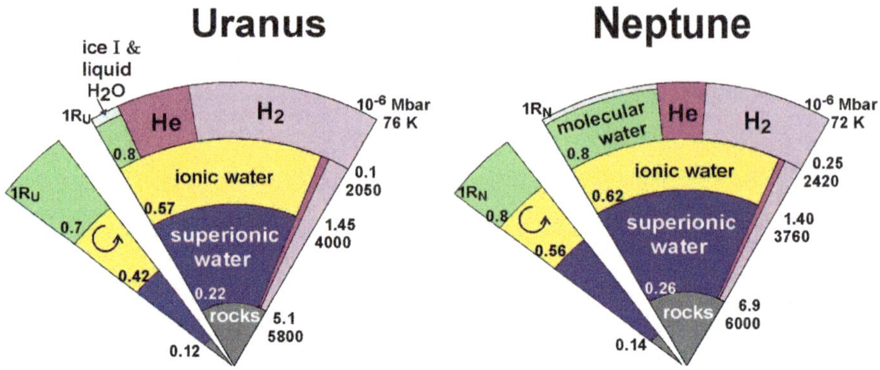

Fig. 17 Three-layer interior models of Uranus and Neptune (a rocky core is surrounded by two adiabatic fluid H-He-H_2O envelopes, larger pie charts on the *right*) [17] compared with predictions from magnetic field simulations (smaller pie charts on the *left*) [189, 190]

Neptune whose radii are only ~35 % of the Jovian one have mean densities of 1.27 g/cm^3 and 1.64 g/cm^3, which implies that they must contain a larger fraction of heavy elements (metals) than Jupiter and Saturn. State-of-the-art interior models usually assume a rocky core surrounded by two adiabatic fluid envelopes composed of hydrogen, helium and water [8, 133, 188]. Water serves as representative for metals or, more specifically, represents the real C-N-O-H mixture inside these planets. This underlines the great need for accurate EOS and conductivity data for this prototypical molecular system or molecular mixtures at WDM conditions, see Sect. 3.4.

Interestingly, predictions for the interior of Uranus and Neptune were derived from three-dimensional magnetic field simulations [189,190]. Their non-dipolar and non-axisymmetric magnetic fields are perhaps generated in a fairly thin layer of at most 0.3 planetary radii surrounding a conducting core that is fluid but stably stratified. In Fig. 17, these results (small pie charts) are compared with three-layer interior models described above [8, 17]. Numbers given on the left are the radius coordinates, numbers on the right denote pressure and temperature values. Below an envelope of helium and molecular hydrogen, water is the main constituent in its ionic phase (yellow) and, for more extreme conditions, in its superionic phase, see Fig. 11. A remarkable agreement of the interior composition of both planets which stem from two different modeling strategies (based solely on gravity data or the shape of the magnetic field) can be stated. However, the conditions for the occurrence of stably stratified layers in the deep interior have to be checked in more detail, e.g. by considering further properties of the superionic phase and possible demixing processes. A detailed analysis of the interior structure of both planets based on new gravity data is performed in Ref. [133].

4.3 Evolution of Giant Planets

After formation in the disk around a young star, the initially warm, luminous and extended planets cool down and contract in a radiative quasi-equilibrium determined by the received radiation flux from the parent star and the emitted radiation of the planet. The received radiation, with a maximum in the visible range, is characterized by the equilibrium temperature T_{eq}, the temperature a black body would have solely exposed to the flux of the star, without internal sources. However, Jupiter and Saturn are observed to emit almost twice the energy they receive from the sun, suggesting intrinsic sources of energy. These sources are addressed via the effective temperature T_{eff}, the temperature of a hypothetical black body emitting the same total radiation as the observed planet does. The evolution of the planet, which is in principal the change of T_{eff} with time, can then be calculated using the cooling equation:

$$4\pi R_p^2 (T_{\mathrm{eff}}^4 - T_{eq}^4) = -\int_0^M dm \frac{T(m,t)ds(m,t)}{dt} . \tag{13}$$

The left-hand side maps the intrinsic radiation source of the planet with radius R_p, since T_{eq} accounts for the influence of the pure solar radiation and T_{eff} that of the reflected solar radiation and the intrinsic planetary radiation sources. Knowing T_{eq} the evolution of T_{eff} is determined by the evolution of the total planetary heat content, given on the right-hand side of the cooling equation. The second law of thermodynamics relates the heat content to the entropy $s = S/m$, which is determined for the planet by T_{eff}, since the layers are isentropic. Integration of the heat content $T(m,t)ds(m,t)$ of each mass shell over the whole planet leads to the total heat content. Its change with time yields the change of T_{eff} with time and hence the evolution of the planet by integrating the cooling equation backward in time, starting from a present model. The age of the solar system and thus the age of the planets is 4.56 billion years (Gyr). Accurate planetary models should reproduce this age for solar system planets in the associated evolution calculations.

If the amount of helium Y_n and heavy elements Z_n as well as the mass M_n of each mass shell (with index n) are assumed to remain constant during time, the evolution is called homogeneous. On the contrary, a change of the particle concentration with time, see [191], leading to a change in Y_n or Z_n, or an extended inhomogeneous region in the planet, see [192], lead to an inhomogeneous evolution of a planet. The homogeneous evolution within one-layer models for Jupiter was first calculated by Hubbard predicting an age of 4–8 Gyr [193] and 5.1 Gyr [193]. Further calculations using the SCvH-EOS [178] with different model assumptions (see references) yield results of 5.2–5.3 Gyr [194], 4.2–5.2 Gyr [195], 4.7 Gyr [196], and 4.66 Gyr [89]. An example cooling curve for Jupiter is shown in Fig. 18. Obviously, homogeneous evolution models are suited to reproduce the correct age of Jupiter, which is also true for Neptune, predicting an age of 4.3–4.8 Gyr [133]. However, similar homogeneous evolution models for Saturn lead to an age of only 2–3 Gyr [194–198], suggesting

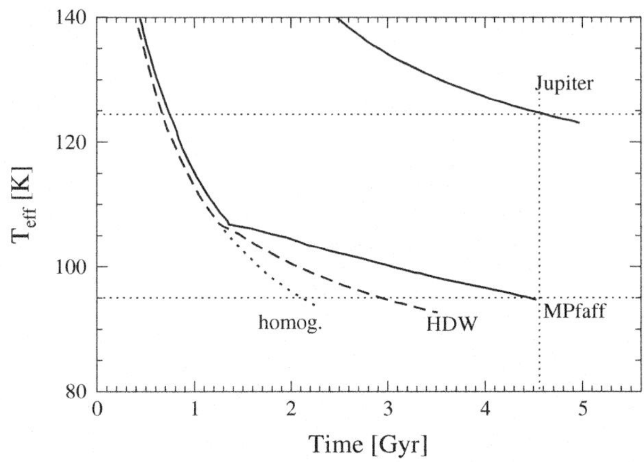

Fig. 18 Homogeneous evolution for Jupiter (*solid*) and Saturn (*dotted*) and inhomogeneous evolution for Saturn with respect to the H-He demixing calculations of Hubbard and DeWitt [202] (*dashed*) and Pfaffenzeller et al. [106] (*solid*) taken from [203]

an additional internal energy source, and of 6.8–10 Gyr [133, 198] in case of Uranus, indicating a retarding effect of ineffective cooling.

The issue of Saturn's short cooling time is closely related to the process of phase separation of hydrogen and helium at high pressures, see Sect. 3.2. First extensive studies on the "helium rain" in Saturn have been performed by Stevenson and Salpeter [191]. In the demixing region, helium is assumed to form droplets with a higher density than their surrounding so that they sink into deeper layers of the planet, thereby releasing gravitational energy and increasing the intrinsic luminosity. There is observational evidence that this process has been occurring in Saturn already for a long time (see Fig. 18) and that it has started in Jupiter as well. For instance, the protosolar He/H-ratio amounts 0.086. In case of Jupiter the Galileo entry probe measured a lower ratio of 0.079 [199], and infrared spectra measurements of Saturn's atmosphere indicate a ratio between 0.055 and 0.08 [200]. In addition, the theoretical isentrope of Saturn above 1 Mbar proceeds well through the theoretical demixing region predicted by Lorenzen et al. [103] and by Morales et al. [94], see Fig. 9. The intersection of the Jovian isentrope with the demixing region indicates that demixing might occur in Jupiter as well. Alternatively, Leconte and Chabrier [201] demonstrated that multi-layer models for Saturn containing compositional gradients and double diffusive convection lead to a considerably slower cooling of the planet.

For illustration, we show here previous results of Fortney and Hubbard [196] in Fig. 18. Homogeneous evolution yields an age of only ~2 Gyr (dotted line). An inhomogeneous evolution using the demixing calculations of Hubbard and DeWitt [202] would lead to an age of ~3 Gyr (dashed line). Assuming that H-He separation is the most important additional intrinsic heat source, they modified the phase diagram of Pfaffenzeller et al. [106] in order to reproduce the correct age of Saturn (solid line). These calculations have been extended to extrasolar giant planets [203], where demixing should occur as well. Evolution calculations for Saturn based on the ab initio H-He EOS of Lorenzen et al. [103] are in progress.

5 Conclusions

In this chapter we have outlined the method of ab initio simulations that is perfectly suited to determine the thermophysical properties of WDM. Various fundamental problems such as the shape of the high-pressure phase diagram (i.e. location of the melting line and of solid-solid transitions, existence of a first-order liquid-liquid phase transition in hydrogen or of a superionic phase in water or ammonia), the consequences of a pressure-driven nonmetal-to-metal transition in molecular systems, and the demixng of hydrogen and helium can be addressed consistently. The corresponding EOS and conductivity data are essential for the development of advanced interior and dynamo models for solar and extrasolar GPs. The ab initio data can be benchmarked by state-of-the-art shock wave and x-ray Thomson scattering experiments, and excellent agreement is usually achieved. Furthermore, new observations for solar (e.g. via the Juno mission to Jupiter) and extrasolar GPs (e.g. via the Kepler and other planned space missions) will substantially extend our so far limited and perhaps not representative data base for planetary physics. We conclude that a better understanding of WDM is a key element for the future development of planetary physics which is fortunately accompanied by the rapid and coincidental progress in ab initio simulations and shock wave experimental technique.

Acknowledgements We thank Mandy Bethkenhagen, Daniel Cebulla, David Ceperley, Michael Desjarlais, Martin French, Sebastien Hamel, Bastian Holst, Jonathan Fortney, Nadine Nettelmann, Robert Püstow, and David Stevenson for many helpful discussions. This work was supported by the Deutsche Forschungsgemeinschaft within the SFB 652, SPP 1385, SPP 1488, and the grant RE 882/11. Support by the North-German Supercomputing Alliance (HLRN) and the ITMZ of the University of Rostock is greatly acknowledged. Pictures of solar planets are taken from NASA/JPL/University of Arizona. This research has made use of the Exoplanet Orbit Database and the Exoplanet Data Explorer at exoplanets.org. We thank the Institute of Pure and Applied Mathematics, UCLA for the organization of the workshop series, the kind hospitality, and the support.

References

1. V. Fortov, I. Iakubov, *Physics of Nonideal Plasma* (Hemisphere Publishing, New York, 1990)
2. Committee on High Energy Density Plasma Physics, National Research Council, *Frontiers in High Energy Density Physics: The X-Games of Contemporary Science* (The National Academies Press, Washington, DC, 2003)
3. R.W. Lee, S.J. Moon, H.K. Chung, W. Rozmus, H.A. Baldis, G. Gregori, R.C. Cauble, O.L. Landen, J.S. Wark, A. Ng, S.J. Rose, J. Opt. Soc. Am. B **20**(4), 770 (2003)
4. S.H. Glenzer, R. Redmer, Rev. Mod. Phys. **81**, 1625 (2009)
5. B. Remington, R. Cavallo, M. Edwards, D.M. Ho, B. Lasinski, K. Lorenz, H. Lorenzana, J. Mcnaney, S. Pollaine, R. Smith, Astrophys. Space Sci. **298**, 235 (2005)
6. R.P. Drake, Physi. Plasmas **16**(5), 055501 (2009)
7. T. Guillot, Science **286**, 72 (1999)
8. J.J. Fortney, N. Nettelmann, Space Sci. Rev. **152**, 423 (2010)

9. S. Udry, N.C. Santos, Ann. Rev. Astron. Astrophys. **45**, 397 (2007)
10. J.J. Fortney, S.H. Glenzer, M. Koenig, B. Militzer, D. Saumon, D. Valencia, Phys. Plasmas **16**(4), 041003 (2009)
11. I. Baraffe, G. Chabrier, T. Barman, Rep. Progr. Phys. **73**, 016901 (2010)
12. N. Haghighipour, Contemporary Phys. **52**(5), 403 (2011)
13. D.J. Stevenson, Phys. Rev. B **12**(10), 3999 (1975)
14. D.J. Stevenson, Annu. Rev. Earth Planet. Sci. **10**, 257 (1982)
15. P.P. Edwards, M.T.J. Lodge, F. Hensel, R. Redmer, Phil. Trans. R. Soc. A **368**, 941 (2010)
16. R. Redmer, B. Holst, in *Metal-to-Nonmetal Transitions. Springer Series in Material Sciences*, vol. 132, ed. by R. Redmer, B. Holst, F. Hensel (Springer, Berlin, 2010), pp. 63–84
17. R. Redmer, T.R. Mattsson, N. Nettelmann, M. French, Icarus **211**, 798 (2011)
18. E.D. Lopez, J.J. Fortney, N. Miller, Astrophys. J. **761**, 59 (2013)
19. H.F. Wilson, M.L. Wong, B. Militzer, Phys. Rev. Lett. **110**, 151102 (2013)
20. D.M. Ceperley, E. Manousakis, J. Chem. Phys. **115**, 10111 (2001)
21. A.E. Mattsson, P.A. Schultz, M.P. Desjarlais, T.R. Mattsson, K. Leung, Model. Simul. Mater. Sci. Eng. **13**, R1 (2005)
22. J. Hafner, J. Comput. Chem. **29**, 2044 (2008)
23. M. Mayor, D. Queloz, Nature **378**, 355 (1995)
24. See http://smsc.cnes.fr/COROT/
25. See http://www.kepler.arc.nasa.gov/
26. J.J. Fortney, M.S. Marley, J.W. Barnes, Astrophys. J. **659**, 1661 (2007)
27. J.J. Fortney, Contrib. Plasma Phys. (2013, accepted)
28. See http://exoplanet.eu/catalog/
29. W. Borucki, et al., Science **327**, 977 (2010)
30. D. Charbonneau, et al., Nature **462**, 891 (2009)
31. N. Nettelmann, J. Fortney, U. Kramm, R. Redmer, Astrophys. J. **733**, 2 (2011)
32. W. Borucki, et al., Astrophys. J. **745**, 120 (2012)
33. J. Lissauer, et al., Nature **470**, 53 (2011)
34. T. Barclay, et al., Nature **494**, 452 (2013)
35. P. Kalas, J. Graham, M. Camplin, Nature **435**, 1067 (2005)
36. J.T. Wright, O. Fakhouri, G.W. Marcy, E. Han, J.A. Feng, Y.and Johnson, D.A. Howard, A. W.and Fischer, J.A. Valenti, J. Anderson, N. Piskunov, Publ. Astron. Soc. Pac. **123**, 412 (2011)
37. G. Kresse, J. Hafner, Phys. Rev. B **47**, 558 (1993)
38. G. Kresse, J. Hafner, Phys. Rev. B **49**, 14251 (1994)
39. G. Kresse, J. Furthmüller, Phys. Rev. B **54**, 11169 (1996)
40. L.A. Collins, I. Kwon, J.D. Kress, N.J. Troullier, D. Lynch, Phys. Rev. E **52**(6), 6202 (1995)
41. M.W.C. Dharma-wardana, F. Perrot, in *Density Functional Theory*, ed. by E.K.U. Gross, R.M. Dreizler (Plenum Press, New York, 1995), p. 625
42. V. Blum, R. Gehrke, F. Hanke, P. Havu, V. Havu, X. Ren, K. Reuter, M. Scheffler, Comput. Phys. Commun. **180**, 2175 (2009)
43. P. Hohenberg, W. Kohn, Phys. Rev. **136**, B864 (1964)
44. N.D. Mermin, Phys. Rev. **137**, A1441 (1965)
45. P. Haas, F. Tran, P. Blaha, Phys. Rev. B **79**, 085104 (2009)
46. W. Kohn, L.J. Sham, Phys. Rev. **140**, A1133 (1965)
47. J.P. Perdew, A. Zunger, Phys. Rev. B **23**, 5048 (1981)
48. J.P. Perdew, Y. Wang, Phys. Rev. B **45**, 13244 (1992)
49. J.P. Perdew, K. Burke, M. Ernzerhof, Phys. Rev. Lett. **77**, 3865 (1996)
50. J. Heyd, G.E. Scuseria, M. Ernzerhof, J. Chem. Phys. **118**, 8207 (2003)
51. J. Heyd, G.E. Scuseria, M. Ernzerhof, J. Chem. Phys. **124**, 219906 (2006)
52. S. Kurth, J.P. Perdew, P. Blaha, Int. J. Quantum Chem. **75**, 889 (1999)
53. J. Heyd, G.E. Scuseria, J. Chem. Phys. **121**, 1187 (2004)
54. K. Hummer, J. Harl, G. Kresse, Phys. Rev. B **80**, 115205 (2009)

55. P. Haas, F. Tran, P. Blaha, L.S. Pedroza, A.J.R. da Silva, M.M. Odashima, K. Capelle, Phys. Rev. B **81**, 125136 (2010)
56. P. Söderlind, A. Gonis, Phys. Rev. B **82**, 033102 (2010)
57. D.R. Hamann, Phys. Rev. B **55**, R10157 (1997)
58. V.N. Staroverov, G.E. Scuseria, J. Tao, J.P. Perdew, J. Chem. Phys. **119**, 12129 (2003)
59. J. Heyd, G.E. Scuseria, J. Chem. Phys. **120**, 7274 (2004)
60. S.V. Faleev, M. van Schilfgaarde, T. Kotani, F. Léonard, M.P. Desjarlais, Phys. Rev. B **74**, 033101 (2006)
61. Y. Zhao, D.G. Truhlar, J. Chem. Phys. **130**, 074103 (2009)
62. M. Städele, J.A. Majewski, P. Vogl, A. Görling, Phys. Rev. Lett. **79**, 2089 (1997)
63. D.R. Hamann, M. Schlüter, C. Chiang, Phys. Rev. Lett. **43**, 1494 (1979)
64. G.B. Bachelet, D.R. Hamann, M. Schlüter, Phys. Rev. B **26**, 4199 (1982)
65. D. Vanderbilt, Phys. Rev. B **41**, 7892 (1990)
66. G. Kresse, D. Joubert, Phys. Rev. B **59**, 1758 (1999)
67. P.E. Blöchl, Phys. Rev. B **50**(24), 17953 (1994)
68. H.J. Monkhorst, J.D. Pack, Phys. Rev. B **13**, 5188 (1976)
69. A. Baldereschi, Phys. Rev. B **7**(12), 5212 (1973)
70. R.P. Feynman, Phys. Rev. **56**, 340 (1939)
71. X. Gonze, B. Amadon, P.M. Anglade, J.M. Beuken, F. Bottin, P. Boulanger, F. Bruneval, D. Caliste, R. Caracas, M. Côté, T. Deutsch, L. Genovese, P. Ghosez, M. Giantomassi, S. Goedecker, D. Hamann, P. Hermet, F. Jollet, G. Jomard, S. Leroux, M. Mancini, S. Mazevet, M. Oliveira, G. Onida, Y. Pouillon, T. Rangel, G.M. Rignanese, D. Sangalli, R. Shaltaf, M. Torrent, M. Verstraete, G. Zerah, J. Zwanziger, Comput. Phys. Commun. **180**, 2582 (2009)
72. X. Gonze, Z. Kristallogr, **220**, 558 (2005)
73. R. Kubo, J. Phys. Soc. Jap. **12**, 570 (1957)
74. R. Kubo, M. Yokota, S. Nakajima, J. Phys. Soc. Jap. **12**, 1203 (1957)
75. D.A. Greenwood, Proc. Phys. Soc. **71**(4), 585 (1958)
76. M.P. Desjarlais, J.D. Kress, L.A. Collins, Phys. Rev. E **66**, 025401(R) (2002)
77. S. Mazevet, M. Torrent, V. Recoules, F. Jollet, High Energy Density Phys. **6**, 84 (2010)
78. B. Holst, M. French, R. Redmer, Phys. Rev. B **83**, 235120 (2011)
79. M. Pozzo, M.P. Desjarlais, D. Alfè, Phys. Rev. B **84**, 054203 (2011)
80. S. Mazevet, M.P. Desjarlais, L.A. Collins, J.D. Kress, N.H. Magee, Phys. Rev. E **71**, 016409 (2005)
81. M. French, T.R. Mattsson, R. Redmer, Phys. Rev. B **82**, 174108 (2010)
82. M. French, S. Hamel, R. Redmer, Phys. Rev. Lett. **107**, 185901 (2011)
83. B. Holst, R. Redmer, M.P. Desjarlais, Phys. Rev. B **77**(18), 184201 (2008)
84. L. Caillabet, S. Mazevet, P. Loubeyre, Phys. Rev. B **83**, 094101 (2011)
85. J.M. McMahon, M.A. Morales, C. Pierleoni, D.M. Ceperley, Rev. Mod. Phys. **84**, 1607 (2012)
86. M.A. Morales, C. Pierleoni, E. Schwegler, D.M. Ceperley, Proc. Natl. Acad. Sci. USA **107**, 12799 (2010)
87. W. Lorenzen, B. Holst, R. Redmer, Phys. Rev. B **82**, 195107 (2010)
88. V. Dzyabura, M. Zaghoo, I. Silvera, PNAS **110**, 8040 (2013)
89. N. Nettelmann, A. Becker, B. Holst, R. Redmer, Astrophys. J. **750**, 52 (2012)
90. I. Tamblyn, S.A. Bonev, Phys. Rev. Lett. **104**(6), 065702 (2010)
91. M.A. Morales, J.M. McMahon, C. Pierleoni, D.M. Ceperley, Phys. Rev. Lett. **110**, 065702 (2013)
92. J. Vorberger, I. Tamblyn, B. Militzer, S.A. Bonev, Phys. Rev. B **75**, 024206 (2007)
93. B. Militzer, Phys. Rev. B **79**, 014202 (2013)
94. M.A. Morales, E. Schwegler, D. Ceperley, C. Pierleoni, S. Hamel, K. Caspersen, Proc. Natl. Acad. Sci. USA **106**(5), 1324 (2009)
95. N. Nettelmann, B. Holst, A. Kietzmann, M. French, R. Redmer, D. Blaschke, Astrophys. J. **683**(2), 1217 (2008)
96. W. Lorenzen, B. Holst, R. Redmer, Phys. Rev. B **84**, 235109 (2011)

97. V.V. Kechin, J. Phys.: Condens. Matter **7**, 531 (1995)
98. R.K. Crawford, W.B. Daniels, J. Chem. Phys. **55**, 5651 (1971)
99. R.L. Mills, D.H. Liebenberg, J.C. Bronson, Phys. Rev. B **21**, 5137 (1980)
100. W.L. Vos, M.G.E. van Hinsberg, J.A. Schouten, Phys. Rev. B **42**, 6106 (1990)
101. F. Datchi, P. Loubeyre, R. Le Toullec, Phys. Rev. B **61**, 6535 (2000)
102. D. Santamaría-Pérez, G.D. Mukherjee, B. Schwager, R. Boehler, Phys. Rev. B **81**, 214101 (2010)
103. W. Lorenzen, B. Holst, R. Redmer, Phys. Rev. Lett. **102**, 115701 (2009)
104. M.A. Morales, S. Hamel, K. Caspersen, E. Schwegler, Phys. Rev. B **87**, 174105 (2013)
105. J.E. Klepeis, K.J. Schafer, T.W. Barbee, M. Ross, Science **254**, 986 (1991)
106. O. Pfaffenzeller, D. Hohl, P. Ballone, Phys. Rev. Lett. **74**, 2599 (1995)
107. P.M. Celliers, G.W. Collins, L.B. DaSilva, D.M. Gold, R. Cauble, R.J. Wallace, M.E. Foord, B.A. Hammel, Phys. Rev. Lett. **84**, 5564 (2000)
108. T. Sano, N. Ozaki, T. Sakaiya, K. Shigemori, M. Ikoma, T. Kimura, K. Miyanishi, T. Endo, A. Shiroshita, H. Takahashi, T. Jitsui, Y. Hori, Y. Hironaka, A. Iwamoto, T. Kadono, M. Nakai, T. Okuchi, K. Otani, K. Shimizu, T. Kondo, R. Kodama, K. Mima, Phys. Rev. B **83**, 054117 (2011)
109. S.T. Weir, A.C. Mitchell, W.J. Nellis, Phys. Rev. Lett. **76**, 1860 (1996)
110. W.J. Nellis, Rep. Prog. Phys. **69**, 1479 (2006)
111. M.D. Knudson, D.L. Hanson, J.E. Bailey, C.A. Hall, J.R. Asay, C. Deeney, Phys. Rev. B **69**, 144209 (2004)
112. M.D. Knudson, M.P. Desjarlais, D.H. Dolan, Science **322**, 1822 (2008)
113. V.E. Fortov, V.Y. Ternovoi, M.V. Zhernokletov, M.A. Mochalov, A.L. Mikhailov, A.S. Filimonov, A.A. Pyalling, V.B. Mintsev, V.K. Gryaznov, I.L. Iosilevskii, J. Exp. Theor. Phys. **97**, 259 (2003)
114. G.V. Boriskov, A.I. Bykov, R.I. Il'Kaev, V.D. Selemir, G.V. Simakov, R.F. Trunin, V.D. Urlin, A.N. Shuikin, W.J. Nellis, Phys. Rev. B **71**, 092104 (2005)
115. M.D. Knudson, D.L. Hanson, J.E. Bailey, C.A. Hall, J.R. Asay, W.W. Anderson, Phys. Rev. Lett. **87**, 225501 (2001)
116. M.D. Knudson, D.L. Hanson, J.E. Bailey, C.A. Hall, J.R. Asay, Phys. Rev. Lett. **90**, 035505 (2003)
117. P. Loubeyre, S. Brygoo, J. Eggert, P.M. Celliers, D.K. Spaulding, J.R. Rygg, T.R. Boehly, G.W. Collins, R. Jeanloz, Phys. Rev. B **86**, 144115 (2012)
118. V.E. Fortov, R.I. Ilkaev, V.A. Arinin, V.V. Burtzev, V.A. Golubev, I.L. Iosilevskiy, V.V. Khrustalev, A.L. Mikhailov, M.A. Mochalov, V.Y. Ternovoi, M.V. Zhernokletov, Phys. Rev. Lett. **99**(18), 185001 (2007)
119. M. Mochalov, R. Il'kaev, V. Fortov, A. Mikhailov, Y. Makarov, V. Arinin, S. Grishechkin, A. Blikov, V. Ogorodnikov, A. Ryzhkov, V. Gryaznov, JETP Lett. **92**, 300 (2010)
120. A. Becker, N. Nettelmann, B. Holst, R. Redmer, Phys. Rev. B **88**, 045122 (2013)
121. B. Holst, R. Redmer, V. Gryaznov, V. Fortov, I. Iosilevskiy, Eur. Phys. J. D **66**, 1 (2012)
122. B. Militzer, Phys. Rev. Lett. **97**, 175501 (2006)
123. W.B. Hubbard, W.J. Nellis, A.C. Mitchell, N.C. Holmes, S.S. Limaye, P.C. MacCandless, Science **253**, 648 (1991)
124. C. Cavazzoni, G.L. Chiarotti, S. Scandolo, E. Tosatti, M. Bernasconi, M. Parrinello, Science **283**, 44 (1999)
125. M. French, T.R. Mattsson, N. Nettelmann, R. Redmer, Phys. Rev. B **79**, 054107 (2009)
126. M.D. Knudson, M.P. Desjarlais, R.W. Lemke, T.R. Mattsson, M. French, N. Nettelmann, R. Redmer, Phys. Rev. Lett. **108**, 091102 (2012)
127. N. Ozaki, T. Kimura, T. Okuchi, M. French, T. Kakeshita, M. Kita, K. Miyanishi, R. Redmer, T. Sano, T. Sano, K. Shimizu, T. Terai, R. Kodama (submitted)
128. A.C. Mitchell, W.J. Nellis, J. Chem. Phys. **76**, 6273 (1982)
129. L. Volkov, et al., JETP Letters **31**, 513 (1980)
130. P.M. Celliers, G.W. Collins, D.G. Hicks, M. Koenig, E. Henry, A. Benuzzi-Mounaix, D. Batani, D.K. Bradley, L.B.D. Silva, R.J. Wallace, S.J. Moon, J.H. Eggert, K.K.M. Lee, L.R.

Benedetti, R. Jeanloz, I.M.N. Dague, B. Marchet, M.R.L. Gloahec, C. Reverdin, J. Pasley, O. Willi, D. Neely, C. Danson, Phys. Plasmas **11**(8), L41 (2004)
131. M. French, R. Redmer, J. Phys. Condens. Matter **21**, 375101 (2009)
132. M. Podurets, et al., Sov. Phys. JETP **35**, 375 (1972)
133. N. Nettelmann, R. Helled, J. Fortney, R. Redmer, Planet. Space Sci. **77**, 143 (2013)
134. H.B. Radousky, A.C. Mitchell, W.J. Nellis, J. Chem. Phys. **93**, 8235 (1990)
135. W.J. Nellis, N.C. Holmes, A.C. Mitchell, D.C. Hamilton, M. Nicol, J. Chem. Phys. **107**(21), 9096 (1997)
136. F. Datchi, S. Ninet, M. Gauthier, A.M. Saitta, B. Canny, F. Decremps, Phys. Rev. B **73**, 174111 (2006)
137. C.J. Pickard, R.J. Needs, Nat. Mater. **7**, 775 (2008)
138. G.I.G. Griffiths, R.J. Needs, C.J. Pickard, Phys. Rev. B **86**, 144102 (2012)
139. S. Ninet, F. Datchi, A.M. Saitta, Phys. Rev. Lett. **108**, 165702 (2012)
140. J.G.O. Ojwang, R.S. McWilliams, X. Ke, A.F. Goncharov, J. Chem. Phys. **137**, 064507 (2012)
141. M. Bethkenhagen, M. French, R. Redmer, J. Chem. Phys. **138**, 234504 (2013)
142. L.R. Benedetti, J.H. Nguyen, W.A. Caldwell, H. Liu, M. Kruger, R. Jeanloz, Science **286**(5437), 100 (1999)
143. G. Gao, A.R. Oganov, Y. Ma, H. Wang, P. Li, Y. Li, T. Iitaka, G. Zou, J. Chem. Phys. **133**(14), 144508 (2010)
144. B.L. Sherman, H.F. Wilson, D. Weeraratne, B. Militzer, Phys. Rev. B **86**, 224113 (2012)
145. R. Chau, S. Hamel, W.J. Nellis, Nat. Commun. **2**, 203 (2011)
146. M.S. Lee, S. Scandolo, Nat. Commun. **2**, 185 (2011)
147. F.D. Stacey, Rep. Prog. Phys. **68**, 341 (2006)
148. A.R. Oganov, G.D. Price, S. Scandolo, Z. Kristallogr. **220**, 531 (2005)
149. M.J. Gillan, D. Alfè, J. Brodholt, L. Vočadlo, G.D. Price, Rep. Prog. Phys. **69**, 2365 (2006)
150. R.J. Hemley, High Press. Res. **30**, 581 (2010)
151. L. Stixrude, N. de Koker, N. Sun, M. Mookherjee, B.B. Karki, Earth Planet. Sci. Lett. **278**, 226 (2009)
152. N. de Koker, L. Stixrude, Geophys. J. Int. **178**, 162 (2009)
153. D.K. Spaulding, R.S. McWilliams, R. Jeanloz, J.H. Eggert, P.M. Celliers, D.G. Hicks, G.W. Collins, R.F. Smith, Phys. Rev. Lett. **108**, 065701 (2012)
154. D.C. Swift, J.H. Eggert, D.G. Hicks, S. Hamel, K. Caspersen, E. Schwegler, G.W. Collins, N. Nettelmann, G.J. Ackland, Astrophys. J. **744**(1), 59 (2012)
155. D. Valencia, M. Ikoma, T. Guillot, N. Nettelmann, Astron. Astrophys. **516**, A20 (2010)
156. L. Stixrude, Phys. Rev. Lett. **108**, 055505 (2012)
157. D. Alfè, M.J. Gillan, G.D. Price, Nature **401**, 462 (1999)
158. E. Sola, D. Alfè, Phys. Rev. Lett. **103**, 078501 (2009)
159. D. Alfè, M. Pozzo, M.P. Desjarlais, Phys. Rev. B **85**, 024102 (2012)
160. N. de Koker, G. Steinle-Neumann, V. Vlcek, Proc. Natl. Acad. Sci. **109**, 470 (2012)
161. D. Valencia, R.J. O'Connell, D. Sasselov, Icarus **181**, 545 (2006)
162. L.A. Rogers, P. Bodenheimer, J.J. Lissauer, S. Seager, Astrophys. J. **738**, 59 (2011)
163. V. Stamenkovic, D. Breuer, T. Spohn, Icarus **216**, 572 (2011)
164. V. Zharkov, A. Makalkin, V. Trubitsyn, Astron. Zh. **51**, 1288 (1974)
165. W.B. Hubbard, M.S. Marley, Icarus **78**, 102 (1989)
166. G. Chabrier, D. Saumon, W.B. Hubbard, J.I. Lunine, Astrophys. J. **391**, 817 (1992)
167. T. Gudkova, V. Zharkov, Planet. Space Sci. **47**, 1201 (1999)
168. T. Guillot, Planet. Space Sci. **47**, 1183 (1999)
169. B. Militzer, W.B. Hubbard, J. Vorberger, I. Tamblyn, S.A. Bonev, Astrophys. J. Lett. **688**, L45 (2008)
170. N. Nettelmann, Astrophys. Space Sci. **336**, 47 (2011)
171. J. Leconte, G. Chabrier, Astron. Astrophys. **540**, A20 (2012)
172. J. Lissauer, Annu. Rev. Astron. Astrophys. **31**, 129 (1993)
173. J. Pollack, O. Hubickyj, P. Bodenheimer, J. Lissauer, M. Podolak, Y. Greenzweig, Icarus **124**, 62 (1996)

174. L. Mayer, T. Quinn, J. Wadsley, J. Stadel, Science **298**, 5599 (2002)
175. A.P. Boss, Astrophys. J. **599**, 577 (2003)
176. R. Helled, T. Guillot, Astrophys. J. **767**, 113 (2013)
177. N. Nettelmann, R. Püstow, R. Redmer, Icarus **225**, 548 (2013)
178. D. Saumon, G. Chabrier, H.M. van Horn, Astrophys. J. Suppl. Ser. **99**, 713 (1995)
179. S. Lyon, J. Johnson, Sesame, Technical report, Los Alamos National Laboratory Equation of State Database. LANL report no. LA-UR-92-3407 (1992)
180. G.I. Kerley, Structures of the planets jupiter and saturn. Kerley Technical Services Research Report KTS04-1, Kerley Technical Services (2004)
181. D. Saumon, T. Guillot, Astrophys. J. **609**, 1170 (2004)
182. B. Militzer, W.B. Hubbard, Astrophys. J. **773**, 95 (2013)
183. J. Liu, P. Goldreich, D. Stevenson, Icarus **196**, 653 (2008)
184. W.J. Nellis, A.C. Mitchell, P.C. McCandless, D.J. Erskine, S.T. Weir, Phys. Rev. Lett. **68**, 2937 (1992)
185. W.J. Nellis, S.T. Weir, A.C. Mitchell, Science **273**(5277), 936 (1996)
186. M. French, A. Becker, W. Lorenzen, N. Nettelmann, M. Bethkenhagen, J. Wicht, R. Redmer, Astrophys. J. Suppl. Ser. **202**, 5 (2012)
187. N. Gómez-Pérez, M. Heimpel, J. Wicht, Phys. Earth Planet. Inter. **181**, 42 (2010)
188. R. Helled, J. Anderson, M. Podolak, G. Schubert, Astrophys. J. **726**, A15 (2011)
189. S. Stanley, J. Bloxham, Nature **428**, 151 (2004)
190. S. Stanley, J. Bloxham, Icarus **184**, 556 (2006)
191. D.J. Stevenson, E.E. Salpeter, Astrophys. J. Suppl. Ser. **35**, 221 (1977)
192. G. Chabrier, I. Baraffe, Astrophys. J. **661**, L81 (2007)
193. W. Hubbard, Astrophys. J. **162**, 687 (1970)
194. D. Saumon, G. Chabrier, Phys. Rev. A **46**, 2084 (1992)
195. T. Guillot, G. Chabrier, D. Gautier, P. Morel, ApJ **450**, 463 (1995)
196. J.J. Fortney, W.B. Hubbard, Icarus **164**, 228 (2003)
197. W. Hubbard, Icarus **30**, 305 (1977)
198. N. Nettelmann, Matter under extreme conditions: modelling giant planets. PhD thesis, University of Rostock (2009)
199. U. von Zahn, D.M. Hunten, G. Lehmacher, J. Geophys. Res. **103**, 0148 (1998)
200. B.J. Conrath, D. Gautier, Icarus **144**, 124 (2000)
201. J. Leconte, G. Chabrier, Nat. Geosci. **6**, 347 (2013)
202. W.B. Hubbard, H.E. DeWitt, Astrophys. J. **290**, 388 (1985)
203. J.J. Fortney, W.B. Hubbard, Astrophys. J. **608**(2, Part 1), 1039 (2004)

Diffusivity of Mixtures in Warm Dense Matter Regime

Tomorr Haxhimali and Robert E. Rudd

Abstract Modeling of ionic diffusion in warm dense plasma mixtures has been of longstanding interest in astrophysics and in Inertial Confinement Fusion. Here we review traditional approaches to calculating plasma diffusion using kinetic theory. We also review earlier classical molecular dynamics (MD) results. We discuss some new results from MD for self and mutual diffusion in a mixture of deuterium and argon at warm dense matter regime. We make use of Yukawa interionic potentials as an effective potential that accounts for the screening effects of the electrons to the ions. We further provide a general description of the Green-Kubo technique to extract the diffusivity of a multicomponent mixture. The description is very general and it can be extended to plasmas.

1 Introduction

Species diffusion is the process in which a net species flux arises in the center of mass frame in response to gradients in the concentration, pressure, or temperature [1]. The species may be atoms or molecules or even larger particles, depending on the system, but for the purpose of this article the species will be different kinds of ions. Species diffusion is often in response to concentration gradients. The species current lowers the free energy by increasing the entropy of mixing, so species diffusion is necessarily a non-equilibrium process. Temperature gradients and pressure gradients may also give rise to gradients in the chemical potential that drive diffusion. These are special cases of species diffusion called thermodiffusion and barodiffusion, and they can be important in plasmas [2–4]. This article focuses

T. Haxhimali • R.E. Rudd (✉)
Lawrence Livermore National Laboratory, Livermore, CA, 94550, USA
e-mail: haxhimali1@llnl.gov; robert.rudd@llnl.gov

F. Graziani et al. (eds.), *Frontiers and Challenges in Warm Dense Matter*, Lecture Notes in Computational Science and Engineering 96, DOI 10.1007/978-3-319-04912-0_9,

primarily on species diffusion driven by concentration gradients, and for simplicity we will refer to it as diffusion.

Diffusion is an important component of mixing. Mixing is a combination of stirring and diffusion, in which stirring is hydrodynamic flow that redistributes initially separated species, and diffusion causes them to mix irreversibly at the atomic level. At high Reynolds numbers vortices can develop and cause stirring on many length scales, and diffusion operates on this complex flow pattern. At low Reynolds numbers, fluid flow is governed by the Stokes equation, so it is reversible. Two initially separated components can be mixed by stirring, and then unmixed by stirring in the reverse direction. Experiments that demonstrate this effect are quite striking [5]. Only a very small amount of mixing persists due to the irreversible diffusion.

Within plasma physics there are well known examples where diffusion is understood to play an important role or where it has been proposed to be important. It is key to the purification of white dwarf atmospheres through the sedimentation of heavy elements [6]. A similar phenomenon is thought to occur in neutron stars. Diffusion can also have an effect in hot spot ignition for inertial confinement fusion (ICF), since it is a mechanism for ablator materials to degrade the fuel [7, 8]. Also, other systems exhibit diffusion of particles interacting through screened Coulomb interactions, such as dust particles in plasmas and colloidal particles suspended in electrolyte [9–12]. Diffusion in these systems spans a large range of regimes. In ICF, the fuel starts cold and at solid density $\sim$1 g/cm^3, is rapidly compressed to warm dense matter conditions, and ultimately while burning should be at temperatures $\sim$1 keV and densities of $\sim$1,000 g/cm^3 [7]. A white dwarf is at a temperature of $\sim$20–1,000 eV and a broad range of densities. The white dwarf is a weakly coupled plasma, whereas the ICF fuel goes from strongly coupled to weakly coupled. Understanding diffusion across this range of conditions is a challenge.

There are several classic pictures of diffusion. In a simplified picture from the kinetic theory of gases, atoms move ballistically on the average of some length λ, the mean free path, before scattering. After scattering, they move ballistically in a different direction. If there is no concentration gradient, there will be just as many atoms coming from one direction as from the opposite direction, so there is no net flux. If there is a concentration gradient, more atoms will come from the region with higher concentration, so the flux will be proportional to the concentration gradient. If the pressure and temperature are uniform, the mass flux $\mathbf{J}$ is related to the gradient in the mass fraction of one species, $c(\mathbf{r})$, according to [13]:

$$\mathbf{J} = -\rho D_{12} \nabla c(\mathbf{r}) \tag{1}$$

where ρ is the mass density. The coefficient of proportionality, D_{12}, is the mutual diffusivity or interdiffusivity. In the simplified ideal gas picture, it is related to the mean free path and mean thermal velocity, v_{th}, by the relation:

$$D_{12} = \frac{1}{3} \lambda v_{th}. \tag{2}$$

Since mass is conserved, the mass flux obeys a continuity equation:

$$\rho\left(\partial_t c + \mathbf{v}\cdot\nabla c(\mathbf{r})\right) = -\nabla\cdot\mathbf{J} \tag{3}$$

Together Eqs. (1) and (3) give rise to the equation:

$$\frac{\partial c}{\partial t} = D_{12}\nabla^2 c(\mathbf{r}) \tag{4}$$

where we have taken $\mathbf{v} = 0$. Equation (1) is known as Fick's first law of diffusion, and Eq. (4) is known as Fick's second law of diffusion. In the simplified kinetic theory, both the mean free path λ and the thermal velocity v_{th} are positive, so the mutual diffusivity is positive. The mass flux acts to decrease the gradients, and the diffusivity increases with increasing thermal velocity and mean free path.

Another familiar case is the diffusion of an interstitial atom in a crystal lattice. The interstitial is an extra atom sitting in a site between the regularly arrayed atoms of the crystal. Interstitials can diffuse, hopping from one interstitial site to another, but there is an energy barrier that must be overcome to hop from site to site. The energy to overcome the barrier is provided by thermal fluctuations. So the atom rattles around in the energy well of one interstitial site until a sufficiently large thermal fluctuation pushes it over the barrier. A similar phenomenon occurs in liquids in which atoms are trapped in the cage of their neighbors for some period of time until a thermal fluctuation allows them to escape the cage. Diffusion of the interstitial atom in a crystal or a caged atom in a liquid is punctuated by hops, just as diffusion in the ideal gas is punctuated by scattering events. After many hops, the distribution function describing the probability of finding the atom a distance r from its starting point is a Gaussian in r whose width increases like the square-root of time, characteristic of diffusion.

These two kinds of diffusive phenomena also occur in plasmas. Weakly coupled plasmas exhibit diffusion like the ideal gas, while strongly coupled plasmas diffuse like the liquid. In plasmas the scattering is due to the Coulomb repulsion between like-charged ions, screened by the electrons. The scattering cross-section that determines the mean free path involves an integral that diverges logarithmically at both short and long range for the pure Coulomb interaction. It is cutoff at long ranges by the Debye length when screening is included, and at short distances by the classical turning point. The resulting kinetic theory making use of Chapman-Enskog theory provides a good description of diffusion in weakly coupled plasmas. Chapman and Cowling derived analytic expressions for diffusion in Coulomb systems [14], and Paquette extended this work to statically screened ions [15].

In the strongly coupled regime, cage effects become important, invalidating the binary collision model upon which the Chapman-Cowling and Paquette models are based. In this regime, molecular dynamics (MD) has been used to capture the many-body effects governing diffusion [16]. There has been a recent finding that effective pair potentials provided a means of extended the realm of validity of a diffusion model based on binary collisions [17]. The MD simulations, valid at strong coupling, become more challenging when the coupling is weak.

We now give a brief review of the parameters of a mixed plasma. We assume a mixture of the ion species $i = 1, 2, \ldots$ with mass and charge A_i and Z_i, respectively. The total number density of ions is given by $n = \sum_i n_i$, where n_i is the number density of the ions of species α. The Wigner-Seitz radius of the system is given by: $a_{WS} = (\frac{4}{3}\pi n)^{-1/3}$. From the charge neutrality of the mixture the number density of the electrons is:

$$n_e = \langle Z \rangle n, \tag{5}$$

where $\langle Z \rangle$ is the average ionization of the mixture. We follow the general definition for the angular brackets $\langle \cdots \rangle$ denoting a number weighted average of a parameter Ξ in a mixture of the type $\langle \Xi \rangle = \sum_i \Xi_i X_i$ where $X_i = n_i/n$ is the mole fraction of species i. The state of the free electrons is determined by the electron number density n_e and temperature T. It is convenient to introduce the dimensionless density parameter $r_s = a_e/a_0$, where $a_e = (\frac{4}{3}\pi n_e)^{-1/3}$ and a_0 is the Bohr radius. The degeneracy of the electrons in the system is determined by the parameter $\Theta \equiv k_B T/E_F$, where $E_F \equiv \frac{\hbar^2}{2m_e}\left(3\pi^2 n_e\right)^{2/3}$ is the Fermi energy and m_e the mass of the electron.

The strength of the ion-ion Coulomb interaction in species i is characterized by the coupling parameter:

$$\Gamma_i = \frac{(Z_i e)^2}{4\pi\epsilon_0 a_i k_B T} = Z_i^{5/3}\langle Z \rangle^{1/3}\Gamma, \tag{6}$$

where $a_i = \frac{Z_i^{1/3}}{\langle Z \rangle^{1/3}} a_{WS}$ is the ion sphere radius introduced by Salpeter [18], and $\Gamma \equiv e^2/(4\pi\epsilon_0 a_{WS} k_B T)$ is the electron-electron coupling parameter. In a multi-component mixed plasma it is useful to introduce an effective coupling [19–23]:

$$\Gamma_{eff} = \sum_i X_i \Gamma_i = \langle Z^{5/3} \rangle \langle Z \rangle^{1/3} \Gamma. \tag{7}$$

An important length scale is also the de Broglie wavelength $\Lambda_i \equiv \left(2\pi\hbar^2/m_i k_B T\right)^{1/2}$, where m_i is the mass of ion of species i.

In the high density and degenerate plasma limit, i.e. $r_s \rightarrow 0$ and $\Theta \rightarrow 0$ respectively, the electrons form a uniform neutralizing background. However, for most conditions the electron background is both polarizable and compressible. In charged fluids the electrons interact not only with their immediate neighbors but with all the other electrons in the system, therefore their motion cannot be decoupled from one another. As a consequence plasmas exhibit a strong collective behavior in the long wavelength limit, $k \rightarrow 0$, where the Fourier transform of the Coulomb potential, $e^2/\epsilon_0 k$, diverges. A manifestation of this behavior is the rearrangement of the electrons around the ions so that the plasma remains locally neutralized.

This leads to an effective ion potential which decays much faster than $1/r$ at large distances. In this work we use the Yukawa potential to describe the ionic effective potential.

To get an insight into the diffusion of dense plasmas we will study a binary ionic mixture immersed in a polarizable background of electrons that adiabatically cloud around the ions. Conceptually, although cumbersome, we can extend the treatment of a binary mixture to a multicomponent mixture. We touch on this topic for the interested reader in the Appendix 1. We denote the mass fraction of component i by $c_i \equiv \rho_i/\rho$, where ρ and ρ_i are the mass density of the total mixture and this component, respectively.

2 General Overview of Diffusivity in Mixtures

Species diffusivity as a response to the concentration gradient is the main contributor to how fluid evolves with time. Other contributions come from pressure and thermal gradients in a way that respects conservation of mass, energy momentum and even number of particles if chemical reactions are ignored. The conservation laws and dissipative fluxes are described by the Navier-Stokes equations for multicomponent fluids. The complete set of these equations is complicated and given in standard texts [13]. Here we consider the binary mixture case and write only equations that are relevant to the diffusion. The conservation equations for mass and momentum and the equation for the species flux are:

$$\frac{\mathfrak{D}\rho}{\mathfrak{D}t} = -\rho\nabla\cdot\mathbf{v}, \tag{8}$$

$$\rho\frac{\mathfrak{D}c}{\mathfrak{D}t} = -\nabla\cdot\mathbf{J} + r, \tag{9}$$

$$\rho\frac{\mathfrak{D}\mathbf{v}}{\mathfrak{D}t} = -\nabla p - \nabla\cdot\tau + \rho\mathbf{g}, \tag{10}$$

$$\mathbf{J} = -\rho D_{12}\left[\nabla c + (k_T/T)\nabla T + (k_p/p)\nabla p\right], \tag{11}$$

where $\mathfrak{D}f/\mathfrak{D}t$ is the comoving derivative (e.g., $\mathfrak{D}\rho/\mathfrak{D}t = \partial_t\rho + \mathbf{v}\cdot\nabla\rho$). In these sets of equations $\mathbf{v}(\mathbf{r})$ is the velocity field, $\rho(\mathbf{r})$ is density and $c(\mathbf{r})$ is the mass fraction. The equations are written for a binary mixture so the indices in the previous definition have been dropped. Also $\mathbf{g}$ is the gravitational acceleration, D_{12} is the interdiffusivity, $k_T D_{12}$ is the thermal diffusion coefficient and $k_P D_{12}$ is the barodiffusion coefficient.

Here τ_{ij} is the energy-momentum tensor related to the stress tensor σ_{ij} by $\tau_{ij} = \rho v_i v_j - \sigma_{ij}$. The rate of production of mass of one species due to chemical reaction, which will be ignored in this study, is denoted by r. The other fields are the internal energy per unit mass E, pressure $\mathbf{p} = -\frac{1}{3}\sigma_{ii}$, and temperature T.

In this work we are interested in Fick's law, which is implied by Eqs. (9) and (11). We consider cases with no temperature, density and pressure gradient. Below we describe the self-diffusion and its relation to the Fickian diffusion.

2.1 Self-Diffusivity

The self-diffusion coefficient $D_\alpha (\alpha = 1,2)$ is related to the random-walk motion of a tagged particle of species α in a mixture on hydrodynamic scales. It can be calculated from the velocity autocorrelation function [1, 24–28]:

$$C_\alpha(t) = \frac{1}{3N_\alpha} \sum_{j=1}^{N_\alpha} \langle \mathbf{v}_j^{(\alpha)}(t) \cdot \mathbf{v}_j^{(\alpha)}(0) \rangle, \tag{12}$$

via a Green-Kubo integral,

$$D_\alpha = \int_0^\infty C_\alpha(t) \mathrm{d}t. \tag{13}$$

In Eq. (12), $\mathbf{v}_j^{(\alpha)}(t)$ is the velocity of particle j of species α at time t. The autocorrelation function C_α depends on the species as well as temperature, pressure, density and composition. The autocorrelation starts at $C_\alpha(0) = k_B T / m_\alpha$ and decreases toward 0 with time. For moderate to strong coupling mixtures the autocorrelation function exhibits an oscillatory decay with time due to the appearance of negative correlation effects. This is a manifestation of the cage effect [29]; where the tagged particle finds itself momentary trapped by its immediate neighbors. At low enough coupling the autocorrelation function decays monotonically; an indication of the absence of the aforementioned many-body effects.

It is straightforward to verify the equivalence of Eq. (13) with the equation relating the mean-squared displacement of a tagged diffusing particle time, in the long time limit. This is given by the well known relation due to Einstein for Brownian motion [30]:

$$D_\alpha = \lim_{t \to \infty} \frac{1}{6t} \langle |\mathbf{r}(t) - \mathbf{r}(0)|^2 \rangle. \tag{14}$$

We can rewrite the Einstein relation (14) in terms of velocity autocorrelation as follows [24, 25]:

$$\langle |\mathbf{r}(t) - \mathbf{r}(0)|^2 \rangle = \left\langle \int_0^t dt' \mathbf{v}(t') \cdot \int_0^t dt'' \mathbf{v}(t'') \right\rangle = 2 \int_0^t dt' \int_0^{t'} dt'' \langle \mathbf{v}(t'') \cdot \mathbf{v}(t') \rangle \tag{15}$$

Equation (15) can be combined with the definition (12) to give:

$$\langle|\mathbf{r}(t)-\mathbf{r}(0)|^2\rangle = 6\int_0^t dt' \int_0^{t'} dt'' C(t'-t''). \tag{16}$$

By changing variables from t' and t'' to t' and $\tau \equiv t' - t''$ and after integration by parts, we find that:

$$\langle|\mathbf{r}(t)-\mathbf{r}(0)|^2\rangle = 6t\int_0^t \left(1-\frac{\tau}{t}\right) C(\tau)d\,\tau. \tag{17}$$

Substitution of Eq. (17) into (14) verifies (13). This equivalence constitutes a simple example of the general *fluctuation−dissipation* relation, also known as Green-Kubo relations [31–34]. It also relates the stochastic process of a random walk in which the mean square displacement of the walker becomes a linear function of time after a sufficiently larger number of collisions and displacements have occurred.

2.2 *Interdiffusion*

Interdiffusion is related to the collective transport of mass driven by concentration gradients [24]. The rate at which concentration fluctuations dissipate is governed by the interdiffusion coefficient D_{12} which linearly relates mass fluxes to gradients in the chemical potential. The latter, in isobaric-isothermal conditions, is linearly proportional to the gradient of mass concentration. This can be easily seen from Eq. (11) by keeping only the first term inside the bracket in the RHS. If we define the center of mass velocity field by:

$$\mathbf{v}(\mathbf{r},t) = \sum_{i=1}^{2} c_i \mathbf{v}_i(\mathbf{r},t), \tag{18}$$

where $\mathbf{v}_i$ is the velocity field of species i, then the mass flux of this species is:

$$\mathbf{J}_i(\mathbf{r},t) \equiv \rho_i(\mathbf{r},t)\left[\mathbf{v}_i(\mathbf{r},t) - \mathbf{v}(\mathbf{r},t)\right], \tag{19}$$

and we define the interdiffusion coefficient by the Fick's law:

$$\mathbf{J}_i(\mathbf{r},t) = -\rho(\mathbf{r},t)D_{12}\nabla c_i(\mathbf{r},t). \tag{20}$$

Using molecular dynamics we can study and compute the interdiffusion by creating a system, like the one in Fig. 1, with an initial sharp interface that creates a very steep composition gradient. From the Fick's law a current will be created

Fig. 1 A non-equilibrium molecular dynamics simulation for binary mixtures of deuterium and argon. In this plot chronologically ordered snapshots of interface broadening between two plasma Ar and D as they mix are shown

that opposes this gradient and with time broadens the interface, and therefore intermixing the different components. This process is very clear in Fig. 1 where we have simulated the mixing process between two plasma of argon (in gold color on right) and deuterium (in silver color on left). In principle, one can compute the interdiffusivity D_{12} by analyzing the evolution of the composition profile $c_i(\mathbf{r}, t)$ which should satisfy the Fick's equation at longtime limit. At the very initial stage when strong gradients are present non-Fickian effects, like diffusion from ballistic motion, occur.

The interdiffusion coefficient like the self-diffusion can be expressed in terms of fluctuations in the system in equilibrium, without any external forces to drive mass or energy flux ([1, 24, 25, 35]), by use of Green-Kubo techniques [31]. In the following we briefly derive the appropriate Green-Kubo formalism by connecting with Fick's law in the low-frequency and long-wavelength fluctuations that correspond to the regime governed by linear hydrodynamics. A more detailed derivation can be found in classical books of Statistical Mechanics [24, 25]. We also provide a general derivation of this technique for multicomponent mixtures in Appendix 1. We consider only mass concentration fluctuations and ignore fluctuations in temperature and pressure. Using the previous notation and the overall mass conservation law, one component should satisfy:

$$\rho(\mathbf{r}, t)\left(\frac{\partial c_1(\mathbf{r}, t)}{\partial t} + \mathbf{v}(\mathbf{r}, t) \cdot \nabla c_1(\mathbf{r}, t)\right) = -\boldsymbol{\nabla} \cdot \mathbf{J}_1(\mathbf{r}, t). \tag{21}$$

Linearizing the above equation with respect to virtual mass fluctuation of species 1, $\delta c_1(\mathbf{r}, t)$ we arrive at the Fick's diffusion equation for this fluctuation:

$$\frac{\partial}{\partial t}\delta c_1(\mathbf{r}, t) = D_{12}\nabla^2 \delta c_1(\mathbf{r}, t). \tag{22}$$

It is very instructive to relate the fluctuations in local mass concentration to the fluctuations in local number densities by the following (cf. Ref. [24], Chap. 8):

$$\delta c(\mathbf{r}, t) = \frac{m_1 m_2}{\rho^2} \left[n_2 \delta n_1(\mathbf{r}, t) - n_1 \delta n_2(\mathbf{r}, t) \right] \tag{23}$$

The real space solution of Eq. (22) in terms of Fourier components is:

$$\delta c_1(\mathbf{k}, t) = \delta c_1(\mathbf{k}, 0) \exp(-D_{12} k^2 t). \tag{24}$$

Ensemble averaging of the correlation of the above fluctuations followed by a Laplace transform in time, results in the following relation:

$$\begin{aligned} \langle \delta c_1(-k, \omega) \delta c_1(k, t = 0) \rangle &\equiv \int_0^\infty e^{i\omega t} \langle \delta c_1(-k, t) \delta c_1(k, t = 0) \rangle \, dt \\ &= \langle | \, \delta c_1(k, t = 0) \, |^2 \rangle \frac{i\omega + D_{12} k^2}{\omega^2 + D_{12}^2 k^4}. \end{aligned} \tag{25}$$

By taking the appropriate hydrodynamic limit (i.e., long wavelength and low frequency limit) and after use of mass conservation and integration by parts [36] we get the following Green-Kubo form for interdiffusion coefficient:

$$D_{12} = \frac{1}{3 \lim\limits_{k \to 0} S_{cc}(k)} \int_0^\infty dt \, \langle \mathbf{j}^c(t) \cdot \mathbf{j}^c(0) \rangle. \tag{26}$$

In the above S_{cc} is the concentration structure factor defined in terms of partial structure factors $S_{ij}(k) = \frac{1}{N} \sum_{\alpha=1}^{N_i} \sum_{\beta=1}^{N_j} \langle \exp i\mathbf{k} \cdot (\mathbf{r}_\alpha - \mathbf{r}_\beta \mathbf{a}) \rangle$, with i, j the species indices, by the following relation due to Bhatia & Thornton [37]:

$$S_{cc}(k) = X_2^2 S_{11}(k) + X_1^2 S_{22}(k) - 2 X_1 X_2 S_{12}(k) \tag{27}$$

The expression (26) for interdiffusivity can simply be written as a product of a thermodynamic factor Φ with a Green-Kubo integrand D_{12}^0:

$$D_{12} = \Phi D_{12}^0 \tag{28}$$

Below we briefly discuss each of these factors.

2.2.1 Thermodynamic Factor

The thermodynamic factor in Eq. (28) is defined by $\Phi \equiv X_1 X_2 / S_{cc}(0)$. In the long wavelength limit all the structure factors can be derived by either thermodynamic

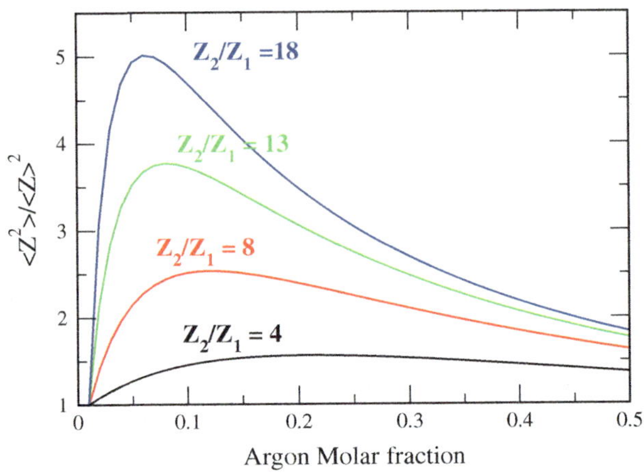

Fig. 2 The thermodynamic factor $\langle Z^2\rangle/\langle Z\rangle^2$ as a function of mole fraction for four different ratios of component charge $Z_2/Z_1 = 4, 8, 13, 18$

fluctuation theory or by calculation in a grand canonical ensemble [38]. The concentration structure factor in isobaric-isothermal conditions is given by:

$$S_{cc}(0) = 1/\left(\frac{\partial^2(\beta G/N)}{\partial X_1^2}\right)_{P,T}. \tag{29}$$

Therefore we can express the thermodynamic factor as:

$$\Phi = X_1 X_2 \left[\frac{\partial^2(\beta G/N)}{\partial X_1^2}\right]_{P,T}, \tag{30}$$

where G is the Gibbs free energy, $\beta \equiv k_B T$ and N is the total number of particles in the system.

In most of the cases for nearly ideal neutral mixtures the thermodynamic factor is very close to unity; however, this may not be true when considering charged plasmas. For weakly coupled plasmas due to the ambipolar electric field of the electrons this factor goes to $\langle Z^2\rangle/\langle Z\rangle^2$, which can be very large at low concentration of an asymmetric mixture [20,35,39,40]. As an example of this effect in Fig. 2 we show the variation of this factor as a function of the mole fraction of the heaviest particle in a binary mixture for four cases of charge ratio of the components, namely, $Z_2/Z_1 = 4, 8, 13, 18$. Note that the asymmetry of the mixture is characterized by this ratio. As it is clear in low mole fraction the higher the asymmetry the further from unity is the thermodynamic factor.

Interestingly, significant deviations of the thermodynamic factor Φ from unity have been recently observed numerically for non-plasma mixture of Al-Ni melts [27].

2.2.2 Green-Kubo: Maxwell-Stefan Diffusivity

The Green-Kubo integrand Eq. (28) is given by:

$$D_{12}^{0} = \frac{1}{3NX_1X_2}\int_0^{\infty} dt\, \langle \mathbf{j}^c(t)\cdot \mathbf{j}^c(0)\rangle, \tag{31}$$

where the term inside the integral is a correlation of interdiffusivity currents:

$$\mathbf{j}^c(t) = X_2\sum_{i=1}^{N_1}\mathbf{v}_i - X_1\sum_{j=1}^{N_2}\mathbf{v}_j = NX_1X_2\left(\bar{v}_1 - \bar{v}_2\right) \tag{32}$$

The above Green-Kubo integrand D_{12}^{0} is also known in literature as Maxwell-Stefan diffusivity [41].

2.3 *Relating Self-Diffusion with Interdiffusion*

The velocity correlation functions in the interdiffusion are collective: only one degree of freedom, the difference between the center of mass velocities for the two species, contributes to the autocorrelation function whereas the self-diffusivity is related to the velocity autocorrelation function of the individual particles. Instead of one degree of freedom, the average is over N_i degrees of freedom. The statistics are much better for calculating self-diffusivity. This difference explains why in molecular dynamics simulations it is difficult to achieve the same numerical precision when computing Maxwell-Stefan diffusivity as compared to self-diffusion coefficient [1,42–44]. Therefore, it is extremely helpful to identify practical relations between Maxwell-Stefan and self-diffusion through different simplifying assumptions.

If we group separately the velocity autocorrelation and cross-correlation functions that appear in Maxwell-Stefan diffusion coefficient (31) and use the $\mathcal{F}$-notation of McCall and Douglass [28,45,46], we have the following:

$$D_{12}^{0} = \left[X_2D_1 + X_1D_2 + X_1X_2\left(\frac{\mathcal{F}_{11}}{X_1^2} + \frac{\mathcal{F}_{22}}{X_2^2} - 2\frac{\mathcal{F}_{12}}{X_1X_2}\right)\right], \tag{33}$$

where the $\mathcal{F}$ -factors are given by:

$$\mathcal{F}_{\alpha\beta} = \frac{1}{3N}\sum_{i=1}^{N_\alpha}\sum_{j\neq i}^{N_\beta}\int_0^{\infty}\langle \mathbf{v}_i(t)\cdot \mathbf{v}_j\rangle dt. \tag{34}$$

If we assume that velocity cross-correlation functions are negligible, i.e.,

$$\frac{\mathcal{F}_{11}}{X_1^2} + \frac{\mathcal{F}_{22}}{X_2^2} - 2\frac{\mathcal{F}_{12}}{X_1X_2} = 0, \tag{35}$$

a simple linear rule that connects Maxwell-Stefan with self-diffusion coefficient in a binary system follows:

$$D_{12}^{0} = X_2 D_1 + X_1 D_2. \tag{36}$$

The above relation is known in the Condensed Matter Community as Darken's equation [46–49]. Similar relation has tacitly been assumed also in the plasma community [22, 23, 35, 42].

A similarity exists between Eq. (36) and the so-called Nernst-Einstein relation, which is an empirical relation that links the electrical conductivity of an ionic system to the self-diffusion coefficients of the cations and anions in the system [50]. In this case the electrical conductivity is related to the same velocity correlation functions as the one that expresses the interdiffusion coefficient.

3 Diffusion Models in Plasmas from Kinetic Theories

Plasmas, whether in astrophysics or in the Inertial Confinement Fusion experiments, span a huge range of the Coulomb coupling parameter Γ (6). In the weakly coupled regime Chapman-Spitzer theories ([14, 51]) have been very successful in describing diffusion coefficient.

In the limit of dilute-gas approximation transport coefficients, including diffusivity, are estimated by considering the Boltzmann equation of kinetic theory. Here the collisions are important in determining the distribution function of the particles. In weakly coupled plasmas ion-ion collisions consist of a series of many small angle binary scattering events. Within these assumptions the Chapman-Enskog theory [14] assumes that the total distribution function of a given species can be expanded as a convergent series of functions with the lowest order corresponding to the equilibrium distribution. Transport coefficients are then calculated from velocity moments of the velocity distribution function. The latter is further expanded in terms of Sonine polynomials (Associated Laguerre Polynomials). Having only the first term in this expansion will correspond to a Maxwell velocity distribution. In this first polynomial approximation following Chapman and Cowling [14], the interdiffusion coefficient is given by:

$$[D_{12}]_1 = \frac{3k_B T}{16 n m_{red} \Omega_{12}^{(11)}}, \tag{37}$$

where T and n are temperature and ion density, respectively, and

$$m_{red} \equiv \frac{m_1 m_2}{m_1 + m_2} \tag{38}$$

is the reduced mass of particles each with mass m_1 and m_2. Here $\Omega_{12}^{(11)}$ is the first collision integral – defined in Appendix 2 – which depends on the exact nature of the interaction of colliding particles. For a pure Coulomb interaction between two particles with charge Z_1 and Z_2 this integral can be derived analytically [14] as:

$$\Omega_{12}^{(11)} = \frac{1}{4\pi\epsilon_0}\sqrt{\frac{\pi}{32}}\frac{Z_1^2 Z_2^2 e^4}{\sqrt{m_{red}}\,(k_B T)^{3/2}}\log\left(1 + 4\pi\epsilon_0 \frac{16\,(k_B T)^2 \lambda_D^2}{Z_1^2 Z_2^2 e^4}\right). \tag{39}$$

The Debye length λ_D is given by:

$$\lambda_D = \sqrt{\frac{\epsilon_0 k_B T}{e^2 \sum X_i Z_i^2}} = \sqrt{\frac{\epsilon_0 k_B T}{e^2 \langle Z_i^2 \rangle}}, \tag{40}$$

where at the last term we have used the molar weighted average convention for the Z^2, with X_i being the mole fraction of species i.

Accounting for a second-order polynomial approximation resulting from deviations from Maxwell distributions provides:

$$[D_{12}]_2 = \frac{[D_{12}]_1}{1 - \Delta}, \tag{41}$$

where Δ is a function of the collision integrals $\Omega_{12}^{(11)}, \Omega_{12}^{(12)}, \Omega_{12}^{(13)}, \Omega_{12}^{(22)}$ and the mole fraction of the mixture. Details of its expression can be found at [15]. Inclusion of terms beyond second polynomial introduce very small correction to the value of diffusivity.

The use of an effective screened Coulomb interaction gives, in principle, a better description of the diffusion in plasmas as it accounts for the collective effects that surround a given charge. The collision integrals for this potential have been numerically calculated and tabulated by Paquette et al. [15]. Using these tabulated collision integrals and a second polynomial expansion we plot in Fig. 3 the diffusivity values for a binary deuterium-argon mixture as a function of the mole fraction, for three different $T = 100$, 500 and 1,000 eV and, ion density $n = 10^{25}/\text{cm}^3$. In all these calculations the ionization of deuterium and argon was kept at values 1 and 13, respectively.

4 Methods

An extremely powerful, albeit simple, tool to calculate diffusion coefficients is molecular dynamics (MD) simulation. MD does not make weak coupling or binary collision approximations, so it allows the study of diffusion with strong coupling. MD simulations follow the motion of atoms (ions) by integrating their equations of

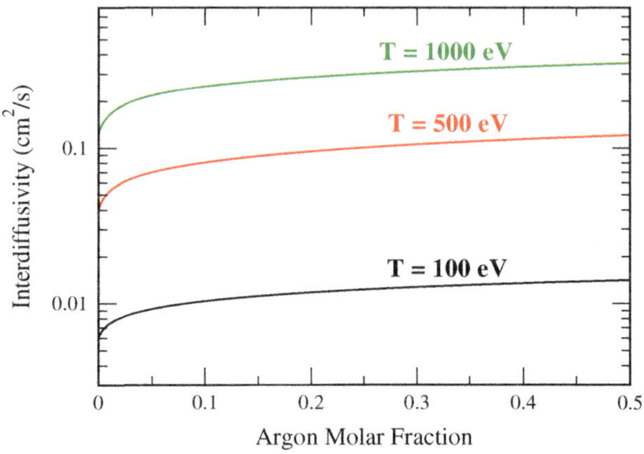

Fig. 3 Interdiffusivity results for a deuterium-argon mixture at three different temperatures with ion density $n = 10^{25}/\,\text{cm}^3$. The values were calculated by using the 2nd polynomial approximation and the tabulated collision integrals for screened Coulomb interaction [15]. In all these calculations the ionization of deuterium and argon was kept at values 1 and 13, respectively

motion while accounting for the pairwise particle-particle interactions [52, 53]. The ions move according to the Newton's Second Law:

$$m\ddot{\mathbf{r}}_i = \sum_{i \neq j}^{N} \mathbf{F}_{i,j}(t). \tag{42}$$

This set of $3N$ coupled ordinary differential equations are integrated explicitly in time using the leap-frog scheme or velocity Verlet [52] algorithm, with a time step that is related to the characteristic time scale of the system. Here N is the total number of particles in the system. In Eq. (42), $\mathbf{F}_{i,j}$ is the force originating from the interaction of "test" particle i with particle j. We study an unconfined system of particles which entails use of periodic boundary conditions (PBC).

We use this general tool to study our plasma system as a mixture of deuterium (D) and argon (Ar). At a temperature $T = 100\,\text{eV}$ we assume that D is fully ionized while Ar partially ionized. The ionization of Ar has been input as a free parameter in order to study its effect. Since the system that we are considering has ions, the interaction force $\mathbf{F}_{i,j}$ among these ions has a Coulombic nature. The presence of free electrons, that act as a neutralizing background, results in an effective ion-ion interaction potential. Within linear response theories [54,55] the interaction between ions is the Yukawa potential:

$$\phi(\mathbf{r}_{ij}) = \frac{Z_i Z_j e^2}{4\pi\epsilon_0 r_{ij}} \exp(-r_{ij} k_D), \tag{43}$$

where $1/k_D$ is the screening length due to the electrons.

In the following we give a succinct description on how this screening arises and how it is related to the thermodynamic conditions. Following the Born-Oppenheimer approximation, the adiabatic separation of motions of heavy and light charged particles leads to an effective ion-ion interactions, or heavy-heavy particle, with the electrons, the light particles, forming a cloud around, and instantaneously following, the ions. To derive this effective potential between the ions we calculate the electron density distribution in space when the ions are embedded in their midst. The potential $\phi(\mathbf{r})$ that gives rise to the effective interaction can be derived from the Poisson equation:

$$\nabla^2\phi(\mathbf{r}) = \frac{e}{\epsilon_0}\left[\sum_i Z_i\delta(\mathbf{r}-\mathbf{r_i}) - n_e\right], \tag{44}$$

with the electron density n_e satisfying the following normalization condition:

$$2\int f(\mathbf{p})\frac{4\pi p^2 dp}{(2\pi\hbar)} = n_e. \tag{45}$$

The factor 2 in front of the integral accounts for the two spin states of the electron: up and down. The electron follow the Fermi-Dirac statistics, therefore $f(\mathbf{p})$ is the electron distribution function in momentum space given by:

$$f(\mathbf{p}) = \frac{1}{\exp\left(\frac{p^2/2m_e-\mu_e}{k_B T}\right)+1}, \tag{46}$$

where μ_e is the chemical potential of the electron.

The presence of this field, or equivalently the ions, varies the electron density as $\Delta n_e(\mathbf{r}) = n_e(\mu_e + e\phi(\mathbf{r})) - n_e(\mu_e)$. Within linear response theory [54] this change is: $\Delta n_e(\mathbf{r}) = -e\frac{\delta n_e}{\delta\mu_e}\phi(\mathbf{r})$. Plugging this into Eq. (44) gives $\phi(\mathbf{r})$ as the Yukawa potential equation (43) with a screening coefficient $k_D = -e\frac{\delta n_e}{\epsilon_0\delta\mu_e}$. This coefficient can be exactly derived in two extreme cases:

- In the limit of non-degenerate electrons $\Theta \to \infty$, i.e., very high T and low density, the electrons satisfy the Maxwell-Boltzmann statistics which results in the Debye-Hückel [56] screening coefficient:

$$k_D = \sqrt{\frac{n_e e^2}{\epsilon_0 k_B T}}. \tag{47}$$

  This corresponds to the classical limit.
- In the fully degenerate limit $\Theta \to 0$, i.e., very low T and high density, which corresponds to an electron gas in the ground state [57], the screening coefficient is given by:

$$k_D = \sqrt{\frac{3}{2}\frac{n_e e^2}{\epsilon_0 E_F}}. \tag{48}$$

At finite temperatures and densities we can use interpolation forms, like the one from Itchimaru [58, 59]

$$\frac{\mu_e}{k_B T} = \frac{3}{2}\ln\Theta + \ln\frac{4}{3\sqrt{\pi}} + \frac{A\Theta^{-b-1} + B\Theta^{-(b+1)/2}}{1 + A\Theta^{-b}}, \tag{49}$$

with A, B and b coefficients given at [59]. A more practical interpolation for numerical studies of transport coefficients, which we adopt for the rest of this work, is the one introduced by Murillo [60]:

$$k_D = \sqrt{\frac{\langle Z\rangle n e^2}{\epsilon_0\sqrt{(k_B T)^2 + \left(\frac{2}{3}E_F\right)^2}}} \tag{50}$$

Note above that we have used Eq. (5) to relate electron density n_e with ion density n and the average charge $\langle Z\rangle$ of the mixture.

The screening length sets the length scale of the system. The system that interacts through Yukawa potential can be described by the dimensionless screening coefficient $\kappa = a_{WS} k_D$ [16] and the effective coupling Γ_{eff}. When the screening coefficient κ is large the Yukawa potential is a short range potential. In this case the MD methods make use of the truncation of the interaction potential therefore limiting the need for the summation of pairwise interactions around a test particle to a region of finite size. In the case of long range interaction, like weak screening Yukawa or Coulomb interaction, such truncation of the potential is not allowed and special techniques, like Ewald [61] summation, have to be used in MD simulations. A more efficient method is to use the so called particle-particle particle-mesh (PPPM) scheme [62, 63], where the interparticle force is partitioned into: (i) a short-range particle-particle force which is to be applied to closely separated ions only, and (ii) a force component that can be calculated on the mesh. In the mesh part of the calculation charged clouds are used instead of point-like particles and their interactions is calculated on a computational mesh taking into account periodic images. This method makes it possible to treat long-range and low-κ Yukawa potentials. For screening values $\kappa \geq 1$, the particle-particle part alone provides sufficient accuracy while being numerically more efficient. In these cases the mesh part need not to be used and the interaction forces are summed for particles situated within a (κ-dependent) cutoff radius around the particle.

The simulations presented here are initiated from a spatially random particle configuration, and particle velocities sampled from a Maxwellian distribution at temperature $T = 100\,\text{eV}$. This random configuration of the particles of each component ensures a spatially uniform mixture as well, suitable for the equilibrium MD. At the initial stage of the run the system is equilibrated to the desired temperature by either (i) rescaling the particle momenta or (ii) by applying Nosé-Hoover [52, 64, 65] thermostats to keep the system initially at a constant number of particles N, volume V and temperature T. On average the system is left to equilibrate for 50,000 timesteps in the NVT ensemble. This is followed by the production run in a microcanonical ensemble with constant N, volume V and total energy E.

The time scale of the system is set by the ionic plasma frequency:

$$\omega_p = \sqrt{\frac{\langle Z \rangle^2 n}{\epsilon_0 \langle m \rangle}}. \tag{51}$$

The timestep in most of the runs was taken $\Delta t \simeq 1/(100\omega_E)$, where ω_E is the Einstein frequency, that physically describes the oscillatory motion of a caged particle in the well potential created by its neighbors. In the limit of no screening $\kappa \to 0$ we have $\omega_E = \omega_p/3$ [16], and for a finite screening we use a fitted form $\omega_E(\kappa) = \omega_p \exp(-0.2\kappa^{1.62})$ [60] to the Ohta and Hamaguchi MD results [16].

To calculate the self-diffusion coefficients of a binary mixed plasma of D and Ar we have performed MD simulations in the following range of conditions: temperature $T = 100\,\text{eV}$, particle number density $n = 10^{23}/\text{cm}^3$ and $10^{25}/\text{cm}^3$, fully ionized deuterium and ionization of Ar $Z_{Ar} = 4, 8, 13$ and 18. The last of Z_{Ar} will correspond to a fully ionized Ar. We take the ionization of Ar as a free parameter; for reference, an average atom model [66] gives the Ar ionization as $Z_{Ar} \simeq 10.5$ at $T = 100\,\text{eV}$, $n = 10^{25}/\text{cm}^3$ and X in the range 0.01–0.5. Specifically, for each of the above conditions we consider mixtures whose Ar mole fraction is: $X = 0.01, 0.05, 0.1, 0.2$ and 0.5. The calculations were performed with enough particles (30,000 $\geq N \geq$ 120,000) over long enough time scales to ensure convergence with insignificant statistical uncertainty.

The autocorrelation function was calculated on the fly as the simulation progressed:

$$C_i(t) = \frac{1}{3N_i} \sum_{\alpha \in i} \mathbf{v}_\alpha(t) \cdot \mathbf{v}_\alpha(0), \tag{52}$$

where i indicates the type of the atom and N_i its numbers. For each atom the initial velocity $\mathbf{v}(0)$ is saved. The simulation was run on parallel supercomputers and so the initial velocity together with other atom attributes has to be communicated when an atom has to move from a spatial domain related to a processor into another. The correlation velocity is saved as a table for a sufficiently large time span. Following Rudd et al. [1], the self-diffusivity was then calculated as an integral over the velocity autocorrelation function by postprocessing:

$$\begin{aligned} D_i &= \int_0^\infty dt C_i(t) \\ &\approx \left[\sum_k a_k \Delta t C_i(t_k) \right] + \frac{t_{max}}{\alpha - 1} C_i(t_{max}), \end{aligned} \tag{53}$$

where the coefficients a_k give the Simpson's rule approximation to the integral. The integral was terminated to account for a power law $t^{-\alpha}$ decay in the correlation function [1]. In particular the long-time tail contribution was included through this term. For the conditions that we consider in our simulations the tail contribution was small (less than 5 %).

5 Results

We have used MD to calculate the velocity autocorrelation functions and self-diffusivities across a broad range of conditions for a binary ion mixture of deuterium and argon in the presence of a polarizable electron background. The ion-ion interaction is the Yukawa potential (screened Coulomb) equation (43), that incorporates the effect of the electrons in the screening coefficient. The latter is calculated using Eq. (50) [60]. In these simulations we have used ionization as an input parameter in order to see its effects along with temperature, density and composition on diffusivity.

Figure 4 shows calculation from MD simulations of the velocity autocorrelation function of deuterium C_D on the left, and of argon C_{Ar} on the right as a function of time. Here different lines correspond to results from mixtures with different Ar mole fraction, namely $X = 0.01, 0.05, 0.1, 0.2$ and 0.5. In these plots the temperature was kept at $T = 100\,\mathrm{eV}$ and the ionization of deuterium and argon at $Z_D = 1$ and $Z_{Ar} = 13$, respectively. As we changed the composition we kept the ion density constant at $n = 10^{25}/\mathrm{cm}^3$ which entails a change of mass density. A change from a 1 to 50 % argon mole fraction, is paralleled with a change in mass density from $\rho = 39.517$ to $348.547\mathrm{g/cm}^3$.

For mixtures with higher argon mole fraction, the screening coefficient κ as well as the couplings, Γ_{eff}, Γ_{Ar} and Γ_D, defined in Eqs. (7) and (6), are increased. For low to moderate coupling values the many-body correlation effects are not very pronounced. As a consequence the autocorrelation function decreases to 0 monotonically with time. On the other hand the autocorrelation function decays faster to 0 when the coupling value is higher.

As the coupling is increased–in this case by increasing the amount of Ar–we notice the formation of structures in the autocorrelation function. This is due to the so-called "cage-effect" [29,67], where the motion of the tagged particle is correlated in a temporary cage of its neighbors. This case is clearly seen at the right of Fig. 4, where the autocorrelation function shows oscillatory characteristics in mixture of 20 and 50 % Ar. The autocorrelation functions are well converged.

The full sets of self-diffusivities are shown in Fig. 5 and Tables 1 and 2. In Fig. 5, we have plotted the calculated self-diffusivities as a function of composition and as a function of Ar ionization Z_{Ar}. The argon mole fraction considered were $X = 0.01, 0.05, 0.1, 0.2$ and 0.5. The value of argon ionizations were $Z_{Ar} = 4, 8, 13$ and fully ionized case $Z_{Ar} = 18$. On the left and the right of Fig. 5 system was kept at ion density $10^{23}/\mathrm{cm}^3$ and $10^{25}/\mathrm{cm}^3$, respectively. In each plot the self-diffusivity

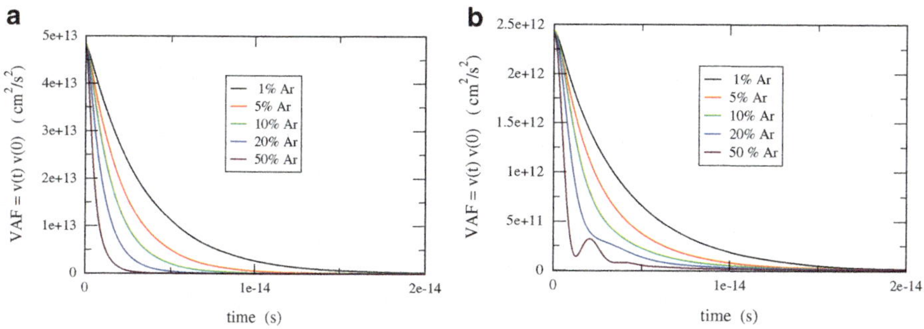

Fig. 4 Velocity autocorrelation function using molecular dynamics as a function of time for five different mole fraction of argon, namely $X = 0.01, 0.05, 0.1, 0.2, 0.5$. In these different mixtures we consider a constant ionization for deuterium and argon at $Z_D = 1$ and $Z_{Ar} = 13$, respectively. These mixtures were kept at temperature $T = 100\,\text{eV}$ and number density $n = 10^{25}/\text{cm}^3$. (**a**) Velocity autocorrelation function of deuterium. (**b**) Velocity autocorrelation function of argon

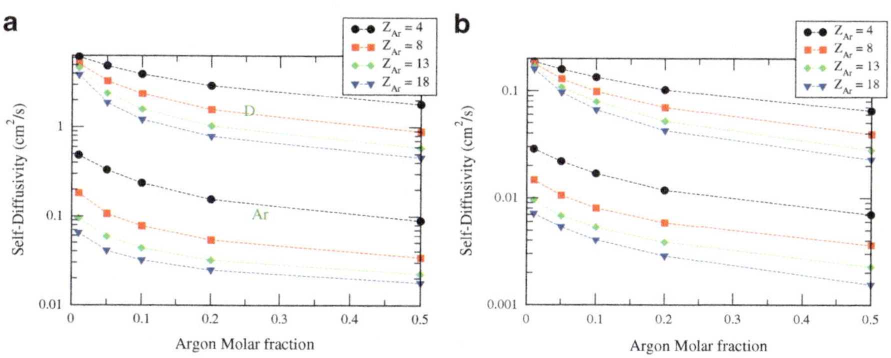

Fig. 5 Self-diffusivity calculation using molecular dynamics for binary mixtures of deuterium and argon. We have considered mixtures of different Ar mole fraction and different ionization level at a given temperature $T = 100\,\text{eV}$. Each color will correspond to a given value of Ar ionization. D is lighter than Ar and as such its self-diffusivity will have higher value. Therefore in each figure the upper band will correspond to values for D and the lower band to Ar. Each line within these bands correspond to different ionization of Ar. (**a**) Ion number density is $n = 10^{23}/\text{cm}^3$. (**b**) Ion number density is $n = 10^{25}/\text{cm}^3$

of Ar and D are plotted together. The values fall in band corresponding to each component, with the upper band corresponding to D. These higher values of D self-diffusion coefficient are related to the fact that its mass is lighter than Ar. In these bands each line corresponds to different Ar ionization Z_{Ar}. As expected increasing Z_{Ar}, increases the couplings Γ_D, Γ_{Ar} of each component, as well as Γ_{eff}, which reduces the self-diffusivity of either component. For exactly the same reason if we compare the values plotted in each graph we see a higher value of self-diffusivity for the less dense case with ion density $n = 10^{23}/\text{cm}^3$ which means

Table 1 Self-diffusion and Darken diffusion values estimated from molecular dynamics for binary mixtures of deuterium and argon. Results correspond to different mixtures at temperature 100 eV and number density $10^{23}/cm^3$. In the second column X denotes Ar molar fraction

Z_{Ar}	X	Γ_{Ar}	Γ_D	Γ_{eff}	κ	ρ (g/cm^3)	D_{Ar} (cm^2/s)	D_D (cm^2/s)	D_{12}^0 (cm^2/s)
4	0.05	1.1379	0.11289	0.16414	0.60921	0.647444	0.32970	4.88611	0.55752
	0.1	1.18537	0.11760	0.22438	0.64763	0.96278	0.23533	3.93434	0.60523
	0.2	1.27032	0.12603	0.35488	0.71826	1.59345	0.15457	2.92675	0.70900
8	0.05	3.81095	0.11909	0.30368	0.65993	0.647444	0.10625	3.30412	0.26615
	0.1	4.11533	0.12860	0.52727	0.74028	0.96278	0.07767	2.36722	0.30663
	0.2	4.61663	0.14427	1.03874	0.87888	1.59345	0.05368	1.56711	0.35637
13	0.05	9.05838	0.12603	0.57264	0.71826	0.647444	0.05858	2.39274	0.17529
	0.1	10.0728	0.14014	1.13341	0.84167	0.96278	0.04348	1.57843	0.19698
	0.2	11.6458	0.16203	2.45879	1.04474	1.59345	0.031786	1.02382	0.23019
18	0.05	16.3537	0.13228	0.94335	0.77213	0.647444	0.04088	1.87104	0.13239
	0.1	18.5501	0.15004	1.99006	0.93185	0.96278	0.03194	1.20874	0.14962
	0.2	21.8295	0.17657	4.50715	1.18681	1.59345	0.02471	0.78650	0.17707

Table 2 Self-diffusion and Darken diffusion values estimated from molecular dynamics for binary mixtures of deuterium and argon. Results correspond to different mixtures at temperature 100 eV and number density $10^{25}/\text{cm}^3$. In the second column X denotes Ar molar fraction

Z_{Ar}	X	Γ_{Ar}	Γ_D	Γ_{eff}	κ	ρ (g/cm^3)	D_{Ar} (cm^2/s)	D_D (cm^2/s)	D_{12}^0 (cm^2/s)
4	0.05	5.28167	0.52400	0.761891	1.04111	64.7444	0.02194	0.15915	0.15228
	0.1	5.50199	0.54586	1.04148	1.07903	96.278	0.01690	0.13433	0.12258
	0.2	5.89628	0.58498	1.64724	1.14247	159.345	0.01177	0.10212	0.08405
8	0.05	17.6888	0.55277	1.40958	1.09063	64.7444	0.01058	0.12940	0.12346
	0.1	19.1017	0.59692	2.4474	1.16079	96.278	0.00804	0.09892	0.08983
	0.2	21.4285	0.66964	4.82142	1.26327	159.345	0.00585	0.06961	0.05685
13	0.05	42.0453	0.58498	2.658	1.14247	64.7444	0.00678	0.10781	0.10275
	0.1	46.7539	0.65049	5.26084	1.23768	96.278	0.00531	0.07889	0.07153
	0.2	54.0551	0.75208	11.4127	1.36437	159.345	0.00384	0.05168	0.04211
18	0.05	75.9072	0.61399	4.37865	1.18619	64.7444	0.00538	0.09728	0.09268
	0.1	86.1021	0.69645	9.23702	1.29768	96.278	0.00406	0.06630	0.06007
	0.2	101.323	0.81957	20.9204	1.43842	159.345	0.00286	0.04241	0.03450

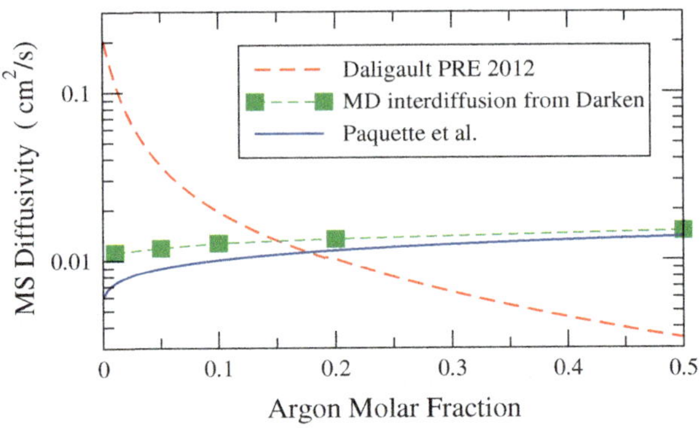

Fig. 6 Maxwell Stefan (MS) diffusivity as a function of Ar mole fraction at a temperature $T = 100\,\text{eV}$, ion density $10^{25}/\text{cm}^3$ and Ar ionization $Z_{Ar} = 13$. We compare MS diffusivity from the darken linear combination (36), with results from kinetic theories from Paquette and Chapman & Cowling [14, 15]. The *long-dashed line* represents an extension to mixtures of the Daligault model [68] for YOCP, following his scheme of extending the OCP model to BIM [23], by use of the mixing rule in Eq. (7)

lower coupling. In both graphs it is seen that for composition with trace element Ar, the self-diffusivity of D is almost unaffected from the ionization of Ar. The variance in self-diffusivity values is mainly due to the mass of Ar. The general trend is for the self-diffusivities to increase with temperature and to decrease with the density, Ar ionization and Ar mole fraction.

For some compositions we put the above results in Tables 1 and 2 together with the corresponding couplings – Γ_D, Γ_{Ar} as well as Γ_{eff}, dimensionless screening κ, mass density ρ in g/cm^3. Also as an initial estimate of the Maxwell-Stefan diffusion equation (33) we assume that velocity cross-correlation terms are negligible and use the Darken relation (36) given by: $D^0_{12} = XD_D + (1-X)D_{Ar}$. In Tables 1 and 2 this value is input at the last column.

At this initial stage it is helpful to compare the Maxwell-Stefan diffusivities with known models. We do this in Fig. 6 for a case with $Z_{Ar} = 13$ and ion density $10^{25}/\text{cm}^3$. Here we compare these results with diffusion from Kinetic Theories [14, 15,51] based on binary collisions. In this figure the results from one of these theories Paquette et al. [15], that assumes a Yukawa potential when calculating the collision integral, is plotted with solid line. Results from our MD simulations combined in the form of Darken relation are shown with green.

Daligault has also done an extensive MD study for One Component Plasma (OCP) [23] as well as for Yukawa One Component Plasma (YOCP) [68], where he validates the kinetic theories at low coupling and extend it by modeling to moderate coupling. In the strong coupling limit he used an empirical law that describes the diffusion process as ions hopping from a cage to another formed by its neighbors. He extended this approach for OCP to binary ionic mixtures (BIM) [23] by use of the Γ_{eff} and the mixed plasma frequency as given in Eq. (51). We followed the same scheme to extend the YOCP model to Yukawa mixtures. In Fig. 6 results from this model are represented with the long-dashed line. Obviously there is discrepancy between the extension of this model with Darken MD. This is due to the mixing rule that we used to extend this model.

In Fig. 6, the agreement with the kinetic theory is fairly good at higher Ar mole fraction, while the discrepancy increases at low Ar mole fraction. This is puzzling, because we would expect the kinetic theories to work very well in the limit of low coupling – low Ar fraction – where the binary collisions are the dominant factor that drives diffusion. This discrepancy might be due to the fact that we are ignoring the cross-correlation terms which might be significant at the low Ar mole fraction. To assess these effects an intense MD study is needed where the Maxwell-Stefan diffusion can be directly computed and then compared against Darken-like relation. We have begun these studies. Also there is a lot of effort on extending kinetic theories to higher coupling [17] through effective potentials, while still retaining the binary collision picture.

6 Discussions

We have reviewed plasma species diffusivity and the various approaches used to calculate it. Kinetic theories, such as the one due to Chapman and Cowling, provide analytic formulas for the diffusivity of weakly coupled mixtures. At stronger coupling the binary collision approximation breaks down and a different technique is needed. MD provides a means of accounting for the many-body effects associated with strong coupling. We have reviewed the self-diffusivity and the Maxwell-Stefan diffusivity. The statistics of the self-diffusivity calculation are much better than those for the Maxwell-Stefan diffusivity, so we have done some comparisons of the Darken expression for the interdiffusivity with the calculated Maxwell-Stefan diffusivity.

Diffusion is one part of mixing. It would be interesting to apply MD to calculate the hydrodynamic (stirring) part of the mixing as well. MD has been used in molten metals to study the mixing in a shear layer forming the Kelvin-Helmholtz instability, capturing both the vortical flow that sweeps arms of one species into the other and then mixes by diffusion [69]. Also, it would be interesting to identify regimes in which the diffusion does not follow Fick's law. Non-Fickian diffusion may result from steep gradients, in which the mole fraction changes significantly within one mean free path. There is still much to be learned about diffusion.

Acknowledgements The authors would like to thank Bill Cabot, Kyle Caspersen, John Castor, Jim Glosli, Frank Graziani, Jeff Greenough, Julie Jackson, A. Bruce Langdon, Paul Miller, Michael Murillo, Dave Richards, Mike Surh, Heather Whitley and all the Cimarron team for their contributions and fruitful discussions. One of us T.H. acknowledges interesting discussion with Jerome Daligault during the IPAM workshop. This work was performed under the auspices of the U.S. Department of Energy by Lawrence Livermore National Laboratory under Contract DE-AC52-07NA27344. This work was funded by the Laboratory Directed Research and Development

Program at LLNL under project tracking code 12-SI-005 and 10-ERD-004. We gratefully acknowledge supercomputer time provided through the Institutional Computing Grand Challenge program at LLNL.

Appendix 1: A General Derivation of the Green-Kubo Relation for the Diffusion of Multicomponent Mixtures

In the following we generalize the derivation due to Hansen and McDonald [24] of the Green-Kubo relation of the interdiffusivity of a binary mixture, Eq. (26), to a more general multicomponent mixture. Similar treatment of diffusion in multicomponent mixtures was given by Zhou and Miller [46, 49] in the geophysical context, and by Wheeler and Newman [70] for the electrolytes. The method although cumbersome is, in principle, straightforward.

Let $c_\alpha(\mathbf{r}, t)$ denote the mass fraction of component α at position $\mathbf{r}$ and time t in a n-component system. The rate of concentration change is linked to the diffusion flux $\mathbf{J}_\alpha(\mathbf{r}, t)$, by the conservation of mass, i.e.,

$$\rho \frac{\partial c_\alpha(\mathbf{r}, t)}{\partial t} + \nabla \cdot \mathbf{J}_\alpha(\mathbf{r}, t) = 0, \tag{54}$$

where ρ is the mass density of the system. There are $(n-1)$ independent equations from (54). The relation for the barycentric frame $\sum_{\alpha=1}^{n} \mathbf{J}_\alpha(\mathbf{r}, t) = 0$ completes the system of these equations. In the linear response regime we can relate diffusion fluxes with the gradient of chemical potentials and through it to mass concentration gradient by the following:

$$\mathbf{J}_\alpha(\mathbf{r}, t) = -\rho \sum_{\beta=1}^{n-1} D_{\alpha\beta} \nabla c_\beta(\mathbf{r}, t), \tag{55}$$

where $D_{\alpha\beta}$ are the mutual diffusion coefficients, which form a $(n-1) \times (n-1)$ matrix in an n-component system. The following general form of Fick's law for a multicomponent mixture [13] is obtained by combining Eq. (54) with Eq. (55)

$$\frac{\partial c_\alpha(\mathbf{r}, t)}{\partial t} = \sum_{\beta=1}^{n-1} D_{\alpha\beta} \nabla^2 c_\beta(\mathbf{r}, t). \tag{56}$$

By Fourier transforming in space and Laplace transforming in time the above diffusion equation becomes:

$$|\tilde{c}_k(\omega)\rangle = \mathscr{D}^{-1} |c_k\rangle, \tag{57}$$

where

$$\mathscr{D} \equiv -i\omega\mathbf{I} + k^2\mathbf{D}, \tag{58}$$

with $\mathbf{I}$ the identity matrix. In Eq. (57) $|\tilde{c}_k(\omega)\rangle$ and $|c_k\rangle$ are $n-1$-dimensional vectors in bra-ket notation, whose αth elements are the Fourier-Laplace transform of $c_\alpha(\mathbf{r},t)$ and the Fourier transform of $c_\alpha(\mathbf{r},0)$, respectively with:

$$\tilde{c}_{\alpha,k}(\omega) = \int_0^\infty dt \int d^3r\, c_\alpha(\mathbf{r},t) \exp(-i\mathbf{k}\cdot\mathbf{r} + \omega t), \tag{59}$$

and

$$c_{\alpha,k} = \int d^3r\, c_\alpha(\mathbf{r},0) \exp(-i\mathbf{k}\cdot\mathbf{r}). \tag{60}$$

From Eq. (57) we can construct an $(n-1)\times(n-1)$ matrix equation relating different correlation concentration factors:

$$\mathscr{D}^{-1} = \mathscr{O}\mathscr{C}^{-1}, \tag{61}$$

where the elements of matrices $\mathscr{O}$ and $\mathscr{C}$ consist of time-dependent and static concentration-correlation functions, respectively, given by:

$$\mathscr{O}_{\alpha\beta} = \langle \tilde{c}_{\alpha,k}(\omega) c_{\beta,-k} \rangle \tag{62}$$

and

$$\mathscr{C}_{\alpha\beta} = \langle c_{\alpha,k} c_{\beta,-k} \rangle, \tag{63}$$

where $\langle \cdots \rangle$ represents statistical ensemble average. Relation (61) can explicitly be written as:

$$\frac{|\mathscr{D}|_{\beta\alpha}}{|\mathscr{D}|} = \frac{1}{|\mathscr{C}|} \sum_{\gamma=1}^{n-1} \mathscr{O}_{\alpha\gamma} |\mathscr{C}|_{\gamma\beta}, \tag{64}$$

where the elements of the inverse of matrix $\mathscr{D}$ are written as $[\mathscr{D}^{-1}]_{\beta\alpha} = |\mathscr{D}|_{\alpha\beta}/|\mathscr{D}|$, with $|\mathscr{D}|_{\alpha\beta}$ the cofactor of $\mathscr{D}_{\alpha\beta}$ in the determinant $|\mathscr{D}|$ [38], and similarly for the concentration correlation matrix $\mathscr{C}$.

Taking the inverse Laplace transform of Eq. (64) gives:

$$-i\omega\frac{|\mathscr{D}|_{\beta\alpha}}{|\mathscr{D}|} - \delta_{\alpha\beta} = \frac{1}{|\mathscr{C}|} \int_0^\infty \frac{d}{dt} \left(\sum_{\gamma=1}^{n-1} \langle c_{\alpha,k}(t) c_{\gamma,-k} \rangle |\mathscr{C}|_{\beta\gamma} \right) \exp(i\omega t)\, dt. \tag{65}$$

In the long wavelength limit the LHS of Eq. (65) is:

$$\lim_{k\to 0}\left(-i\omega\frac{|\mathscr{D}|_{\beta\alpha}}{|\mathscr{D}|}-\delta_{\alpha\beta}\right)=\frac{D_{\alpha\beta}k^2}{i\omega}, \tag{66}$$

from which the following expression $D_{\alpha\beta}$ is obtained:

$$D_{\alpha\beta}=\lim_{\omega\to 0}\lim_{k\to 0}\frac{i\omega}{k^2|\mathscr{C}|}\int_0^\infty\frac{d}{dt}\left(\sum_{\gamma=1}^{n-1}\langle c_{\alpha,k}(t)c_{\gamma,-k}\rangle|\mathscr{C}|_{\beta\gamma}\right)\exp(i\omega t)\,dt. \tag{67}$$

Equation (67) is a generalization of a multicomponent mixture to the one derived by Hansen and McDonald [24]. In the above relation it is essential taking the limit of $k \to 0$ before $\omega \to 0$. By use of the mass conservation relation in Fourier space $\rho\frac{\partial c_{\alpha,k}(t)}{\partial t} = -i\mathbf{k}\cdot\mathbf{J}_{\alpha,k}(t)$, in the above equation we can replace the time dependent concentration structure factor by the corresponding diffusion flux correlation functions. After an integration by parts and use of the time-translational invariance property $\langle\mathbf{J}_{1,k}(t)c_{1,-k}\rangle = \langle\mathbf{J}_{1,k}c_{1,-k}(-t)\rangle$, we have:

$$D_{\alpha\beta}=\lim_{\omega\to 0}\lim_{k\to 0}\frac{1}{3\rho}\left(\sum_{\gamma=1}^{n-1}\frac{|\mathscr{C}|_{\beta\gamma}}{|\mathscr{C}|}\int_0^\infty\langle\mathbf{J}_{\alpha,k}(t)\cdot\mathbf{J}_{\gamma,-k}\rangle\exp(i\omega t)\,dt\right) \tag{68}$$

from which we get:

$$D_{\alpha\beta}=\frac{1}{3\rho}\left(\sum_{\gamma=1}^{n-1}\frac{|\mathscr{C}_0|_{\beta\gamma}}{|\mathscr{C}_0|}\int_0^\infty\langle\mathbf{J}_{\alpha,k\to 0}(t)\cdot\mathbf{J}_{\gamma,k\to 0}\rangle\,dt\right). \tag{69}$$

Here the elements of matrix $\mathscr{C}_0$ are given by $|\mathscr{C}_0|_{\alpha\beta} \equiv \langle c^\alpha c^\beta\rangle$. This are the long wavelength limit concentration correlation functions. In binary mixtures they are directly related to the concentration structure factor S_{cc} defined by Eq. (27). Equation (69) is written as a product of two matrices, one whose elements are related to composition-composition structure factor, and the other with elements constructed from Green-Kubo (Maxwell-Stefan) interdiffusion correlation terms. It is straightforward to derive the two component interdiffusion coefficient (26) as a special case of this multicomponent relation (69). In a ternary mixture matrix relation (69) simplifies to:

$$\begin{bmatrix} D_{11} & D_{12}\\ D_{21} & D_{22}\end{bmatrix}=\frac{1}{3\rho|C_0|}\begin{bmatrix}\langle(c_2)^2\rangle & -\langle c_1c_2\rangle\\ -\langle c_1c_2\rangle & \langle(c_1)^2\rangle\end{bmatrix}$$
$$\times\begin{bmatrix}\int_0^\infty\langle\mathbf{J}_1(t)\cdot\mathbf{J}_1\rangle\,dt & \int_0^\infty\langle\mathbf{J}_1(t)\cdot\mathbf{J}_2\rangle\,dt\\ \int_0^\infty\langle\mathbf{J}_2(t)\cdot\mathbf{J}_1\rangle\,dt & \int_0^\infty\langle\mathbf{J}_2(t)\cdot\mathbf{J}_2\rangle\,dt\end{bmatrix}. \tag{70}$$

Here

$$|C_0| = \langle (c_1)^2\rangle \langle (c_2)^2\rangle - (\langle c_1 c_2\rangle)^2 \tag{71}$$

is the determinant of matrix $\mathbf{C_0}$. In analogy with the binary mixing case we can define a thermodynamic factor matrix Φ such that:

$$\begin{bmatrix} \Phi_{11} & \Phi_{12} \\ \Phi_{21} & \Phi_{22} \end{bmatrix} = \frac{1}{3\rho|C_0|}\begin{bmatrix} \langle (c_2)^2\rangle & -\langle c_1 c_2\rangle \\ -\langle c_1 c_2\rangle & \langle (c_1)^2\rangle \end{bmatrix}. \tag{72}$$

The Green-Kubo form presented in Eq. (70) provides a simple form that can be used to study ternary mixtures through equilibrium MD.

Appendix 2: Collision Integrals

The collision integrals $\Omega_{\alpha\beta}^{(ij)}$ between two particles α and β are related to the total cross section after integrating over a Maxwellian velocity distribution. They are given by [14, 15]:

$$\Omega_{\alpha\beta}^{(ij)} = \left(\frac{k_B T}{2\pi m_{red}}^{1/2}\right)\int_0^\infty e^{-g^2} g^{2j+3}\phi_{\alpha\beta}^{(i)}\, dg, \tag{73}$$

where g is a dimensionless velocity, $m_{red} \equiv m_\alpha m_\beta/(m_\alpha + m_\beta)$ is the reduced mass, and $\phi_{\alpha\beta}^{(i)}$ are the collisions cross section for a given energy:

$$\phi_{\alpha\beta}^{(i)} = 2\pi \int_0^\infty (1 - \cos^i \chi_{\alpha\beta}) b\, db, \tag{74}$$

with the integration over the impact parameter b.

In Eq. (74) $\chi_{\alpha\beta}$ is the scattering angle given:

$$\chi_{\alpha\beta} = \pi - 2\int_{r_{\alpha\beta}^{min}}^\infty \frac{b\, dr}{r^2\left[1 - \frac{b^2}{r^2} - \frac{V_{\alpha\beta}(r)}{g^2 k_B T}\right]^{1/2}}. \tag{75}$$

Here $V_{\alpha\beta}(r)$ is the interaction potential between the particle of type α and β, and $r_{\alpha\beta}^{min}$ is the distance of the closest approach between the particle which is the root of the following equation:

$$1 - \frac{b^2}{\left(r_{\alpha\beta}^{min}\right)^2} - \frac{V_{\alpha\beta}\left(r_{\alpha\beta}^{min}\right)}{g^2 k_B T} = 0. \tag{76}$$

References

1. R.E. Rudd, W.H. Cabot, K.J. Caspersen, J.A. Greenough, D.F. Richards, F.H. Streitz, P.L. Miller, Phys. Rev. E **85**, 031202 (2012)
2. P. Amendt, O.L. Landen, H.F. Robey, C.K. Li, R.D. Petrasso, Phys. Rev. Lett. **105**, 115005 (2010)
3. P. Amendt, S. Wilks, C. Bellei, C.K. Li, R.D. Petrasso, Phys. Plasma **18**, 056308 (2011)
4. P. Amendt, C. Bellei, S. Wilks, Phys. Rev. Lett. **109**, 075002 (2012)
5. D.B. Dusenbery, *Living at Micro Scale* (Harvard University Press, Cambridge, 2009)
6. G. Michaud, Astrophys. J. **160**, 641 (1970)
7. J. Lindl, Phys. Plasmas **2**, 3933 (1995)
8. J. Nuckolls, L. Wood, A. Thiessen, G. Zimmerman, Nature **239**, 139 (1972)
9. S. Hamaguchi, R.T. Farouki, J. Chem. Phys. **101**, 9876 (1994)
10. R.T. Farouki, S. Hamaguchi, J. Chem. Phys. **101**, 9885 (1994)
11. S. Hamaguchi, R.T. Farouki, H.E. Dubin, J. Chem. Phys. **105**, 7641 (1996)
12. R. Farouki, S. Hamaguchi, J. Comput. Phys. **115**(2), 276 (1994)
13. R. Bird, W. Stewart, E. Lightfoot, *Transport Phenomena*, Wiley International Edition (Wiley, New York, 2007)
14. S. Chapman, T. Cowling, *The Mathematical Theory of Non-uniform Gases: An Account of the Kinetic Theory of Viscosity, Thermal Conduction and Diffusion in Gases*. Cambridge Mathematical Library (Cambridge University Press, Cambridge, 1970)
15. C. Paquette, C. Pelletier, G. Fontaine, G. Michaud, Astrophys. J. Suppl. Ser. **61**, 177 (1986)
16. H. Ohta, S. Hamaguchi, Phys. Plasmas **7**, 4506 (2000)
17. S.D. Baalrud, J. Daligault, Phys. Rev. Lett. **110**, 235001 (2013)
18. E.E. Salpeter, Australian. J. Phys. **7**, 373 (1954)
19. J.P. Hansen, G.M. Torrie, P. Vieillefosse, Phys. Rev. A **16**, 2153 (1977)
20. S. Bastea, Phys. Rev. E **71**, 056405 (2005)
21. Y. Rosenfeld, Phys. Rev. E **47**, 2676 (1993)
22. Y. Rosenfeld, E. Nardi, Z. Zinamon, Phys. Rev. Lett. **75**, 2490 (1995)
23. J. Daligault, Phys. Rev. Lett. **108**, 225004 (2012)
24. J. Hansen, I. McDonald, *Theory of Simple Liquids* (Elsevier, New York, 2006)
25. N. March, M. Tosi, *Atomic Dynamics in Liquids*. Dover Books on Physics and Chemistry (Dover, New York, 1991)
26. D. McQuarrie, *Statistical Mechanics* (HarperCollinsPublishers, New York, 1976)
27. J. Horbach, S.K. Das, A. Griesche, M.P. Macht, G. Frohberg, A. Meyer, Phys. Rev. B **75**, 174304 (2007)
28. J. Trullàs, J.A. Padró, Phys. Rev. E **50**, 1162 (1994)
29. J. Daligault, Phys. Rev. Lett. **96**, 065003 (2006)
30. A. Einstein, Annalen der Physik **322**(8), 549 (1905)
31. R. Kubo, J. Phys. Soc. Jpn. **12**(6), 570 (1957)
32. R. Kubo, M. Yokota, S. Nakajima, J. Phys. Soc. Jpn. **12**(11), 1203 (1957)
33. R. Zwanzig, J. Chem. Phys. **40**, 2527 (1964)
34. R. Zwanzig, Annu. Rev. Phys. Chem. **16**, 67 (1965)
35. D.B. Boercker, E.L. Pollock, Phys. Rev. A **36**, 1779 (1987)
36. C. Cohen, J.W.H. Sutherland, J.M. Deutch, Phys. Chem. Liq. **2**, 213 (1971)
37. A.B. Bhatia, D.E. Thornton, Phys. Rev. B **2**, 3004 (1970)
38. J.G. Kirkwood, F.P. Buff, J. Chem. Phys. **19**, 774 (1951)
39. J.P. Hansen, G.M. Torrie, P. Vieillefosse, Phys. Rev. A **16**, 2153 (1977)
40. D. Léger, C. Deutsch, Phys. Rev. A **37**, 4916 (1988)
41. J. Stefan, Akad. Wiss. Wien **Abt. II**, 63 (1871)
42. J.P. Hansen, F. Joly, I.R. McDonald, Physica **132A**, 4722 (1985)
43. M. Schoen, C. Hoheisel, Mol. Phys. **52**, 1029 (1984)
44. D.L. Jolly, R.J. Bearman, Mol. Phys. **41**, 137 (1980)

45. D.W. McCall, D.C. Douglass, J. Phys. Chem. **71**, 987 (1967)
46. Y. Zhou, G.H. Miller, Phys. Rev. E **53**, 1587 (1996)
47. L.S. Darken, Trans. AIME **175**, 184 (1948)
48. D.C. Douglass, H.L. Frisch, J. Phys. Chem. **73**, 3039 (1969)
49. Y. Zhou, G.H. Miller, J. Phys. Chem. **100**, 5516 (1996)
50. J.P. Hansen, I.R. McDonald, Phys. Rev. A **11**, 2111 (1975)
51. L. Spitzer, *Physics of Fully Ionized Gases*. Interscience Tracts on Physics and Astronomy (Interscience Publishers, New York, 1962)
52. D. Frenkel, B. Smit, *Understanding Molecular Simulation: From Algorithms to Applications*. Computational Science (Elsevier, Burlington, 2001)
53. G. Ciccotti, D. Frenkel, I. McDonald, *Simulation of Liquids and Solids: Molecular Dynamics and Monte Carlo Methods in Statistical Mechanics*. North-Holland Personal Library (North-Holland, New York, 1987)
54. N. Ashcroft, N. Mermin, *Solid State Physics*. Science: Physics (Saunders College, Fort Worth, 1976)
55. B. Firey, N.W. Ashcroft, Phys. Rev. A **15**, 2072 (1977)
56. P. Debye, E. Hückel, Phys. Z **24**, 185 (1923)
57. D. Pines, P. Nozieres, *The Theory of Quantum Fluids* (Addison-Wesley, Redwood, 1990)
58. S. Itchimaru, *Statistical Plasma Physics* (Addison-Wesley, Reading, 1991)
59. G. Gregori, S.H. Glenzer, W. Rozmus, R.W. Lee, O.L. Landen, Phys. Rev. E **67**, 026412 (2003)
60. M.S. Murillo, High Energy Density Phys. **4**(1–2), 49 (2008)
61. P.P. Ewald, Ann. Phys. **64**, 253 (1921)
62. M. Desrno, C. Holm, J. Chem. Phys. **109**, 7678 (1998)
63. E.L. Pollock, J. Glosli, Comp. Phys. Comm. **95**, 93 (1996)
64. S. Nosé, J. Chem. Phys. **81**, 511 (1984)
65. W.G. Hoover, Phys. Rev. A **31**, 1695 (1985)
66. R.P. Feynman, N. Metropolis, E. Teller, Phys. Rev. **75**, 1561 (1949).
67. M. Vergeles, G. Szamel, J. Chem. Phys. **110**, 3009 (1999)
68. J. Daligault, Phys. Rev. E **86**, 047401 (2012)
69. J. Glosli, K. Caspersen, D. Richards, R. Rudd, F. Streitz, J. Gunnels, in Proceedings of the Supercomputing 2007 (SC07), Reno, Nov 2007. UCRL-CONF-230679 pp. 276–287 (2007)
70. D.R. Wheeler, J. Newman, J. Phys. Chem. B **108**, 18353 (2004)

A Review of Wave Packet Molecular Dynamics

Paul E. Grabowski

Abstract Warm dense matter systems created in the laboratory are highly dynamical. In such cases electron dynamics is often needed to accurately simulate the evolution and properties of the system. Large systems force one to make simple approximations enabling computational feasibility. Wave packet molecular dynamics (WPMD) provides a simple framework for simulating time-dependent quantum plasmas. Here, this method is reviewed. The different variants of WPMD are shown and compared and their validity is discussed.

1 Introduction

The creation of warm dense matter in the laboratory is a dynamic process. A large amount of energy is delivered to a target in a short period of time. Several different methods for delivering this energy have been developed, including exploding wire [1–3], laser foil [4–19], inertial confinement fusion [20, 21], ion beam [22–27], Z machine [28–31], free electron laser [32–35], and laser induced shock [36–41] experiments. Understanding energy exchange processes is crucial to the proper interpretation of these experiments. Lasers and ion beams deposit most of their energy in the electrons while shocks impart most of their energy to the ions. Energy absorption by the electrons in the former cases leads to a time-dependent light induced ionization and scattering [42, 43] or a stopping power problem [44, 45], respectively, while all of these cases produce electron-ion temperature relaxation [46]. Furthermore, electrons transport their energy via possibly non-equilibrium electrical and thermal conduction. Recent efforts [42, 43] to measure electron and

P.E. Grabowski (✉)
University of California, Irvine, CA, USA
e-mail: paul.grabowski@uci.edu

F. Graziani et al. (eds.), *Frontiers and Challenges in Warm Dense Matter*, Lecture Notes in Computational Science and Engineering 96,
DOI 10.1007/978-3-319-04912-0_10,

ion dynamics at attosecond and picosecond timescales reinforce the need for a short-timescale simulation capability. It is now possible to directly image electron density [47] and such efforts can validate any theoretical predictions.

Ideally, one should calculate a numerically converged solution to the many-body time-dependent Schrödinger equation for the electron wave function in the external potential due to the time-dependent electron-ion interactions. Tens or a few hundred degrees of freedom can be evolved with the efficient multi-configuration time-dependent Hartree (MCTDH) method [48, 49]. Big simulations are needed to solve non-equilibrium systems for statistical reasons, to resolve gradients, and to resolve mean free paths of the particles in the system. Despite making significant progress in reducing computational effort for many-body quantum problems, MCTDH still scales exponentially with the number of degrees of freedom and cannot handle the system sizes needed for these systems. The obvious simplification would be to use time-dependent density functional theory [50] for which one would only need to evolve a three-dimensional density. However, very little is known about the accuracy of commonly used functionals for non-equilibrium high energy density systems. Finite computational resources then necessitate simple models that capture only the requisite quantum mechanics.

Wave packet molecular dynamics (WPMD) is a simple time-dependent quantum mechanical method with a rich underlying mathematical structure [51]. Here, WPMD is defined as the joint propagation of classical ions, represented as point particles, and quantum electrons, represented as wave packets. The wave packets are single electron states, localized in both position and momentum, and are combined via a Hartree product or Slater determinant to form the many-body wave function. Several groups have used WPMD to calculate a diverse set of observables, including equations of state [52–57], the collision rate [58], electrical conductivity [54, 56], the diffusion coefficient [59], the dynamic structure factor [60], stopping power [61, 62], and shock Hugoniot curves [53, 55, 56, 63, 64]. In this chapter, the different derivations of the equations of motion for the wave packets are summarized in Sect. 2. The many different variations in wave packet forms used, antisymmetrization approximations, and interpretation of variational parameters are reviewed in Sect. 3 and an outlook is given in Sect. 4.

2 Theoretical Basis

2.1 *Ehrenfest's Theorem*

Quantum mechanics is infinitely more complicated than classical mechanics. Quantum dynamics happens within the full Hilbert space of the system while classical dynamics are described by a path through a finite-dimensional phase space. The quantum mechanical state does not correspond to any classical quantity such as position or momentum. We can, however, describe this state by its expectation values of moments of these quantities. Ehrenfest [65] was the first to show how these moments evolve with his famous theorem,

$$\frac{d}{dt}\left\langle\hat{A}\right\rangle = \frac{1}{i\hbar}\left\langle\left[\hat{A}, \hat{H}\right]\right\rangle + \left\langle\frac{\partial\hat{A}}{\partial t}\right\rangle, \tag{1}$$

where $\hat{A}$ is an arbitrary operator and $\hat{H}$ is the Hamiltonian operator, typically given by

$$\hat{H} = \sum_i \frac{\hat{p}_i^2}{2m_i} + \sum_{i<j} V_{ij}(\hat{\mathbf{r}}_i, \hat{\mathbf{r}}_j). \tag{2}$$

Here, the summations are over the particles in the system, V_{ij} is the interaction potential energy between particles i and j, and $\hat{\mathbf{p}}_i$, $\hat{\mathbf{r}}_i$, and m_i are the momentum operator, position operator, and mass of particle i, respectively. It is easy to show that

$$\frac{d}{dt}\langle\hat{\mathbf{r}}_i\rangle = \frac{\langle\hat{\mathbf{p}}_i\rangle}{m_i} \tag{3}$$

$$\frac{d}{dt}\langle\hat{\mathbf{p}}_i\rangle = -\sum_{j\neq i}\left\langle\nabla_i V_{ij}\right\rangle \tag{4}$$

$$\frac{d}{dt}\left\langle\hat{r}_i^2\right\rangle = \frac{\langle\{\hat{\mathbf{r}}_i, \hat{\mathbf{p}}_i\}\rangle}{m_i} \tag{5}$$

$$\frac{d}{dt}\left\langle\hat{p}_i^2\right\rangle = \frac{1}{i\hbar}\sum_{j\neq i}\left\langle[\hat{p}_i^2, V_{ij}]\right\rangle, \tag{6}$$

where ∇_i is the gradient with respect to $\mathbf{r}_i$, the anti-commutator is defined as $\{\hat{\mathbf{r}}_i, \hat{\mathbf{p}}_i\} = \hat{\mathbf{r}}_i \cdot \hat{\mathbf{p}}_i + \hat{\mathbf{p}}_i \cdot \hat{\mathbf{r}}_i$, and we have suppressed the dependence of V_{ij} on the position operators. These equations are only a few of the infinite number of equations needed to describe complete quantum dynamics. If V_{ij} is a smooth C^∞ function, then all of the time derivatives of $\langle\hat{\mathbf{r}}_i^n\hat{\mathbf{p}}_j^m\rangle$ for any i, j, n, and m will be a function of these same moments. So Ehrenfest's theorem leads to an infinite hierarchy of equations. These equations can be truncated at finite values of n and m with the help of a closure. For example, a common choice of restricted wave function in WPMD is the isotropic Gaussian for each quantum particle,

$$\varphi_G(\mathbf{r}_i) = \left(\frac{3}{2\pi\sigma^2}\right)^{3/4}\exp\left[-\left(\frac{3}{4\sigma^2} - \frac{ip_\sigma}{2\hbar\sigma}\right)|\mathbf{r} - \mathbf{r}_i|^2 + \frac{i\mathbf{p}\cdot(\mathbf{r} - \mathbf{r}_i)}{\hbar}\right], \tag{7}$$

where $\mathbf{r}_i$ is the position coordinate, $\mathbf{r} = \langle\hat{\mathbf{r}}_i\rangle$ is its expectation value, $\mathbf{p} = \langle\hat{\mathbf{p}}_i\rangle$ is the expectation value of momentum, $\sigma = \sqrt{\langle\hat{r}_i^2\rangle - |\langle\hat{\mathbf{r}}_i\rangle|^2}$ is the uncertainty in position, and p_σ is its conjugate momentum. Within the Hartree approximation, Eqs. (3–6) imply

$$\dot{\mathbf{r}} = \frac{\mathbf{p}}{m_i} \tag{8}$$

$$\dot{\mathbf{p}} = -\sum_{j\neq i} \langle \nabla_i V_{ij} \rangle \tag{9}$$

$$\dot{\sigma} = \frac{p_\sigma}{m_i} \tag{10}$$

$$\dot{p}_\sigma = \frac{9\hbar^2}{4m_i\sigma^3} - \frac{1}{\sigma}\sum_{j\neq i} \langle (\hat{\mathbf{r}}_i - \mathbf{r}) \cdot \nabla_i V_{ij} \rangle . \tag{11}$$

These equations of motion predict that the center of the wave packet moves at the expectation value of velocity. This velocity changes according to the expectation value of the force. The width of the wave packet changes at a rate of p_σ/m_i, and its conjugate momentum evolves so as to satisfy the Heisenberg uncertainty principal and minimize the expectation value of the potential energy.

2.2 Local Harmonic Approximation

Heller's original formulation [66] of time-dependent semiclassical wave packet dynamics approached the problem differently. Instead of making a wave function ansatz, he expanded the potential energy to second order in the distance to the expectation value of the configuration of the system.

$$V(\mathbf{X}) \approx V(\tilde{\mathbf{X}}) + \nabla_X V(\tilde{\mathbf{X}}) \cdot (\mathbf{X} - \tilde{\mathbf{X}}) + \frac{1}{2} \nabla_X \otimes \nabla_X V(\tilde{\mathbf{X}}) : (\mathbf{X} - \tilde{\mathbf{X}}) \otimes (\mathbf{X} - \tilde{\mathbf{X}}), \tag{12}$$

where $\mathbf{X} = \{\mathbf{x}_1, \ldots, \mathbf{x}_N\}$ is the set of positions of all N particles in the system, $\tilde{\mathbf{X}} = \langle \mathbf{X} \rangle$ is its expectation value, $\otimes$ is the tensor product, and : indicates the contraction of the indices of the rank-two tensors on either side of it. The system is then a $3N$ dimensional harmonic oscillator with characteristic frequencies that depend directly on $\tilde{\mathbf{X}}$ and so indirectly on time. The further simplification of neglecting the inter-particle terms in the last term of Eq. (12) and the particle statistics leads to an anisotropic formulation of WPMD. The potential energy has the reduced form:

$$V(\mathbf{X}) \approx \sum_i^N V_i(\tilde{\mathbf{x}}_i) + \nabla_{\mathbf{x}_i} V(\tilde{\mathbf{x}}_i) \cdot (\mathbf{x}_i - \tilde{\mathbf{x}}_i) + \frac{1}{2} \nabla_{\mathbf{x}_i} \otimes \nabla_{\mathbf{x}_i} V(\tilde{\mathbf{x}}_i) : (\mathbf{x}_i - \tilde{\mathbf{x}}_i) \otimes (\mathbf{x}_i - \tilde{\mathbf{x}}_i), \tag{13}$$

where $\tilde{\mathbf{x}}_i = \langle \mathbf{x}_i \rangle$, and the approximate many-body wave function is

$$\psi(\mathbf{X}, t) \approx \prod_i^N \left(\frac{\det \Sigma_i}{\pi} \right)^{1/4} e^{-(\mathbf{x}_i - \mathbf{r}_i)^T \cdot (\Sigma_i + i\Pi_i) \cdot (\mathbf{x}_i - \mathbf{r}_i) + i\mathbf{p}_i \cdot (\mathbf{x}_i - \mathbf{r}_i)/\hbar + i\xi_i}, \tag{14}$$

where Σ_i and Π_i, $\mathbf{r}_i$ and $\mathbf{p}_i$, and ξ_i are time-dependent tensors, vectors, and scalars, respectively, which depend on the local value of potential energy and its derivatives. The main issue in applying this formulation to plasma physics is that the Coulomb potential is singular. So the expansion (12), is divergent near the singularities.

2.3 Time Dependent Variational Principle

One variational principle, proposed by McLachlan [67] and used in Refs. [59, 68], is to minimize

$$I(\psi, \theta) = \int |i\hbar\theta - \hat{H}\psi|^2 dV \tag{15}$$

with respect to $\theta = \partial\psi/\partial t$, where the integration is performed over all of configuration space.

The Dirac-Frenkel time-dependent variational principle (TDVP) [69] leads to a rigorous approximation of the time-dependent Schrödinger equation (TDSE) with a given variational ansatz. With this method the residual of the TDSE is minimized over a given subspace of states $|\psi\rangle$,

$$\delta \int_{t_i}^{t_f} \left\langle \psi \left| i\hbar \frac{\partial}{\partial t} - \hat{H} \right| \psi \right\rangle dt = 0, \tag{16}$$

where t_i and t_f are the initial and final times of the integration, and $\hat{H}$ is the Hamiltonian. If the state $|\psi\rangle$ is allowed to vary throughout a Hilbert space that includes the solution, the TDSE will be exactly solved. Otherwise, the error in the state grows linearly with time over short times [70]. This variational principle is equivalent to the McLachlan approach for the most general form of a Gaussian wave packet [71].

A variational state $|\mathbf{q}\rangle$ can be parametrized by a vector of complex time-dependent variational parameters,

$$\mathbf{q} = \{q_1, q_2, \ldots, q_{N_v}\}. \tag{17}$$

The variational parameters follow the equations of motion [70]:

$$\mathrm{i}\mathbf{N}\dot{\mathbf{q}} = \frac{\partial\langle\hat{H}\rangle}{\partial\mathbf{q}^*}, \quad -\mathrm{i}\mathbf{N}\dot{\mathbf{q}}^* = \frac{\partial\langle\hat{H}\rangle}{\partial\mathbf{q}}, \tag{18}$$

where $\langle\hat{H}\rangle = \langle\psi|\hat{H}|\psi\rangle$ and * denotes the complex conjugate. The Hermitian norm matrix is defined by [70]:

$$N_{ab} = \frac{\partial}{\partial q_a^*}\frac{\partial}{\partial q_b}\ln\langle\mathbf{q}|\mathbf{q}\rangle. \tag{19}$$

Note, Eq. (18) are time reversed forms of each other; so models derived from the TDVP preserve time reversal symmetry. For special choices of the variational form and parameters, the matrix $\mathbf{N}$ reduces to a trivially-inverted matrix and canonical positions and momenta can be defined that make the equations of motion have a Hamilton form in N_v dimensions (see for example Ref. [72]):

$$\dot{\boldsymbol{\rho}} = \frac{\partial\langle\hat{H}\rangle}{\partial\boldsymbol{\pi}}, \quad \dot{\boldsymbol{\pi}} = -\frac{\partial\langle\hat{H}\rangle}{\partial\boldsymbol{\rho}}. \tag{20}$$

In spite of the persuasiveness of this form, it has to be noted that $\boldsymbol{\rho}$ and $\boldsymbol{\pi}$ are variational parameters inextricably tied to a particular variational wave function that should not be mistaken for classical positions and momenta if the quantum nature of the method is to be preserved. Using the TDVP with a small number of parameters requires physical intuition as to the form of the wave function. It must be flexible enough to give reasonable observables as well as numerically convenient and capable of representing the desired initial state.

3 Usage

3.1 Alternate Wave Packet Forms

The standard isotropic Gaussian form (Eq. 7) for the wave packets in WPMD was mainly chosen for its mathematical simplicity. Expectation values of the kinetic energy and the potential energy for many different types of interactions (including the Coulomb potential) can be analytically calculated. This becomes a great utility in many-body dynamics, for which extensive calculations limit computationally-feasible system sizes. However, there is much to be desired physically from a variational form that this simple wave packet lacks. Well known asymptotic behaviors both near and far from nuclei are incorrectly modeled by Gaussians. The wave packets are unable to breakup and properly share their density with all the nuclei to produce an accurate representation of a free state, nor can an isotropic

Gaussian accurately form bonds, but simple bonding can occur [57]. Furthermore, there is the issue of reconciling a periodic system with an aperiodic wave function. Various authors have made attempts to improve all of these shortcomings. Their trial wave functions are listed below:

- Anisotropic Gaussian [51]

$$\varphi_{ag}(\mathbf{x}, t) = \left(\frac{\det \Sigma}{\pi}\right)^{1/4} e^{-(\mathbf{x}-\mathbf{r})^T \cdot (\Sigma + i\Pi)\cdot(\mathbf{x}-\mathbf{r}) + i\mathbf{p}\cdot(\mathbf{x}-\mathbf{r})/\hbar + i\xi}, \tag{21}$$

  where Σ and Π are real symmetric matrices and Σ is positive definite. This form allows the Gaussians to evolve differently in each dimension. It is most useful for systems which have electron densities around each ion that are not isotropic, that is, when bonds are important.
- Hermite Gaussian Wave Packet [73]

$$\varphi_{hg}(\mathbf{x}, t) = \sum_{\{n_x, n_y, n_z\} \in A} c_{n_x n_y n_z} \varphi_{hg}^{n_x n_y n_z}(\mathbf{x}, t), \tag{22}$$

  where the Hermite-Gauss functions φ_{hg}^{ijk} are

$$\varphi_{hg}^{n_x n_y n_z} = \prod_{j=x,y,z} h_{n_j}(\sqrt{\omega_j}(r_j - x_j)) \exp[-(\omega_j/2 + \mathrm{i}\gamma_j)(r_j - x_j)^2 + \mathrm{i}p_j(r_j - x_j)], \tag{23}$$

  the h_{n_j} are the normalized Hermite polynomials of degree n_j, A is the set of allowed triplets, and ω_j, r_j, γ_j, and p_j are variational parameters. The set A can be restricted to reduce the numerical effort in inverting the overlap matrix; otherwise, in the limit that A includes all possible triplets, the Hermite Gaussian wave packet becomes exact for one-electron problems.
- Split Wave Packet [74, 75]

$$\varphi_s(\mathbf{x}, t) \propto \sum_{\alpha=1}^{M} c_\alpha \varphi_\alpha(\mathbf{x}, t), \tag{24}$$

  where each wave packet (φ_α) has the same form as Eq. (7), and the variational parameters ($\mathbf{r}$, $\mathbf{p}$, σ, and p_σ) take on different values and evolve independently for each wave packet. This form represents a single electron wave function by M Gaussians with mixing coefficients c_α. It allows the wave packet to breakup, following the behavior observed in Ref. [75].
- Periodic Wave Packet [52, 53, 63]

$$\varphi_p(\mathbf{x}, t) \propto \sum_{\mathbf{n}} \varphi_G(\mathbf{x} - \mathbf{n}L), \tag{25}$$

  where $\mathbf{n}$ is a lattice index for the three-dimensional periodic system and L is the length of the periodic box. This form is more consistent with periodic boundary conditions used in bulk plasma simulations. It causes a large wave packet to be even more weakly interacting with the rest of the system and so be more likely to have divergent width.

- Periodic-Bloch Wave Packet [54, 56]

$$\varphi_{pb}(\mathbf{x}, t) = e^{i\mathbf{q}\cdot\mathbf{x}/\hbar}\varphi_p(\mathbf{x}), \tag{26}$$

where $\mathbf{q}$ is the Bloch momentum.

- Bound-Free Wave Packets [76]

$$\varphi_b(\mathbf{x}, t) = \frac{1}{\sqrt{\pi a_0^3}} e^{-|\mathbf{x}-\mathbf{r}_I|/a_0}, \tag{27}$$

where a_0 is the Bohr radius and $\mathbf{r}_I$ is the position of an ion,

$$\varphi_f(\mathbf{x}, t) = \left(\frac{3}{2\pi\sigma^2}\right)^{3/4} e^{3(\mathbf{x}-\mathbf{r})^2/4\sigma^2 + i\mathbf{p}\cdot(\mathbf{x}-\mathbf{r})/\hbar}, \tag{28}$$

and the width σ is fixed. Ebeling and Militzer used both of these forms to represent bound and free states. The electrons were allowed to transition between these states according to known cross sections. Of course, the exact bound and free states can differ significantly from φ_b and φ_f.

- Self Similar Wave Packet [77]

$$\varphi_{ss}(\mathbf{x}, t) = \sqrt{s^{-3}\rho_0\left(\frac{\mathbf{x}-\mathbf{r}}{s}\right)} e^{i[\mathbf{p}\cdot(\mathbf{x}-\mathbf{r}) + p_s(\mathbf{x}-\mathbf{r})^2/2s]/\hbar + \phi}, \tag{29}$$

where s, $\mathbf{r}$, $\mathbf{p}$ and p_s are time-dependent variational parameters and

$$\rho_0(\mathbf{x}) = \begin{cases} \left(\frac{1}{\pi G^2}\right)^{3/2} e^{-\mathbf{x}^2/G^2} & \text{Gaussian} \\ \frac{1}{8\pi E} e^{-\mathbf{x}/E} & \text{Exponential} \end{cases} \tag{30}$$

Murillo and Timmermans compared the relative accuracy of the Gaussian and exponential forms of ρ_0 in calculating the ground state energies of hydrogenic systems, helium-like ions, and the hydrogen molecule. The exponential wave packet performed better for single atoms, but the Gaussian wave packet better reproduced the hydrogen molecular bond. It is unclear which is better for time dependent systems for which $\mathbf{r}$ can become displaced from the positions of the ions or centers of bonds.

- Harmonically Constrained Wave Packet [78]

$$\varphi_h(\mathbf{x}, t) = \varphi_G(\mathbf{x}, t) e^{id(\mathbf{x}-\mathbf{r})^4/\sigma^2}, \tag{31}$$

where d is an adjustable parameter which controls how large width σ can get. The consequences of fixing d are discussed in Sect. 3.2.

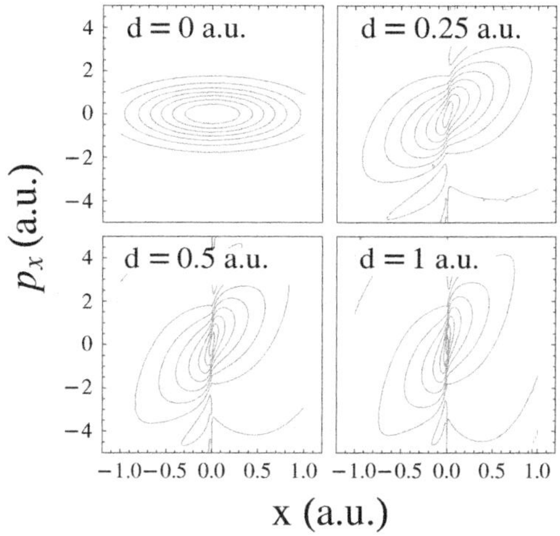

Fig. 1 Wigner density cross section of a three-dimensional Gaussian wave function with extra phase factor as in Eq. (31) for different values of d. The expectation values of position and momentum are zero, the width σ is one, and its conjugate momentum p_σ is zero

3.2 *Width Constraints*

It has been observed [52, 53, 58, 63, 78–81] that at high enough temperatures that the mean width of the wave packets increases without bound unless the equations of motion are altered. A diverging width leads to a uniform electron density and diminished electron-ion and electron-electron interactions, preventing electron equilibration or the correct measurement of transport quantities.

The most common method of preventing width spreading was introduced in Ref. [79] and continued by Refs. [52, 53, 58, 63, 80, 81]. A harmonic constraint is added to the energy expectation value in an ad hoc fashion,

$$H_{Harm} = a\sigma^2, \tag{32}$$

where a is an adjustable parameter setting the mean width of the electrons. Alternatively, Ebeling and coworkers [78] were able to derive such an expression by changing the variational wave function to Eq. (31), giving a slightly different harmonic constraint,

$$H'_{Harm} = \frac{20d}{3m}\left(p_\sigma \sigma + \frac{14}{3}d\sigma^2\right). \tag{33}$$

Changing the variational wave function to Eq. (31) strongly affects the properties of the wave function. To visualize the change, the Wigner density

$$f_W(\mathbf{R}, \mathbf{P}, t) = \frac{1}{(2\pi\hbar)^3}\int \varphi_h^*\left(\mathbf{R} + \frac{\mathbf{s}}{2}\right)\varphi_h\left(\mathbf{R} - \frac{\mathbf{s}}{2}\right)e^{i\mathbf{P}\cdot\mathbf{s}/\hbar}d^3s \tag{34}$$

was calculated at $\mathbf{R} = \{x, 0, 0\}$ and $\mathbf{P} = \{p_x, 0, 0\}$ for $\mathbf{r} = 0$, $\mathbf{p} = 0$, $p_\sigma = 0$, and $\sigma = 1\ a.u.$ for several different values of d and shown in Fig. 1. When $d \neq 0$ the Wigner density is twisted and develops a discontinuity in its derivatives near $x = 0$.

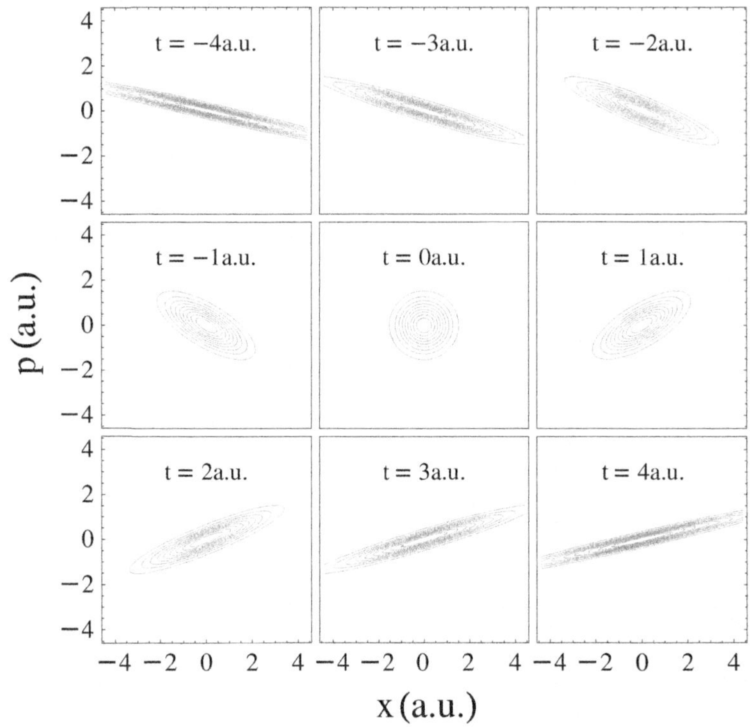

Fig. 2 Wigner density of a one-dimensional Gaussian wave function propagating in vacuum. The expectation values of position and momentum are zero at all times

Morozov and Valuev [58] proposed two alternative constraints. They introduced periodic boundary conditions for the width coordinate (equivalent to a reflecting boundary condition at some maximum width). This change allowed the wave packets to equilibrate to a width distribution, but the equilibrated density was too constant. They also invented an energy-based constraint, in which the wave packets experienced a confining potential only if their interaction potential energy with the nearest ion were above a given threshold. By tuning this threshold they could obtain different results for the collision frequency. It is unlikely that every dynamical quantity will obtain the correct value at the same threshold energy.

The underlying assumption of the harmonic constraint is that a diverging wave packet width is unphysical. However, it is physical for wave packet widths to diverge, but they should localize near ions at the same time. In fact, the simplest system (a single electron wave packet in a vacuum) has a well known exact solution which exhibits wave packet width divergence. Isocontours of the Wigner density are shown for the non-interacting wave packet in Fig. 2 at a series of different times. The time coordinate is defined so that the wave packets reaches a minimum uncertainty configuration at $t = 0$. At negative times the uncertainty in position is decreasing and at positive times it is increasing. The momentum uncertainty remains constant for this special case of the non-interacting wave packet because momentum eigenstates are also energy eigenstates. This behavior is generic to all non-interacting particles. At late times, they will always have a diverging width and will never reach minimum uncertainty again.

Grabowski et al. [75] showed that this divergence also occurs in their model plasma, made of a single dynamic electron propagating through a fixed periodic system of statically screened protons. This simple system could be propagated exactly and it was observed that the simple wave packet variational form actually underestimates the spreading, and its main failure was its inability to individually scatter off each proton, which would have created the necessary fluctuations in density.

3.3 Antisymmetrization

The most computationally expensive part of a wave-function-based quantum calculation is the accurate treatment of Fermi statistics. Such a treatment requires a fully antisymmetrized many-body wave function, which becomes especially cumbersome when the single particle states are not orthogonal to each other as in WPMD. The totally antisymmetric wave function is

$$\psi(\mathbf{x}_1, \ldots, \mathbf{x}_N, t) = n\hat{A}\prod_i \varphi_i(\mathbf{x}_i, t)\chi_i, \tag{35}$$

where $\{\mathbf{x}_1, \ldots, \mathbf{x}_N\}$ are the coordinates of N electrons, t is time, χ_i is a Pauli spinor, n is a normalization constant, φ_i is ith wave packet, here of the form Eq. (7), $\hat{A}$ is the antisymmetrization operator,

$$\hat{A} = \prod_{i<j}(1 - \hat{\epsilon}_{ij}), \tag{36}$$

and $\hat{\epsilon}_{ij}$ is the exchange operator, which exchanges the positions and spins of particles i and j.

The expectation value of the Hamiltonian is needed to calculate the equations of motion of the variational parameters (see Eq. (18)). The kinetic and potential expectation values for the state (35) are [70]

$$\langle\hat{T}\rangle = \sum_{k,l=1}^{N} \langle\varphi_k\chi_k|\hat{t}|\varphi_l\chi_l\rangle\mathcal{O}_{lk} \tag{37}$$

$$\langle\hat{V}_{ei}\rangle = \sum_{k,l=1}^{N} \langle\varphi_k\chi_k|\hat{v}_{ei}|\varphi_l\chi_l\rangle\mathcal{O}_{lk} \tag{38}$$

$$\langle\hat{V}_{ee}\rangle = \sum_{k,l,m,n=1}^{N} \langle\varphi_k\chi_k\varphi_l\chi_l|\hat{v}_{ee}|\varphi_m\chi_m\varphi_n\chi_n\rangle(\mathcal{O}_{mk}\mathcal{O}_{nl} - \mathcal{O}_{ml}\mathcal{O}_{nk}), \tag{39}$$

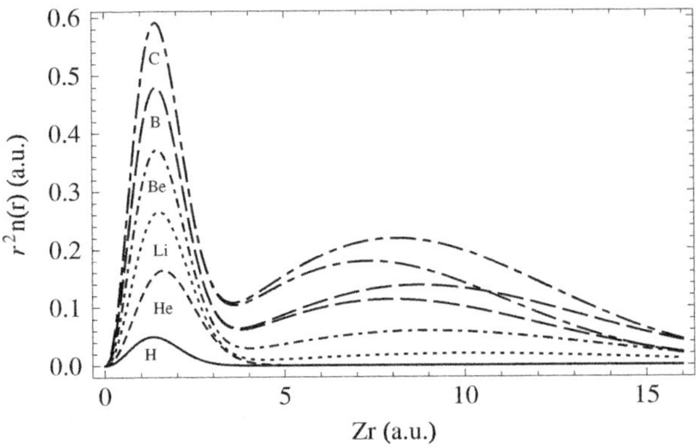

Fig. 3 Radial density of fully antisymmetrized variational wave functions for the lowest six elements of the periodic table: hydrogen (*solid*), helium (*dashed*), lithium (*dotted*), beryllium (*dot-dashed*), boron (*long dashed*), and carbon (*long dot-dashed*). Two curves are shown for boron and carbon to indicate anisotropy. These are the maximum and minimum densities found at each value of r

where $\hat{t}$ is the single-body kinetic energy operator, $\hat{v}_{ei}$ and $\hat{v}_{ee}$ are the two-body electron-ion and electron-electron potential energy operators, respectively, and $\mathcal{O}_{lk}$ is the inverse of the overlap matrix,

$$(\mathcal{O}^{-1})_{kl} = \langle \varphi_k \chi_k | \varphi_l \chi_l \rangle. \tag{40}$$

It is immediately apparent from Eq. (39) that one must complete an order N^4 operation per time step as well as deal with the possibility that the overlap matrix can become nearly singular, which explains why such calculations have mainly been limited to systems of finite size, such as the modeling of nucleons in a nucleus [70].

Using Eqs. (37–39), the energy and density of atoms can be calculated from the Raleigh-Ritz variational principle. The electron densities of the first six atoms on the periodic table are shown in Fig. 3 using individual electron wave packets of the form (7). Even with such a simple variational form, the correct shell structure is produced. Hydrogen has a single s orbital, helium has two s orbitals of the same size and opposite spin, lithium has two inner s orbitals and one outer, and beryllium has two inner and two outer s orbitals. The structure is more interesting for boron and carbon, in which pairs of wave packets are displaced on opposite sides of the nucleus in order to form effective p orbitals. So the boron electron density has a maximum along the z axis and a minimum in the x-y plane, while the carbon atom has maxima along the x and y axes and a minimum on the z axis. Of course, the detailed structure is wrong; the cusp condition is not satisfied and the density falls off too fast at infinity. However, having shell structure means antisymmetrized WPMD includes a simple version of bound-free transitions.

Due to the expense of full antisymmetrization, several different approximations have been made in the literature. The simplest approximation is the Hartree approximation,

$$\hat{A} \approx \hat{A}_1 = 1, \tag{41}$$

used in Refs. [58, 60, 78]. This approximation is valid at temperatures much above the Fermi energy. Unfortunately, this is also the regime in which wave packet spreading is the biggest issue, so width constraints were used in all of those references.

Klakow et al. [82, 83] published the first WPMD results for plasmas. They and others [79, 80, 84] used the pairwise antisymmetrization approximation

$$\hat{A} \approx \hat{A}_2 = 1 - \sum_{i<j} \hat{\epsilon}_{ij}, \tag{42}$$

which includes only two-body exchange. This approximation holds when all of the wave packets are well separated in phase space. That is, the distance between their expectation values of position and momentum is big compared to their uncertainties.

Another approximation [52–54, 56, 63] is to use the Hartree form while calculating the expectation value of the electron-electron potential energy, but use full antisymmetrization for the kinetic energy,

$$\langle V_{ee} \rangle \approx \sum_{k,l=1}^{N} \langle \varphi_k \chi_k \varphi_l \chi_l | \hat{v}_{ee} | \varphi_k \chi_k \varphi_l \chi_l \rangle. \tag{43}$$

Such a scheme allows simulations at temperatures up to 30,000 K before wave packet divergence becomes an issue [58].

The simplest computational way of including exchange effects is through the empirical electron Force Field (eFF) model [55, 57, 64]. Energy expectation values are calculated with the Hartree approximation, but then corrected with a Pauli potential,

$$E_{Pauli} = \sum_{m_{s,i}=m_{s,j}} E(\uparrow\uparrow)_{ij} + \sum_{m_{s,i}\neq m_{s,j}} E(\uparrow\downarrow)_{ij}, \tag{44}$$

$$E(\uparrow\uparrow)_{ij} = \left(\frac{S_{ij}^2}{1 - S_{ij}^2} + (1-\rho)\frac{S_{ij}^2}{1 + S_{ij}^2} \right) \Delta T_{ij}, \tag{45}$$

$$E(\uparrow\uparrow)_{ij} = \frac{\rho S_{ij}^2}{1 + S_{ij}^2} \Delta T_{ij}, \tag{46}$$

$$\Delta T_{ij} = \frac{3}{2}(\bar{\sigma}_i^{-2} + \bar{\sigma}_j^{-2}) - \frac{2[3(\bar{\sigma}_i^2 + \bar{\sigma}_j^2) - 2\bar{r}_{ij}^2]}{(\bar{\sigma}_i^2 + \bar{\sigma}_j^2)^2}, \tag{47}$$

$$S_{ij} = \left(\frac{2}{\bar{\sigma}_i/\bar{\sigma}_j + \bar{\sigma}_j/\bar{\sigma}_i} \right)^{3/2} \exp\left(-\frac{\bar{r}_{ij}^2}{\bar{\sigma}_i^2 + \bar{\sigma}_j^2} \right), \tag{48}$$

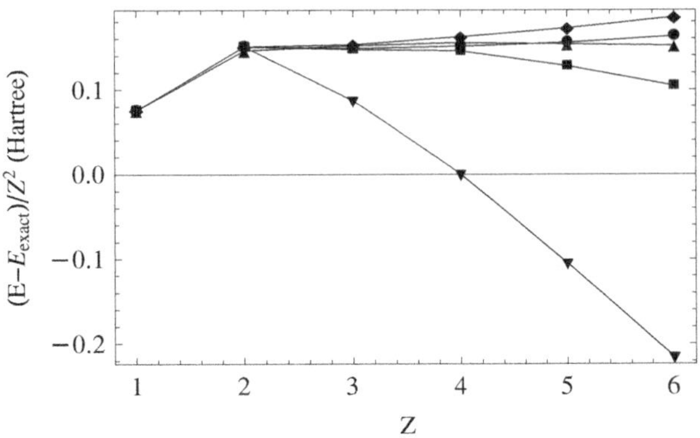

Fig. 4 Error in variational energy as a function of Z for different choices of approximation to the energy expectation value: Hartree (*upside down triangles*), antisymmetrized kinetic energy (*squares*), eFF (*triangles*), fully antisymmetrized (*circles*), and two-body exchange (*diamonds*)

where $\bar{\sigma}_i = c_1\sigma_i$, $\bar{r}_{ij} = c_2|\mathbf{r}_i - \mathbf{r}_j|$, and the empirical parameters $\rho = -0.2$, $c_1 = 0.9$ and $c_2 = 1.125$ are set by fitting to accurate molecular properties. The form of E_{Pauli} is motivated by terms which appear with two-body exchange.

Figure 4 shows the error in energy of the lowest six atoms on the periodic table using all of these antisymmetrization methods. Full antisymmetrization makes an error roughly equal to $0.15Z^2$ a.u. This error is due to the lack of inclusion of electron correlation and from the simple single electron variational form. Since the approximations made are with respect to the fully antisymmetrized Gaussian wave packet calculation, their accuracy should be determined by how close they are to that result. In order of accuracy, the approximations are eFF, two-body exchange, antisymmetrized kinetic energy, and the Hartree approximation.

3.4 Interpretation

The theoretical basis of WPMD is quantum mechanical. However, the variational parameters describing the electrons' wave function and the energy expectation value are often interpreted as if they are classical quantities. The latter quantity is continuous with a lower bound, necessarily greater than the true ground state due to the restriction of the Hilbert space to the subspace spanned by the Gaussian wave packets. There is confusion in the literature as to why this energy is not quantized for bound systems, which is taken to be a problem of the model [76]. However, intermediate energy expectation values can easily be reached by creating time-dependent states, which are mixtures of the states with the discrete energy eigenvalues, which is exactly the case here.

Another confusion arises with respect to the partition function. If the energy expectation value is taken to be classical and the width taken to be a classical fourth degree of freedom for each electron, then the classical partition function is divergent unless a width constraint is added. However, this is caused by the approximation

$$e^{-\beta\hat{H}} \approx e^{-\beta\langle\hat{H}\rangle}, \tag{49}$$

where β is the inverse temperature. In fact, exact thermodynamic properties can be calculated in any basis, including Gaussians. The interested reader is referred to Refs. [70] and [85] for accurate quantum treatments.

Within the plasma physics community dynamical quantities and transport coefficients have been calculated by interpreting the position and momentum expectation values as classical quantities and then using classical formulas [54, 56, 58, 60]. This choice along with the width constraints turns the WPMD model into an effective classical system meant to mimic properties of quantum mechanics; it is no longer an ab initio quantum model. Tuning this classical model to known quantities may still lead to sensical results.

4 Outlook

The dream of having a simple and accurate time-dependent method capable of accurately simulating WDM systems has not been fully realized. To model high temperature systems, of order or greater than the Fermi energy, either an unphysical width constraint must be used or the system develops an unphysical homogeneous density. At lower temperatures, the eFF variant of WPMD has made the greatest strides having been applied to such diverse calculations as stopping power [57], equation of state [55], lithium cluster-slab impact [64], and the shock Hugoniot curves of liquid hydrogen and deuterium [55, 57, 64]. However, the empirical parameters in this model make it difficult to quantify errors a priori.

The greatest errors in WPMD stem from its simple variational form. Improvements for high temperature systems should focus on more accurate representations of delocalized free states that have higher densities near ions. The drive for a better representation of low temperature systems should focus on improving the ability of the wave packets to form bonds by allowing anisotropies.

Acknowledgements The author would like to thank Michael S. Murillo for mentorship while he was learning about WPMD and writing this review, John Benage for helpful discussions and background information, Ronald Redmer for useful suggestions on improving the content of this review, and Frank Graziani, the organizing committee, and the staff of the Institute for Pure and Applied Mathematics for organizing the workshop: Computational Challenges in Warm Dense Matter.

This work was mostly written during the time the author was an employee of the Los Alamos National Security, LLC. (LANS), operator of the Los Alamos National Laboratory under Contract No. DE-AC52-06NA25396 with the U.S. Department of Energy and funded by the Laboratory Directed Research and Development Program at LLNL under project tracking code 09-SI-011.

References

1. A.W. DeSilva, J.D. Katsouros, Phys. Rev. E **57**, 5945 (1998)
2. J.F. Benage, W.R. Shanahan, M.S. Murillo, Phys. Rev. Lett. **83**, 2953 (1999)

3. A.W. DeSilva, G.B. Vunni, Phys. Rev. E **83**, 037402 (2011)
4. R.C. Malone, R.L. McCrory, R.L. Morse, Phys. Rev. Lett. **34**, 721 (1975)
5. B.H. Ripin, P.G. Burkhalter, F.C. Young, J.M. McMahon, D.G. Colombant, S.E. Bodner, R.R. Whitlock, D.J. Nagel, D.J. Johnson, N.K. Winsor, C.M. Dozier, R.D. Bleach, J.A. Stamper, E.A. McLean, Phys. Rev. Lett. **34**, 1313 (1975)
6. W.C. Mead, R.A. Haas, W.L. Kruer, D.W. Phillion, H.N. Kornblum, J.D. Lindl, D.R. MacQuigg, V.C. Rupert, Phys. Rev. Lett. **37**, 489 (1976)
7. B. Yaakobi, T.C. Bristow, Phys. Rev. Lett. **38**, 350 (1977)
8. J.P. Anthes, M.A. Gusinow, M.K. Matzen, Phys. Rev. Lett. **41**, 1300 (1978)
9. R. Decoste, S.E. Bodner, B.H. Ripin, E.A. McLean, S.P. Obenschain, C.M. Armstrong, Phys. Rev. Lett. **42**, 1673 (1979)
10. A. Ng, D. Parfeniuk, P. Celliers, L. DaSilva, R.M. More, Y.T. Lee, Phys. Rev. Lett. **57**, 1595 (1986)
11. K. Widmann, T. Ao, M.E. Foord, D.F. Price, A.D. Ellis, P.T. Springer, A. Ng, Phys. Rev. Lett. **92**, 125002 (2004)
12. Y. Ping, D. Hanson, I. Koslow, T. Ogitsu, D. Prendergast, E. Schwegler, G. Collins, A. Ng, Phys. Rev. Lett. **96**, 255003 (2006)
13. Y. Ping, A. Correa, T. Ogitsu, E. Draeger, E. Schwegler, T. Ao, K. Widmann, D. Price, E. Lee, H. Tam, P. Springer, D. Hanson, I. Koslow, D. Prendergast, G. Collins, A. Ng, High Energy Density Phys. **6**(2), 246 (2010)
14. U. Zastrau, P. Audebert, V. Bernshtam, E. Brambrink, T. Kämpfer, E. Kroupp, R. Loetzsch, Y. Maron, Y. Ralchenko, H. Reinholz, G. Röpke, A. Sengebusch, E. Stambulchik, I. Uschmann, L. Weingarten, E. Förster, Phys. Rev. E **81**, 026406 (2010)
15. U. Zastrau, A. Sengebusch, P. Audebert, E. Brambrink, R. Faustlin, T. Kampfer, E. Kroupp, R. Loetzsch, Y. Maron, H. Reinholz, G. Rpke, E. Stambulchik, I. Uschmann, E. Frster, High Energy Density Phys. **7**(2), 47 (2011)
16. B. Rus, T. Mocek, M. Kozlov, J. Polan, P. Homer, M. Fajardo, M. Foord, H. Chung, S. Moon, R. Lee, High Energy Density Phys. **7**(1), 11 (2011)
17. N.A. Inogamov, V.V. Zhakhovsky, S.I. Ashitkov, V.A. Khokhlov, V.V. Shepelev, P.S. Komarov, A.V. Ovchinnikov, D.S. Sitnikov, Y.V. Petrov, M.B. Agranat, S.I. Anisimov, V.E. Fortov, Contrib. Plasma Phys. **51**(4), 367 (2011)
18. J. Li, J. Zhou, T. Ogitsu, Y. Ping, W.D. Ware, J. Cao, High Energy Density Phys. **8**(3), 298 (2012)
19. T. Ogitsu, Y. Ping, A. Correa, B. ick Cho, P. Heimann, E. Schwegler, J. Cao, G.W. Collins, High Energy Density Phys. **8**(3), 303 (2012)
20. J. Nuckolls, L. Wood, A. Thiessen, G. Zimmerman, Nature **239**(5368), 139 (1972)
21. J.S. Clarke, H.N. Fisher, R.J. Mason, Phys. Rev. Lett. **30**, 89 (1973)
22. K.A. Flippo, E. d'Humieres, S.A. Gaillard, J. Rassuchine, D.C. Gautier, M. Schollmeier, F. Nurnberg, J.L. Kline, J. Adams, B. Albright, M. Bakeman, K. Harres, R.P. Johnson, G. Korgan, S. Letzring, S. Malekos, N. Renard-LeGalloudec, Y. Sentoku, T. Shimada, M. Roth, T.E. Cowan, J.C. Fernandez, B.M. Hegelich, Phys. Plasmas (1994-present) **15**(5), 056709 (2008)
23. D.H.H. Hoffmann, A. Blazevic, N.A. Tahir, S. Udrea, D. Varaentsov, Y. Zhao, Int. J. Mod. Phys. E Nucl. Phys. **18**(2), 381 (2009)
24. A. Lvy, F. Dorchies, M. Harmand, C. Fourment, S. Hulin, O. Peyrusse, J.J. Santos, P. Antici, P. Audebert, J. Fuchs, L. Lancia, A. Mancic, M. Nakatsutsumi, S. Mazevet, V. Recoules, P. Renaudin, S. Fourmaux, Plasma Phys. Control. Fusion **51**(12), 124021 (2009)
25. F.M. Bieniosek, E. Henestroza, S. Lidia, P.A. Ni, Rev. Sci. Instrum. **81**(10) (2010)
26. Y. Zhao, Z. Hu, R. Cheng, Y. Wang, H. Peng, A. Golubev, X. Zhang, X. Lu, D. Zhang, X. Zhou, X. Wang, G. Xu, J. Ren, Y. Li, Y. Lei, Y. Sun, J. Zhao, T. Wang, Y. Wang, G. Xiao, Laser Part. Beams **30**, 679 (2012)
27. M. Gauthier, S.N. Chen, A. Levy, P. Audebert, C. Blancard, T. Ceccotti, M. Cerchez, D. Doria, V. Floquet, E. Lamour, C. Peth, L. Romagnani, J.P. Rozet, M. Scheinder, R. Shepherd, T. Toncian, D. Vernhet, O. Willi, M. Borghesi, G. Faussurier, J. Fuchs, Phys. Rev. Lett. **110**, 135003 (2013)

28. M.D. Knudson, R.W. Lemke, D.B. Hayes, C.A. Hall, C. Deeney, J.R. Asay, J. Appl. Phys. **94**(7), 4420 (2003)
29. M.D. Knudson, D.L. Hanson, J.E. Bailey, C.A. Hall, J.R. Asay, C. Deeney, Phys. Rev. B **69**, 144209 (2004)
30. M.D. Knudson, M.P. Desjarlais, Phys. Rev. Lett. **103**, 225501 (2009)
31. S. Root, K.R. Cochrane, J.H. Carpenter, T.R. Mattsson, Phys. Rev. B **87**, 224102 (2013)
32. C. Bostedt, H.N. Chapman, J.T. Costello, J.R.C. Lopez-Urrutia, S. Dusterer, S.W. Epp, J. Feldhaus, A. Fohlisch, M. Meyer, T. Moller, R. Moshammer, M. Richter, K. Sokolowski-Tinten, A. Sorokin, K. Tiedtke, J. Ullrich, W. Wurth, Nucl. Instrum. Methods Phys. Res. Sect. A Accel. Spectrometers Detect. Assoc. Equip. **601**(1–2), 108 (2009)
33. M. Altarelli, Crystallogr. Rep. **55**(7), 1145 (2010)
34. M. Altarelli, Nucl. Instrum. Methods Phys. Res. Sect. B Beam Interact. Mater. At. **269**(24), 2845 (2011)
35. S.P. Hau-Riege, Phys. Rev. E **87**, 053102 (2013)
36. C.G.M. van Kessel, R. Sigel, Phys. Rev. Lett. **33**, 1020 (1974)
37. L.R. Veeser, J.C. Solem, Phys. Rev. Lett. **40**, 1391 (1978)
38. R.J. Trainor, J.W. Shaner, J.M. Auerbach, N.C. Holmes, Phys. Rev. Lett. **42**, 1154 (1979)
39. F. Cottet, J.P. Romain, R. Fabbro, B. Faral, Phys. Rev. Lett. **52**, 1884 (1984)
40. A. Ng, D. Parfeniuk, L. DaSilva, Phys. Rev. Lett. **54**, 2604 (1985)
41. N.L. Kugland, G. Gregori, S. Bandyopadhyay, C.M. Brenner, C.R.D. Brown, C. Constantin, S.H. Glenzer, F.Y. Khattak, A.L. Kritcher, C. Niemann, A. Otten, J. Pasley, A. Pelka, M. Roth, C. Spindloe, D. Riley, Phys. Rev. E **80**, 066406 (2009)
42. S. Pabst, Eur. Phys. J. Spec. Top. **221**(1), 1 (2013)
43. B. Ziaja, Z. Jurek, N. Medvedev, R. Thiele, S. Toleikis, High Energy Density Phys. **9**(3), 462 (2013)
44. D. Hoffmann, V. Fortov, M. Kuster, V. Mintsev, B. Sharkov, N. Tahir, S. Udrea, D. Varentsov, K. Weyrich, Astrophys. Space Sci. **322**(1–4), 167 (2009)
45. F.R. Graziani, V.S. Batista, L.X. Benedict, J.I. Castor, H. Chen, S.N. Chen, C.A. Fichtl, J.N. Glosli, P.E. Grabowski, A.T. Graf, S.P. Hau-Riege, A.U. Hazi, S.A. Khairallah, L. Krauss, A.B. Langdon, R.A. London, A. Markmann, M.S. Murillo, D.F. Richards, H.A. Scott, R. Shepherd, L.G. Stanton, F.H. Streitz, M.P. Surh, J.C. Weisheit, H.D. Whitley, High Energy Density Phys. **8**(1), 105 (2012)
46. L.X. Benedict, M.P. Surh, J.I. Castor, S.A. Khairallah, H.D. Whitley, D.F. Richards, J.N. Glosli, M.S. Murillo, C.R. Scullard, P.E. Grabowski, D. Michta, F.R. Graziani, Phys. Rev. E **86**, 046406 (2012)
47. G. Dixit, J.M. Slowik, R. Santra, Phys. Rev. Lett. **110**, 137403 (2013)
48. H. Wang, M. Thoss, J. Chem. Phys. **119**(3), 1289 (2003)
49. U. Manthe, J. Chem. Phys. **128**(16), 164116 (2008)
50. E. Runge, E.K.U. Gross, Phys. Rev. Lett. **52**, 997 (1984)
51. R.G. Littlejohn, Phys. Rep. **138**(4–5), 193 (1986)
52. M. Knaup, P.G. Reinhard, C. Toepffer, G. Zwicknagel, Comput. Phys. Commun. **147**(1–2), 202 (2002). Proceedings of the Europhysics Conference on Computational Physics Computational Modeling and Simulation of Complex Systems
53. M. Knaup, P.G. Reinhard, C. Toepffer, G. Zwicknagel, J. Phys. A Math. Gen. **36**(22), 6165 (2003)
54. B. Jakob, P.G. Reinhard, C. Toepffer, G. Zwicknagel, Phys. Rev. E **76**, 036406 (2007)
55. J.T. Su, W.A. Goddard, Phys. Rev. Lett. **99**, 185003 (2007)
56. B. Jakob, P.G. Reinhard, C. Toepffer, G. Zwicknagel, J. Phys. A Math. Theor. **42**(21), 214055 (2009)
57. J.T. Su, W.A. Goddard, J. Chem. Phys. **131**(24), 244501 (2009)
58. I.V. Morozov, I.A. Valuev, J. Phys. A Math. Theor. **42**(21), 214044 (2009)
59. K. Singer, W. Smith, Mol. Phys. **57**(4), 761 (1986)
60. G. Zwicknagel, T. Pschiwul, J. Phys. A Math. Gen. **39**(17), 4359 (2006)

61. D. Klakow, M. Knaup, H. Matuszok, P.G. Reinhard, C. Toepffer, Fusion Eng. Des. **32–33**, 533 (1996). Proceedings of the Seventh International Symposium on Heavy Ion Inertial Fusion
62. G. Zwicknagel, C. Toepffer, P.G. Reinhard, Laser Part. Beams **13**(02), 311 (1995)
63. M. Knaup, P.G. Reinhard, C. Toepffer, Contrib. Plasma Phys. **41**(2–3), 159 (2001)
64. A. Jaramillo-Botero, J. Su, A. Qi, W.A. Goddard, J. Comput. Chem. **32**(3), 497 (2011)
65. P. Ehrenfest, Z. fur Phys. **45**(7–8), 455 (1927)
66. E.J. Heller, J. Chem. Phys. **62**(4), 1544 (1975)
67. A. McLachlan, Mol. Phys. **8**(1), 39 (1964)
68. N. Corbin, K. Singer, Mol. Phys. **46**(3), 671 (1982)
69. P. Dirac, *The Principles of Quantum Mechanics*. Oxford Science Publications (Oxford University Press, New York 1981)
70. H. Feldmeier, J. Schnack, Rev. Mod. Phys. **72**, 655 (2000)
71. J. Broeckhove, L. Lathouwers, E. Kesteloot, P.V. Leuven, Chem. Phys. Lett. **149**(5–6), 547 (1988)
72. A.K. Kerman, S.E. Koonin, Ann. Phys. **100**(1–2), 332 (1976)
73. A. Lenglet, G. Maynard, Nucl. Instrum. Methods Phys. Res. Sect. A Accel. Spectrometers Detect. Assoc. Equip. **577**(1–2), 343 (2007)
74. I.V. Morozov, I.A. Valuev, Contrib. Plasma Phys. **52**(2), 140 (2012)
75. P.E. Grabowski, A. Markmann, I.V. Morozov, I.A. Valuev, C.A. Fichtl, D.F. Richards, V.S. Batista, F.R. Graziani, M.S. Murillo, Phys. Rev. E **87**, 063104 (2013)
76. W. Ebeling, B. Militzer, Phys. Lett. A **226**(5), 298 (1997)
77. M. Murillo, E. Timmermans, Contrib. Plasma Phys. **43**(5–6), 333 (2003)
78. W. Ebeling, A. Filinov, M. Bonitz, V. Filinov, T. Pohl, J. Phys. A Math. Gen. **39**(17), 4309 (2006)
79. M. Knaup, P.G. Reinhard, C. Toepffer, Contrib. Plasma Phys. **39**(1–2), 57 (1999)
80. M. Knaup, G. Zwicknagel, P.G. Reinhard, C. Toepffer, J. Phys. IV France **10**, 307 (2000)
81. M. Knaup, G. Zwicknagel, P.G. Reinhard, C. Toepffer, Nucl. Instrum. Methods Phys. Res. Sect. A Accel. Spectrometers Detect. Assoc. Equip. **464**(1–3), 267 (2001)
82. D. Klakow, C. Toepffer, P.G. Reinhard, J. Chem. Phys. **101**(12), 10766 (1994)
83. D. Klakow, C. Toepffer, P.G. Reinhard, Phys. Lett. A **192**(1), 55 (1994)
84. A. Lenglet, G. Maynard, Y.K. Kurilenkov, J. Phys. A Math. Gen. **39**(17), 4671 (2006)
85. B. Militzer, E.L. Pollock, Phys. Rev. E **61**, 3470 (2000)

Editorial Policy

1. Volumes in the following three categories will be published in LNCSE:

i) Research monographs
ii) Tutorials
iii) Conference proceedings

Those considering a book which might be suitable for the series are strongly advised to contact the publisher or the series editors at an early stage.

2. Categories i) and ii). Tutorials are lecture notes typically arising via summer schools or similar events, which are used to teach graduate students. These categories will be emphasized by Lecture Notes in Computational Science and Engineering. **Submissions by interdisciplinary teams of authors are encouraged.** The goal is to report new developments – quickly, informally, and in a way that will make them accessible to non-specialists. In the evaluation of submissions timeliness of the work is an important criterion. Texts should be well-rounded, well-written and reasonably self-contained. In most cases the work will contain results of others as well as those of the author(s). In each case the author(s) should provide sufficient motivation, examples, and applications. In this respect, Ph.D. theses will usually be deemed unsuitable for the Lecture Notes series. Proposals for volumes in these categories should be submitted either to one of the series editors or to Springer-Verlag, Heidelberg, and will be refereed. A provisional judgement on the acceptability of a project can be based on partial information about the work: a detailed outline describing the contents of each chapter, the estimated length, a bibliography, and one or two sample chapters – or a first draft. A final decision whether to accept will rest on an evaluation of the completed work which should include

- at least 100 pages of text;
- a table of contents;
- an informative introduction perhaps with some historical remarks which should be accessible to readers unfamiliar with the topic treated;
- a subject index.

3. Category iii). Conference proceedings will be considered for publication provided that they are both of exceptional interest and devoted to a single topic. One (or more) expert participants will act as the scientific editor(s) of the volume. They select the papers which are suitable for inclusion and have them individually refereed as for a journal. Papers not closely related to the central topic are to be excluded. Organizers should contact the Editor for CSE at Springer at the planning stage, see *Addresses* below.

In exceptional cases some other multi-author-volumes may be considered in this category.

4. Only works in English will be considered. For evaluation purposes, manuscripts may be submitted in print or electronic form, in the latter case, preferably as pdf- or zipped ps-files. Authors are requested to use the LaTeX style files available from Springer at http://www.springer.com/authors/book+authors/helpdesk?SGWID=0-1723113-12-971304-0 (Click on Templates → LaTeX → monographs or contributed books).

For categories ii) and iii) we strongly recommend that all contributions in a volume be written in the same LaTeX version, preferably LaTeX2e. Electronic material can be included if appropriate. Please contact the publisher.

Careful preparation of the manuscripts will help keep production time short besides ensuring satisfactory appearance of the finished book in print and online.

5. The following terms and conditions hold. Categories i), ii) and iii):

Authors receive 50 free copies of their book. No royalty is paid.
Volume editors receive a total of 50 free copies of their volume to be shared with authors, but no royalties.

Authors and volume editors are entitled to a discount of 33.3 % on the price of Springer books purchased for their personal use, if ordering directly from Springer.

6. Springer secures the copyright for each volume.

Addresses:

Timothy J. Barth
NASA Ames Research Center
NAS Division
Moffett Field, CA 94035, USA
barth@nas.nasa.gov

Michael Griebel
Institut für Numerische Simulation
der Universität Bonn
Wegelerstr. 6
53115 Bonn, Germany
griebel@ins.uni-bonn.de

David E. Keyes
Mathematical and Computer Sciences
and Engineering
King Abdullah University of Science
and Technology
P.O. Box 55455
Jeddah 21534, Saudi Arabia
david.keyes@kaust.edu.sa

and

Department of Applied Physics
and Applied Mathematics
Columbia University
500 W. 120 th Street
New York, NY 10027, USA
kd2112@columbia.edu

Risto M. Nieminen
Department of Applied Physics
Aalto University School of Science
and Technology
00076 Aalto, Finland
risto.nieminen@aalto.fi

Dirk Roose
Department of Computer Science
Katholieke Universiteit Leuven
Celestijnenlaan 200A
3001 Leuven-Heverlee, Belgium
dirk.roose@cs.kuleuven.be

Tamar Schlick
Department of Chemistry
and Courant Institute
of Mathematical Sciences
New York University
251 Mercer Street
New York, NY 10012, USA
schlick@nyu.edu

Editor for Computational Science
and Engineering at Springer:
Martin Peters
Springer-Verlag
Mathematics Editorial IV
Tiergartenstrasse 17
69121 Heidelberg, Germany
martin.peters@springer.com

Lecture Notes in Computational Science and Engineering

1. D. Funaro, *Spectral Elements for Transport-Dominated Equations.*
2. H.P. Langtangen, *Computational Partial Differential Equations.* Numerical Methods and Diffpack Programming.
3. W. Hackbusch, G. Wittum (eds.), *Multigrid Methods V.*
4. P. Deuflhard, J. Hermans, B. Leimkuhler, A.E. Mark, S. Reich, R.D. Skeel (eds.), *Computational Molecular Dynamics: Challenges, Methods, Ideas.*
5. D. Kröner, M. Ohlberger, C. Rohde (eds.), *An Introduction to Recent Developments in Theory and Numerics for Conservation Laws.*
6. S. Turek, *Efficient Solvers for Incompressible Flow Problems.* An Algorithmic and Computational Approach.
7. R. von Schwerin, ***M**ulti **B**ody **S**ystem **SIM**ulation.* Numerical Methods, Algorithms, and Software.
8. H.-J. Bungartz, F. Durst, C. Zenger (eds.), *High Performance Scientific and Engineering Computing.*
9. T.J. Barth, H. Deconinck (eds.), *High-Order Methods for Computational Physics.*
10. H.P. Langtangen, A.M. Bruaset, E. Quak (eds.), *Advances in Software Tools for Scientific Computing.*
11. B. Cockburn, G.E. Karniadakis, C.-W. Shu (eds.), *Discontinuous Galerkin Methods.* Theory, Computation and Applications.
12. U. van Rienen, *Numerical Methods in Computational Electrodynamics.* Linear Systems in Practical Applications.
13. B. Engquist, L. Johnsson, M. Hammill, F. Short (eds.), *Simulation and Visualization on the Grid.*
14. E. Dick, K. Riemslagh, J. Vierendeels (eds.), *Multigrid Methods VI.*
15. A. Frommer, T. Lippert, B. Medeke, K. Schilling (eds.), *Numerical Challenges in Lattice Quantum Chromodynamics.*
16. J. Lang, *Adaptive Multilevel Solution of Nonlinear Parabolic PDE Systems.* Theory, Algorithm, and Applications.
17. B.I. Wohlmuth, *Discretization Methods and Iterative Solvers Based on Domain Decomposition.*
18. U. van Rienen, M. Günther, D. Hecht (eds.), *Scientific Computing in Electrical Engineering.*
19. I. Babuška, P.G. Ciarlet, T. Miyoshi (eds.), *Mathematical Modeling and Numerical Simulation in Continuum Mechanics.*
20. T.J. Barth, T. Chan, R. Haimes (eds.), *Multiscale and Multiresolution Methods.* Theory and Applications.
21. M. Breuer, F. Durst, C. Zenger (eds.), *High Performance Scientific and Engineering Computing.*
22. K. Urban, *Wavelets in Numerical Simulation.* Problem Adapted Construction and Applications.

23. L.F. Pavarino, A. Toselli (eds.), *Recent Developments in Domain Decomposition Methods.*

24. T. Schlick, H.H. Gan (eds.), *Computational Methods for Macromolecules: Challenges and Applications.*

25. T.J. Barth, H. Deconinck (eds.), *Error Estimation and Adaptive Discretization Methods in Computational Fluid Dynamics.*

26. M. Griebel, M.A. Schweitzer (eds.), *Meshfree Methods for Partial Differential Equations.*

27. S. Müller, *Adaptive Multiscale Schemes for Conservation Laws.*

28. C. Carstensen, S. Funken, W. Hackbusch, R.H.W. Hoppe, P. Monk (eds.), *Computational Electromagnetics.*

29. M.A. Schweitzer, *A Parallel Multilevel Partition of Unity Method for Elliptic Partial Differential Equations.*

30. T. Biegler, O. Ghattas, M. Heinkenschloss, B. van Bloemen Waanders (eds.), *Large-Scale PDE-Constrained Optimization.*

31. M. Ainsworth, P. Davies, D. Duncan, P. Martin, B. Rynne (eds.), *Topics in Computational Wave Propagation.* Direct and Inverse Problems.

32. H. Emmerich, B. Nestler, M. Schreckenberg (eds.), *Interface and Transport Dynamics.* Computational Modelling.

33. H.P. Langtangen, A. Tveito (eds.), *Advanced Topics in Computational Partial Differential Equations.* Numerical Methods and Diffpack Programming.

34. V. John, *Large Eddy Simulation of Turbulent Incompressible Flows.* Analytical and Numerical Results for a Class of LES Models.

35. E. Bänsch (ed.), *Challenges in Scientific Computing - CISC 2002.*

36. B.N. Khoromskij, G. Wittum, *Numerical Solution of Elliptic Differential Equations by Reduction to the Interface.*

37. A. Iske, *Multiresolution Methods in Scattered Data Modelling.*

38. S.-I. Niculescu, K. Gu (eds.), *Advances in Time-Delay Systems.*

39. S. Attinger, P. Koumoutsakos (eds.), *Multiscale Modelling and Simulation.*

40. R. Kornhuber, R. Hoppe, J. Périaux, O. Pironneau, O. Wildlund, J. Xu (eds.), *Domain Decomposition Methods in Science and Engineering.*

41. T. Plewa, T. Linde, V.G. Weirs (eds.), *Adaptive Mesh Refinement – Theory and Applications.*

42. A. Schmidt, K.G. Siebert, *Design of Adaptive Finite Element Software.* The Finite Element Toolbox ALBERTA.

43. M. Griebel, M.A. Schweitzer (eds.), *Meshfree Methods for Partial Differential Equations II.*

44. B. Engquist, P. Lötstedt, O. Runborg (eds.), *Multiscale Methods in Science and Engineering.*

45. P. Benner, V. Mehrmann, D.C. Sorensen (eds.), *Dimension Reduction of Large-Scale Systems.*

46. D. Kressner, *Numerical Methods for General and Structured Eigenvalue Problems.*

47. A. Boriçi, A. Frommer, B. Joó, A. Kennedy, B. Pendleton (eds.), *QCD and Numerical Analysis III.*

48. F. Graziani (ed.), *Computational Methods in Transport.*

49. B. Leimkuhler, C. Chipot, R. Elber, A. Laaksonen, A. Mark, T. Schlick, C. Schütte, R. Skeel (eds.), *New Algorithms for Macromolecular Simulation.*

50. M. Bücker, G. Corliss, P. Hovland, U. Naumann, B. Norris (eds.), *Automatic Differentiation: Applications, Theory, and Implementations.*

51. A.M. Bruaset, A. Tveito (eds.), *Numerical Solution of Partial Differential Equations on Parallel Computers.*

52. K.H. Hoffmann, A. Meyer (eds.), *Parallel Algorithms and Cluster Computing.*

53. H.-J. Bungartz, M. Schäfer (eds.), *Fluid-Structure Interaction.*

54. J. Behrens, *Adaptive Atmospheric Modeling.*

55. O. Widlund, D. Keyes (eds.), *Domain Decomposition Methods in Science and Engineering XVI.*

56. S. Kassinos, C. Langer, G. Iaccarino, P. Moin (eds.), *Complex Effects in Large Eddy Simulations.*

57. M. Griebel, M.A Schweitzer (eds.), *Meshfree Methods for Partial Differential Equations III.*

58. A.N. Gorban, B. Kégl, D.C. Wunsch, A. Zinovyev (eds.), *Principal Manifolds for Data Visualization and Dimension Reduction.*

59. H. Ammari (ed.), *Modeling and Computations in Electromagnetics: A Volume Dedicated to Jean-Claude Nédélec.*

60. U. Langer, M. Discacciati, D. Keyes, O. Widlund, W. Zulehner (eds.), *Domain Decomposition Methods in Science and Engineering XVII.*

61. T. Mathew, *Domain Decomposition Methods for the Numerical Solution of Partial Differential Equations.*

62. F. Graziani (ed.), *Computational Methods in Transport: Verification and Validation.*

63. M. Bebendorf, *Hierarchical Matrices.* A Means to Efficiently Solve Elliptic Boundary Value Problems.

64. C.H. Bischof, H.M. Bücker, P. Hovland, U. Naumann, J. Utke (eds.), *Advances in Automatic Differentiation.*

65. M. Griebel, M.A. Schweitzer (eds.), *Meshfree Methods for Partial Differential Equations IV.*

66. B. Engquist, P. Lötstedt, O. Runborg (eds.), *Multiscale Modeling and Simulation in Science.*

67. I.H. Tuncer, Ü. Gülcat, D.R. Emerson, K. Matsuno (eds.), *Parallel Computational Fluid Dynamics 2007.*

68. S. Yip, T. Diaz de la Rubia (eds.), *Scientific Modeling and Simulations.*

69. A. Hegarty, N. Kopteva, E. O'Riordan, M. Stynes (eds.), *BAIL* 2008 – *Boundary and Interior Layers.*

70. M. Bercovier, M.J. Gander, R. Kornhuber, O. Widlund (eds.), *Domain Decomposition Methods in Science and Engineering XVIII.*

71. B. Koren, C. Vuik (eds.), *Advanced Computational Methods in Science and Engineering.*

72. M. Peters (ed.), *Computational Fluid Dynamics for Sport Simulation.*

73. H.-J. Bungartz, M. Mehl, M. Schäfer (eds.), *Fluid Structure Interaction II - Modelling, Simulation, Optimization.*

74. D. Tromeur-Dervout, G. Brenner, D.R. Emerson, J. Erhel (eds.), *Parallel Computational Fluid Dynamics 2008.*

75. A.N. Gorban, D. Roose (eds.), *Coping with Complexity: Model Reduction and Data Analysis.*

76. J.S. Hesthaven, E.M. Rønquist (eds.), *Spectral and High Order Methods for Partial Differential Equations.*

77. M. Holtz, *Sparse Grid Quadrature in High Dimensions with Applications in Finance and Insurance.*

78. Y. Huang, R. Kornhuber, O.Widlund, J. Xu (eds.), *Domain Decomposition Methods in Science and Engineering XIX.*

79. M. Griebel, M.A. Schweitzer (eds.), *Meshfree Methods for Partial Differential Equations V.*

80. P.H. Lauritzen, C. Jablonowski, M.A. Taylor, R.D. Nair (eds.), *Numerical Techniques for Global Atmospheric Models.*

81. C. Clavero, J.L. Gracia, F.J. Lisbona (eds.), *BAIL 2010 – Boundary and Interior Layers, Computational and Asymptotic Methods.*

82. B. Engquist, O. Runborg, Y.R. Tsai (eds.), *Numerical Analysis and Multiscale Computations.*

83. I.G. Graham, T.Y. Hou, O. Lakkis, R. Scheichl (eds.), *Numerical Analysis of Multiscale Problems.*

84. A. Logg, K.-A. Mardal, G. Wells (eds.), *Automated Solution of Differential Equations by the Finite Element Method.*

85. J. Blowey, M. Jensen (eds.), *Frontiers in Numerical Analysis - Durham 2010.*

86. O. Kolditz, U.-J. Gorke, H. Shao, W. Wang (eds.), *Thermo-Hydro-Mechanical-Chemical Processes in Fractured Porous Media - Benchmarks and Examples.*

87. S. Forth, P. Hovland, E. Phipps, J. Utke, A. Walther (eds.), *Recent Advances in Algorithmic Differentiation.*

88. J. Garcke, M. Griebel (eds.), *Sparse Grids and Applications.*

89. M. Griebel, M.A. Schweitzer (eds.), *Meshfree Methods for Partial Differential Equations VI.*

90. C. Pechstein, *Finite and Boundary Element Tearing and Interconnecting Solvers for Multiscale Problems.*

91. R. Bank, M. Holst, O. Widlund, J. Xu (eds.), *Domain Decomposition Methods in Science and Engineering XX.*

92. H. Bijl, D. Lucor, S. Mishra, C. Schwab (eds.), *Uncertainty Quantification in Computational Fluid Dynamics.*

93. M. Bader, H.-J. Bungartz, T. Weinzierl (eds.), *Advanced Computing.*

94. M. Ehrhardt, T. Koprucki (eds.), *Advanced Mathematical Models and Numerical Techniques for Multi-Band Effective Mass Approximations.*

95. M. Azaïez, H.E. Fekih, J.S. Hesthaven (eds.), *Spectral and High Order Methods for Partial Differential Equations ICOSAHOM 2012.*

96. M.P. Desjarlais et al. (eds.), *Frontiers and Challenges in Warm Dense Matter.*

For further information on these books please have a look at our mathematics catalogue at the following URL: `www.springer.com/series/3527`

Monographs in Computational Science and Engineering

1. J. Sundnes, G.T. Lines, X. Cai, B.F. Nielsen, K.-A. Mardal, A. Tveito, *Computing the Electrical Activity in the Heart.*

For further information on this book, please have a look at our mathematics catalogue at the following URL: www.springer.com/series/7417

Texts in Computational Science and Engineering

1. H. P. Langtangen, *Computational Partial Differential Equations.* Numerical Methods and Diffpack Programming. 2nd Edition
2. A. Quarteroni, F. Saleri, P. Gervasio, *Scientific Computing with MATLAB and Octave.* 4th Edition
3. H. P. Langtangen, *Python Scripting for Computational Science.* 3rd Edition
4. H. Gardner, G. Manduchi, *Design Patterns for e-Science.*
5. M. Griebel, S. Knapek, G. Zumbusch, *Numerical Simulation in Molecular Dynamics.*
6. H. P. Langtangen, *A Primer on Scientific Programming with Python.* 4th Edition
7. A. Tveito, H. P. Langtangen, B. F. Nielsen, X. Cai, *Elements of Scientific Computing.*
8. B. Gustafsson, *Fundamentals of Scientific Computing.*
9. M. Bader, *Space-Filling Curves.*
10. M. Larson, F. Bengzon, *The Finite Element Method: Theory, Implementation and Applications.*
11. W. Gander, M. Gander, F. Kwok, *Scientific Computing: An Introduction using Maple and MATLAB.*

For further information on these books please have a look at our mathematics catalogue at the following URL: www.springer.com/series/5151

CPSIA information can be obtained
at www.ICGtesting.com
Printed in the USA
BVOW05*0726110118
504982BV00012B/24/P

9 783319 049113